# American Imperial Pastoral

# American Imperial Pastoral

## *The Architecture of US Colonialism in the Philippines*

REBECCA TINIO MCKENNA

*The University of Chicago Press Chicago and London*

The University of Chicago Press, Chicago 60637
The University of Chicago Press, Ltd., London

Printed in the United States of America
26 25 24 23 22 21 20 19 18 17 1 2 3 4 5

ISBN-13: 978-0-226-41776-9 (cloth)
ISBN-13: 978-0-226-41793-6 (e-book)
DOI: 10.7208/chicago/9780226417936.001.0001

Library of Congress Cataloging-in-Publication Data
Names: McKenna, Rebecca Tinio, author.
Title: American imperial pastoral : the architecture of US colonialism in the Philippines / Rebecca Tinio McKenna.
Description: Chicago ; London : The University of Chicago Press, 2017. | Includes bibliographical references and index.
Identifiers: LCCN 2016025206 | ISBN 9780226417769 (cloth : alk. paper) | ISBN 9780226417936 (e-book)
Subjects: LCSH: Baguio (Philippines)—History. | Phillippines—Colonization. | Philippines—Relations—United States. | United States—Relations—Philippines. | Baguio (Philippines)—Ethnic relations. | Igorot (Philippine people)—Philippines—Benguet (Province)—History. | City planning—Philippines—Baguio—History. | Burnham, Daniel Hudson, 1846–1912. | Philippines—History—1898–1946.
Classification: LCC DS689.B2 M35 2017 | DDC 959.9/1—dc 23 LC record available at https://lccn.loc.gov/2016025206

⊗ This paper meets the requirements of ANSI/NISO Z39.48-1992 (Permanence of Paper).

*For Jake and Delaney Lundberg*
*and*
*Aïda McKenna*
*and*
*In memory of Thomas McKenna*

# Contents

# Acknowledgments

This project began at Yale University, where I had the benefit of working with a dream-team committee of historians. Jean-Christophe Agnew has shown me a model of the intellectual life; his generosity as a teacher and scholar is an inspiration. If I enable my students half as much as he has enabled me, I will count myself a success. John Mack Faragher has been a kind mentor, and through his courses (and holiday parties), he has built intellectual communities that have endured long past Yale. The scholarship and teaching of Mary Lui and Seth Fein have shaped mine, and I am grateful for the time, encouragement, and friendship they have given me. At Yale, I had the good fortune to work with other historians whose scholarship left an imprint on me; not least of these are Dolores Hayden, John Demos, David Blight, Steven Stoll, Kevin Repp, and Cynthia Russett. My classmates were my teachers, too, and I thank especially Dara Orenstein, Kate Unterman, Lisa Pinley Covert, Julia Irwin, Barry Muchnick, Miriam Posner, and members of the Asian American Studies Workshop, all of whom read pieces of my project. I remain indebted to Dara, for hours upon hours of conversation about everything under the sun. Not least of these subjects was Baguio and US imperialism. My experience at Yale would not have been the same without her, Erin Wood, Melissa Stuckey, Brian Fobi, and Jeremi Szaniawski.

I found many more teachers when I embarked on research and presentations of my work well beyond New Haven. I thank Delfin Tolentino, Erlyn Ruth Alcantara, and Oscar Campomanes, among many others in the Philippines, who kindly welcomed me, shared their work, and offered comment on and encouragement of my own. Members of the extended Tinio family graciously opened their Baguio home to me. I am especially grateful to Martin Tinio, Jr., who taught me about Tinio family history and the political and cultural history of the Philippines. In Manila, Cesar Enrique Aguinaldo Virata, Ernie Zarate, the Alampays, the Jesenas, the Reicherts, and the Serratos were the warmest hosts and teachers. Thank you to the great many librarians and archivists from Manila to Cambridge who helped me find my way to sources. And thanks go to Scott Zillmer at Terra Carta; he produced a map of the greater Baguio region for me.

I am grateful to the moderators, co-panelists, and audience members at conferences whose comments and questions have nudged me further, especially Paul Kramer, Julie Greene, Daniel Immerwahr, Christopher Capozzola, Anne Foster, and Mae Ngai. I thank Margaret Garb and the two blind readers of my manuscript who showed tremendous generosity of time, offering me indispensable comment and critique; this project would have looked quite different without them. I'm immensely grateful to Timothy Mennel at the University of Chicago Press; he has been an ideal editor and ally through the whole process. Editorial associate Rachel Kelly has answered all my questions, large and small, and kept me on track; Caterina MacLean has checked that I've dotted my i's and crossed my t's. Any errors and oversights that remain in the book are mine alone.

I could not be more fortunate in the friends, colleagues, and mentors I have found in the Department of History at the University of Notre Dame, most especially Catherine Cangany, John Deak, Paul Ocobock, Gail Bederman, Annie Coleman, Jon Coleman, Dan Graff, Karen Graubart, Patrick Griffin, Alex Martin, Richard Pierce, Jason Ruiz, Robert Sullivan, and Julia Thomas. I also thank Notre Dame's Institute for Scholarship in the Liberal Arts for a grant in support of this project.

I am grateful to my extended family: Henry Hof and Rosalia and Karina Tinio Hof, Stephanie Tayengco, Francis McKenna, and all the rest of the Tinios and McKennas; Bruce and Delaney and the whole of the Lundberg clan; and my loyal Barnard College pals. I'm humbled by their years of unstinting support and encouragement. I thank my immediate family for the same. Jake Lundberg is a devoted teacher and brilliant writer whose zest for history and intellectual curiosity have been renewing mine since our days sitting across each other at Sterling Library. With him near, cheering me on, distant hopes—like finishing this book—have felt within reach. Jake inspires me to seek a full life, one that we now share with Delaney Tinio Lundberg. Countless times, while I sat at my computer, biting my lip, she took my hand, leading me to play. More often than not, this was just what I needed. She is our most wonderful gift. I thank my mother Aïda Tinio

McKenna. Were it not for her labor and love in caring for Delaney every day, I'd still be writing. In some ways, this project is an effort to understand my mom and my dad, Thomas McKenna, and the historical currents and ironies that made possible their meeting decades ago. Their examples of love, wisdom, and faith that most things are possible continue to see me through.

# Introduction

In 1904, urban planner Daniel Burnham boarded a steamship in San Francisco and set off for the Philippines. He brought along a stellar resume. Work on the Columbian Exposition, the 1893 World's Fair held in Chicago, had cast Burnham in the national spotlight and helped him secure plum projects such as the redesign of the Washington, DC Mall. Now, at the call of William Howard Taft, then secretary of war, Burnham journeyed across the Pacific to help polish the Philippines, a recent acquisition from Spain, into the "Pearl of the Orient." Two projects would occupy him: improvements to the Manila capital and a plan for a colonial mountain resort nearly 5,000 feet above sea level that came to be called Baguio.[1] The latter, a stretch of pastureland for indigenous peoples' wealth in cattle, offered Americans a retreat from the torrid heat and politics of Manila.[2] Designing the resort in the City Beautiful style with impressive arrangements of colonial buildings, a long promenade, and grassy parks, Burnham and those who executed his plans would make Baguio a home away from home for Americans, a place unlike any other on the archipelago. One visitor described it as Uncle Sam's "beautiful summer home . . . for his official family."[3]

To William Howard Taft, the mountain region recalled the Adirondacks or Wyoming in the summer.[4] Others saw in Baguio a town leaps ahead of many American cities. Its "parks should serve as fitting examples for our smaller towns here at home," wrote one visitor in 1913. "It may be conservatively asserted that no American community anywhere of equal size can equal, let alone excel, her drives and plazas,"

0.1 Map of northern Luzon; borders and place names are contemporary. Prepared by Scott Zillmer, Terra Carta.

he noted.[5] To the protagonist of Filipino novelist Carlos Bulosan's *America Is in the Heart* (1946), by the 1920s, the retreat had become "a small city in the heart of tall mountains where the weather is always temperate. . . . The roads are asphalt and the most modern and beautiful in the Philippines. The houses and theaters are built in Western fashion. Tall pine trees cover the mountains and at night one can hear the leaves singing in the slight wind from the deep canyons beyond the city."[6] Rather than follow Burnham's wake back to San Francisco, colonialists could ascend the Cordillera Central mountains, stem the tide of tropical neurasthenia, or what was called Philippinitis on the archipelago, and be renewed for the task of the Philippines's "benevolent assimilation" amid the pines and the modern amenities of Baguio. US officials would use this mountain resort in all its transcendent and modern glory as the summer capital of the Philippines—a green, commanding peak from which to exercise power. If Burnham's neoclassical White City at the 1893 World's Fair had suggested US imperial ambitions, Baguio, a model American town in the upper elevations of the Pacific archipelago, declared that the United States was an empire indeed.[7]

This is a story about the making of this American retreat, the transformation of Philippine pasture into American pastoral, and what its design, construction, and use can tell us about the literal and figurative architecture of US imperialism, its soaring ambitions, the colonial practices that buttressed them, and the challenges these met on the ground. Defining the particular formations of US imperial power has vexed generations of historians. As a recent anthology puts the problem: "Among the colonial empires that once ruled the globe, the United States was an elusive, even paradoxical power. All the usual imperial labels that attach so readily to Great Britain or France seem to require qualification when applied to America."[8] The United States' territorial acquisitions following the Spanish-American War have seemed anomalous in the history of an otherwise "elusive" empire.[9] By the treaty that ended the conflict, the United States took the Philippine Islands for $20 million and seized Puerto Rico and Guam. The United States also claimed authority to intervene in Cuban affairs and leasing rights for a naval base at Guantánamo Bay through the Platt Amendment of 1901. In the Philippines, the first nationalist revolution in Asia, which began against Spain in 1896, gave way to the Philippine-American War once Filipinos recognized that US Navy warships were not docked in Manila Bay to aid them in the work of liberation. While retention of the Philippines raised significant debate in the United States, pro-annexationists won the day.

Some like Theodore Roosevelt envisioned colonialism in the Philippines as an opportunity to pursue the "strenuous life," a tonic to that peril of modern, bourgeois life, overcivilization.[10] He also justified the occupation of the Philippines by referring to the history of the US West: if Americans were "morally bound to abandon the Philippines," Roosevelt challenged his weak-kneed, anti-imperialist opponents, "we were also morally bound to abandon Arizona to the Apaches."[11] For others, especially those rattled by the effects of the 1893 economic depression and seeking an outlet for the supposed overproduction of American goods, the Philippines held out the hope of better economic futures. It seemed to bring the coveted markets of China within reach of a greater United States. Speaking in 1900 at the start of the US occupation of the Philippines, Indiana Senator Albert Beveridge warned that "More and more Europe will manufacture what it needs, secure from its colonies what it consumes." "Where shall we turn for consumers of our surplus?" he asked. Beveridge offered an answer: "Geography." "China is our natural customer. . . . Most future wars will be conflicts for commerce. The power that rules the Pacific, therefore, is the power that rules the world. And, with the Philippines, the power will be the American republic."[12] Beveridge's fellow Republican, Massachusetts Senator Henry Cabot Lodge, shared his views; Lodge believed the capital-port of Manila to be "the prize and the pearl of the East." "In our hands," he pledged, "it will become one of the greatest distributing points, one of the richest emporiums of the world's commerce. Rich in itself, with all its fertile islands behind it, it will keep open to us the markets of China and enable American enterprise and intelligence to take a master share in all the trade of the Orient."[13]

These visions of a lucrative trade in the Far East drew inspiration from the past just as they looked forward. They hearkened back to the Spanish use of Manila as an entrepôt for galleon ships traveling between Mexico and China, exchanging silver for precious goods like "spices, ivory, porcelain, lacquerware, and processed silk."[14] The year 1571, when Manila was founded, can be held as the "birth of global trade"; the port-city made possible "substantial, direct, and continuous trade between America and Asia for the first time in history."[15] About fifty tons of silver passed through Manila, which roughly corresponds to what Portugal and the Dutch and English East India Companies were together exporting to Asia in the seventeenth century.[16] By the end of the sixteenth century, this trade had created a "golden age" in Manila.[17] Until opposition from rival ports in Seville and Cadiz won limits to shipping between it and Acapulco,[18] Manila functioned as "the foremost colonial capital in Asia" and "an emporium of Oriental trade."[19] If Beveridge's and Cabot's aspirations call to mind that Spanish mercantile past, they also referenced visions native to the United

States; they extended Thomas Hart Benton's mid-nineteenth-century scheme for a national road connecting the Mississippi River and the Pacific Ocean. Such a highway, Benton imagined, could provide "passage to India" and fulfill centuries-old dreams of a trade route to the Orient.[20]

Yet the commercial expansion that Beveridge promoted has also been understood to mark a new form of American imperialism. Rather than acquire large swaths of land for settler colonialism, in the late nineteenth century and into the twentieth, the state pursued market expansion. Like other European powers at the time, pro-imperialists sought to pry open doors for free trade in the East. Historian Walter LaFeber reminds us that President William McKinley first demanded from Spain only the port of Manila. McKinley's interest, like Beveridge's and Cabot's, was "to use these holdings [like the Philippines] as a means to acquire markets for the glut of goods pouring out of highly mechanized factories and farms."[21] While the Philippines might itself serve as an outlet for surplus American goods and capital, the Manila port would facilitate American steamships' journeys to the China market.[22] Trade with China would solve the crisis of overproduction, the problem that many had cited as cause of the 1893 depression, and retention of the Philippines could thus help to diffuse labor radicalism at home.

The motive behind the 1898 acquisitions foreshadowed a characteristic of US power in the world in the coming decades: the "dissociation of economic from territorial expansion."[23] As geographer Neil Smith has written, the American imperial form that developed through the twentieth century exercised "power in the first place through the more abstract geography of the world market rather than through direct political control of territory."[24] This American-styled "globalism," he argues, "represents a long-term strategic rebuttal of European colonialism and anticolonial movements alike."[25] According to the logic of this imperial formation, territory was less an end and signifier of supremacy than a means. In 1963, LaFeber called it "new empire."

The distinction suggested by "new" is valuable; it emphasizes the difference between a state's acquisition of land for settlement or raw material extraction and a state's accumulation of parcels of territory as bases, ports, and canals to facilitate commercial expansion. LaFeber's work built on decades of scholarship that had considered the relationship between empire and commercial markets, or, as John Gallagher and Ronald Robinson put it in 1953, "the imperialism of free trade." The two historians defined "informal empire," a concept intended to distinguish between "degrees" of control and kinds of "expansion": from annexation and "political possession" of a region (or "formal" empire) to economic integration without recourse to "paramount power" (or "informal empire"). Gallagher and Robinson were disputing the view that mid-Victorians' pursuit of free trade differed substantially from late Victorians' annexations of territory; they contested the notion of a "mid-Victorian 'indifference' and a late-Victorian 'enthusiasm' for empire."[26] Both, they argued, pursued imperial relations. This was LaFeber's point: while the conquests of 1898 signaled something new, they still constituted imperial expansion, one in formation or "preparation" for several decades.[27]

Historians have been reconsidering the terms "formal" and "informal," and some have found them wanting. Ann Laura Stoler has argued that "'indirect rule' and 'informal empire' are unhelpful euphemisms." They have come to imply that older empires were "securely bounded" and "firmly entrenched," as those blocks of color across European imperial maps may suggest, and they prevent us from recognizing that all "imperial formations" are "states of becoming rather than being, macropolities in constant formation." Terms like "formal" and "informal" tend to render anything but direct, territorial control as "aberrant, quasi-empires; exceptional cases; peripheral forms," when arguably all imperial formations subsist on the creation of exceptions.[28] Further, as Paul Kramer has noted, "[T]he category of 'informal empire' abstracted the relationship between capitalist social relations and state power."[29] Meanwhile, if one follows the distinction wherein imperialism denotes a "concept" and state policy, and colonialism, a "practice," "an activity on the periphery, economically driven,"[30] then we might say that the concept of "new" empire privileges imperial motives and aspirations while obscuring colonial peoples and their ambitions.[31] In effect, the notion of a new, informal empire posed in contradistinction to some older, more formal form risks perpetuating a sense of US imperial power as dematerialized, even "invisible" and "anticolonial," when that was far from the experience on the ground.[32]

With that historiographical projection before us, we journey to Baguio, by 1903 an enclave of America, or a "government reservation," as administrators deemed it, in one node of the United States' emergent market empire. I ask readers to dwell on and in this colonial place—to go up a road, around a town site, inside a marketplace, and into a private home and country club—to confront the seeming abstraction of market making and the making of an "elusive" empire with the formal aspects of US rule and the labor of building it.[33] Beyond tracking the largely forgotten Pacific crossings of famed Progressives like Daniel Burnham, William Howard Taft, and Gifford Pinchot,[34] this story serves as a reminder that in trying to make Manila into "the prize and pearl of the East" and ease the way for "free" trade, Americans undertook the conquest and transformation of land, work that was premised on and that advanced the creation of social difference. US colonial administrators seized native lands as a base of power, expropriated labor for building roads, raised marketplaces to manage trade and people, and produced social and cultural capital in the hill station's Western-style homes, in its country club, and across its golf greens. In the process, they generated capital and subjects of rule. This story renders visible the work of capital and subject formation through enclosure, a process by which things held in common—land and national resources, ideas, and even time—"become cordoned off, either literally or figuratively, for the exclusive use of property owners, states, and corporate entities."[35] It shows how Americans attempted to integrate relations of buying and selling, production and consumption, across the seas, through acts of separation and division in and through particular places.

## Cordillera Pasture

We can take Baguio, this particular place—a colonial enclave, by definition, limited in its territorial take—as emblematic of "new empire" acquisitions. Americans' "government reservation" at Baguio initially amounted to about five square miles of land and would grow to cover about twenty-one by 1907.[36] In 1905, this was one of thirty-nine "civil reservations" in the Philippines claimed by the colonial government; all but five were intended for lighthouses. To these could be added the twenty-five military reservations also held by the government by that year.[37] These were small parcels of land, islands of power, from which the United States sought to exercise authority in the Philippines and launch a sphere of influence in Asia. Baguio stands as representative in another sense. The acts of dispossession that underwrote its creation as an American capital would apply elsewhere across the archipelago; they exemplify the United States' dispossession of Philippine sovereignty and its enclosure of a political future.

In other respects, this colonial place stands distinct from much of the Philippines in its geography and climatic features and in its history. Baguio sits in the present-day province of Benguet, which rests in the southern portion of the Philippines' imposing Cordillera Central, a mountain range that rises north of the Manila capital and Luzon's central plains. Stretching across nearly one-sixth of the island of Luzon and some 7,000 miles of land,[38] the Cordillera region is bounded on the east by the Cagayan Valley, known in the nineteenth century for its success in tobacco cultivation, and on the west by the South China Sea. The mountain chain runs almost two hundred miles north to south, and its peaks reach as high as

8,000 feet.[39] As this topography may suggest, the region Americans chose for their retreat necessarily had a past unlike that of the lowlands. In one of its first reports, the Philippine Commission, originally a group of five American men appointed by the president with executive and legislative powers in the colony, described those natives of Benguet—just over 20,000 in 1903—as "a friendly, harmless tribe," a description they surely would not have applied to Filipino revolutionaries of the lowlands.[40] Most of the "tribe" were likely Nabaloi-speaking Ibaloi, who had used "Baguio," a variation on the Ibaloi word *bagyu*, a "submerged slimy waterplant," to refer to a valley between current day Baguio and the nearby town of La Trinidad. By the mid-nineteenth century, the Spanish had identified "Baguio" as a *ranchería*, a settlement of Ibaloi in the valley. Americans would come to use "Baguio" to signify the whole of the US government reservation.[41] They gathered a series of distinct Ibaloi places like Campeo, Orengao, Cavalijureza, and Kafagway—largely grazing land for the Ibaloi peoples' cattle—into one.[42]

Prior to that, the Spanish had consolidated the peoples inhabiting the Cordillera under a single name: *Igorrote* or Igorot. Today, these ethnolinguistic groups dwell across the northern Luzon provinces of Benguet, Abra, Apayao, Kalinga, Ifugao, and the Mountain Province. At the time of the 1903 census, they were estimated to number 183,319. It seems that these peoples had no collective name for themselves.[43] "Igorot" derives from the Tagalog root *golot*, or mountain chain, and the prefix *in-*, or "dweller in" or "people of," a meaning that historian William Henry Scott notes became lost over generations of colonial use.[44] He writes that its trajectory "nicely parallels the decline of the English word, heathen, from its original meaning of 'dweller on the heath' to its present meaning of 'pagan.'"[45] In the pages that follow, I will use *Igorot* despite its weaknesses. In referring, literally, to mountain dwellers, the term does capture an important historical reality, and as Scott points out, the Tagalog term is at least relatively indigenous in origin.[46] And further, we know that the meanings and associations attached to "Igorot" have changed in time; they need not be fixed by imperial powers but can be remade through use. William Henry Scott, for one, wrote with admiration for the Igorots, reminding us that they occupy a unique place in the history of the Philippines: the Spanish did not manage to discipline them into colonial subjects until later in the nineteenth century. The Spanish subjected lowlanders to a process of Christianization and Hispanicization, and they certainly pursued converts and caches of gold in the mountains. But Igorots managed to eke out a measure of autonomy, one enabled by the Cordillera and its treasures.

By the time Americans arrived on the archipelago, neither these Igorots nor the "Moros" of the south were incorporated into the notion of "Filipino." During Spanish colonialism, "Filipino" referred to one "belonging to, located in, [or] originating from Las Islas Filipinas" or to a person born in the Philippines but of Spanish extraction. Revolution would change this; around the turn of the twentieth century, the term would take on its present usage as an "indigenous nationality-ethnicity."[47] These new Filipinos would seek an independent Philippines, and they assumed the Spanish division of Philippine peoples, with *ilustrado* elites not recognizing "any quality of national belonging among their putatively less civilized upland animist [Igorot] and southern Muslim neighbors."[48] In the course of their administration of the islands, Americans would further this division, assigning the Moros and "pagans" alike to the category of "non-Christian tribes." Colonialists would administer regions where these groups predominated separately, referring to them as "special provinces." In turn, Baguio, which gained a city charter in 1909, became a special province within a special province. It joined the provincial governments, municipal governments, the city of Manila, and the Department of Mindanao and Sulu as one of five "units" that the colonial government recognized; it administered the city of Baguio directly.[49]

## Imperial Pastoral

Upon his return from visits to Baguio and Manila, Daniel Burnham wrote a friend of his eye-opening trip to the East and of his "Baguio scheme"; it "begins to warrant a hope of something unusual among cities," he believed.[50] He was right in one way. A 1913 article on Baguio for the *Overland Monthly* began by noting that "Summer capitals have long been the rage everywhere, except in the United States."[51] But in another respect, in creating a mountain retreat, Americans were only following their Spanish predecessors. In an early report on conditions in the Philippines, commissioners referenced a Spanish study on Benguet as the prospective home to a health resort. The Spanish had envisioned a mountain sanitarium that might satisfy local demand for such a retreat, generate revenue by competing with similar resorts throughout Asia, and, with the influx of these visitors, provide civilizing influences on the Igorots. When the Philippine Commission recommended a sanitarium at the future site of Baguio, then, they drew on Spanish prospects for the area. Arguably, they also issued a salute to the French and British. The French were known to ascend to Da Lat in the Central Highlands of Vietnam. The English in India enjoyed Darjeeling and Shimla, the latter made famous by Rudyard

Kipling and his stories from "the hills." These "hill stations," established between 1815 and 1870, served purposes similar to Baguio's: they offered cooler sites for colonial health care and recreation and military quarters, too.[52] Private capital led the development of these towns in colonial India, and employees of the British East India Company were their first Anglo settlers. These agents "acted as often without as with the encouragement and support of the government." Later in the nineteenth century, however, the development of hill stations like Shimla achieved the status of "state policy."[53]

In the Philippines, discharged American soldiers who had learned of Cordillera gold were among the first to clamor up the mountains, but ultimately, it was the US colonial government that paved the way to the hill station. And its representatives did so with some good reason. When Burnham and his American hosts looked across Manila and beyond, they observed the collateral damage of war and even the traces of an ongoing insurgency. Although the US military declared the war to pacify Filipino revolutionaries complete in 1902, insurgents challenged US sovereignty well beyond that year. "Bandits," "robbers," and "brigands" undermined US authority across the archipelago through the first decade of rule. This gave the mountains a special appeal. In their 1901 report, commissioners wrote that "It is hard to see how a hostile force of any size could hope successfully to invade" Benguet. Pursuers would lack a supply base; "friendly Igorrotes"—probably the Ibaloi—would alert US troops to any incursions; and further, "[t]he scattering of Igorrote houses would afford no protection to outsiders."[54]

Early visitors to the Philippines also believed that Baguio's cooler climate was key to the health of white American workers. Here, soldiers' wounds were thought to heal more quickly. The colony's American workers and their families could enjoy refreshing surroundings and some rest and relaxation, too, and without costly trips to the mainland. One colonial official had written of "the administrative loss to the islands from the tendency of our best men to stay at home to be near their children when at school or because their wives cannot live in Manila." "[T]he immense expense incurred in shipping families to and fro" made the hill station a budget saver.[55] The price of transporting troops between the Philippines and the United States was alone said to reach $5 million in an editorial that one commissioner neatly clipped and filed away.[56] "In the end," wrote another Baguio advocate, "trips into the mountains of Benguet province . . . got to be the thing in lieu of expensive, time-consuming jaunts abroad to China, Siam, Japan, or the Straits Settlements between February and June of each year, the period when Manila sizzles and sears."[57] Baguio was understood as essential, then, to both the security and the health and efficiency of the colonial regime. Building a hill station was a capital investment serving the reproduction of colonial labor. While the Spanish had made such a secure zone for themselves in Manila called Intramuros, a haven "between the walls," Americans chose to dwell across the capital city, taking their refuge high into the mountains.[58] These were the circumstances under which colonialists fashioned Baguio as a retreat.

Through the first two decades of the twentieth century, across Ibaloi lands, Americans would create a landscape that included a governor general's mansion, city hall, government office complex, camp for American teachers, a hospital, a military reservation for US soldiers, and a school for training officers of the Philippine Constabulary, the colonial police force.[59] They built a famous "zig-zag" road that wound up the mountains and carried visitors to the Burnham-designed town, a marketplace, a country club, private homes like that of Philippine Commissioner and eventual Governor General William Cameron Forbes, the subjects of the chapters that follow. To build this Baguio, Americans appropriated Ibaloi meadow, transforming what had been fodder for cattle into sites for recreation and landscape views to offset colonialists' nostalgia and gird them for the colonial occupation. Ibaloi pasture became grassy parks, a polo field, a golf green tended by Igorot caddies, flower and vegetable gardens of the colony's new headmen. In this place, a City Beautiful nestled in the wilds of the sublime Cordillera, colonialists could follow Tityrus of Virgil's classic pastoral and "enjoy the best of both worlds—the sophisticated order of art and the simple spontaneity of nature."[60] Baguio would be a place of "peace, leisure, and economic sufficiency," as Leo Marx defined the classical ideal in *The Machine in the Garden* (1964).[61]

Indeed, for many visitors, the town was no mere "government reservation." It was a great American garden miraculously blossoming in the Orient. "I find it hard to describe this mountain capital," wrote Frank Carpenter, a well-traveled American journalist. "The whole scene has the effect of a great landscape garden designed by Jehovah and developed by man," he wrote in 1925.[62] Another Baguio booster wrote of the "regeneration" to be had by overworked colonial clerks in the land of the pines: "Go, and help your clerks go. It will make better men of you and them. In the shadow of those giant mountains and 'Neath the wide canopy of a heaven which seems measurably nearer one drops his mean and petty thoughts and his heart throbs close to Nature's own. It is a regeneration."[63] Building the hill station was an attempt at escape from the heat, from exhaustion and overwork, and arguably from the realities of the US occupation, or the "Philippine tangle," as William James so aptly put it.[64] It was a retreat into nature or, rather, nature reengineered as a "great landscape garden." This was Baguio as imperial pastoral.

I refer to the pastoral not as the genre of literature noted for "its conventionalized imagery of happy, sportive shepherds, shepherdesses, and flocks."[65] Rather, I invoke a sensibility or a "metaphor" for a simple and romantic vision of "country life."[66] In our case, that country was a valley in the Cordillera. In *The Country and the City* (1973), Raymond Williams argues that "the contrast of country and city is one of the major forms in which we become conscious of a central part of our experience and of the crises of our society." He identifies as the root of these crises the "minority ownership" of the means of production, whether organized by feudalism or capitalism. Preceding this conclusion is an exploration of the various incarnations of idealizations of the country from Marvell's poetry to Henry James's fiction. Williams's objective is not to define or fix the pastoral form but to observe its transformations. He asks not what the pastoral *is* but rather what it *does*—its cultural labor—at particular moments in time. Some versions idealized romantic love; others, rural virtue or economy; still others, the country's "way of life as a whole" and "the metaphorical but also the actual retreat" into it. Varied as they may be, these transformations did move in an overarching "direction," one "in the interest of a new kind of society: that of a developing agrarian capitalism." Pastorals especially since the Renaissance, Williams observes, demonstrate a break from the classical pastoral of Virgil, which is animated by a "contrast" "between the pleasures of rural settlement and the threat of loss or eviction." In the classical pastoral, the return of soldiers from imperial war threatens farmers with the loss of their lands; these lands are to be confiscated and rewarded to Octavius's soldiers on their return from war. During the Renaissance, this conflict at the heart of the classical pastoral becomes "excised." "[S]elected images stand as themselves: not in a living but in an enameled world," a change that renders the form "highly artificial and abstracted." This is the pastoral of landscape painting and the country-house novel.[67]

The idealizations communicated through these later pastoral forms tend to present a contrast between the city and a "happier past" or more "innocent" place signified by the country. This contrast presents a critique of a present time and place but one that displays a "retrospective radicalism." For, as Williams argues, the contrast is ultimately a fictional one. Not only does it obscure the relations of power and dominance that prevailed in that supposed "happier" feudal or post-feudal past, but it also cloaks the relations that link country and city, the connections "between the processes of rural exploitation" and "the register of that exploitation, in the law courts, the money markets, the political power and the conspicuous expenditure of the city."[68]

Williams considers primarily English literature and its context, but the framework he presents resonates with Leo Marx's analysis of nineteenth-century American literature in *The Machine in the Garden* (1964). And rightly so. The writings of Hawthorne, Emerson, and Thoreau emerged in the context of the "take-off" of industrial capitalism and its attendant social and cultural dislocations, and Marx identifies a common pastoral theme in their work: "a variation upon the contrast between two worlds, one identified with rural peace and simplicity, the other with urban power and sophistication." These American writers, Marx argues, employed the pastoral to make sense of and cope with their tumultuous circumstances. He suggests more broadly that "[T]he yearning for a simpler, more harmonious style of life, an existence 'closer to nature,'" which some of these works exhibit, "is the psychic root of all pastoralism."[69]

To the extent that the pastoral has been tapped to reckon with perceived social, economic, and political conflicts around land and labor, it is conceivable that one could find distinct but still similar invocations of "country life" outside the Anglo world. Chinese rulers since the beginnings of the Chinese state enjoyed retreats beyond royal palaces where appreciation of nonhuman nature was the primary occupation. Some of these royal demesnes became "country houses called the 'suburban pleasure palaces'" and would inspire members of the Chinese aristocracy to build their own private gardens.[70] Back in the West, Meredith Martin explores a pastoral formation in French history: the *ancien regime*'s pleasure dairies from the time of Catherine de Medici to Marie Antoinette. These, "[l]ike other forms of early modern European pastoral art and literature," "offered an idealized representation of rural life that embodied the desires, alleviated the anxieties, and certified the authority of the ruling class in a time of great social and political upheaval," she writes. "Pleasure dairies enabled the crown and the nobility to project an image of Arcadian peace and prosperity," Martin argues, "and to profess an enduring devotion to the land, while also playing with new forms of courtly refinement, leisure, and display—thus merging old and new forms of the aristocratic self."[71] Chandra Mukerji has written more narrowly of Louis XIV's construction of the gardens of Versailles and the movement of court life there. She interprets the construction of these gardens in the context of the Sun King's efforts at state building. He sought, she argues, "an uncontested and identifiable political *location* for accumulating and exercising power: a territorial state." Versailles became that site; for visitors, its gardens stood as exhibits of "French capacities to use the countryside as a political resource for power." These gardens also "helped to naturalize the imposition of political order on the countryside by making it seem continuous with nature's own forms and forces."[72]

Enlisting nature to obscure exercises of power is no stranger to our own times. Louise Mozingo has written of a "pastoral capitalism" expressed by American corporations' suburban campuses of the mid-twentieth century. She writes that "[l]ike many suburban dwellers before them (and many executives were themselves suburban home owners), the corporate managers were propelled to the pastoral urban periphery by a distaste for the sensory and social realities of industrial production, the noise and congestion of dense urban cores, and class and ethnic conflicts."[73] The suburb as pastoral, incidentally, was not far from one colonial administrator's image of Baguio's development; he promised comers that "one needed only to construct driveways and go to work with a lawn-mower."[74]

These various manifestations of the pastoral structure evidence Raymond Williams's argument that the form presents itself as a "deep desire for stability" that has "served to cover and to evade the actual and bitter contradictions of the time."[75] For this reason, it is not surprising that the pastoral should be deployed and retooled in the imperial context. Bringing to bear the world-systems theory of capitalist development, Williams argues, in fact, that "one of the last models of 'city and country' is the system we now know as imperialism."[76] In structural terms, the colony does share with the country a shift toward "minority ownership" of land—a history of enclosure and dispossession.[77] This is the context of anthropologist Rosalind Morris's invocation of the "imperial pastoral" in an article that turns on the work of a British teacher stationed in British Malaya. While interned in Singapore in 1942, this man began a translation of Virgil's *Georgics*, an endeavor consistent with many pastorals: they are often "written from the perspective of exile—including the exile of the imperial representative in colonized lands." From the life and writings of this ordinary colonialist, Morris considers the "tendency to anachronism," which, she argues, ran through the British regime in Malaya. "This imperial pastoralism," she writes, "with its idylls of rusticity and patriarchal authority, often in the form of monarchism, is invariably opposed to revolution and may be thought of, precisely, as the (fetishistic) disavowal of revolution—in Europe and elsewhere." This version of the pastoral "rested on the conflation of the ideas of native and nature, and it worked to hierarchize the populations of colonized territories in a falsely temporalized sense, with rurality and peasantry assuming the place of the primordial." The pastoral ideal "constru[ed] nativeness as a primordial connection to the land, but one that could be distinguished from both ownership and title." The ideal thus rationalized the development of the colonial rubber industry, which entailed the dispossession of lands from Malays.[78]

Morris emphasizes the discursive and temporal dimensions of the pastoral form. Through Baguio, this study attends to the discursive and also material construction of the pastoral in the creation of the built environment and in the process of colonial place making—in assigning meaning to that built environment.[79] The forces animating our subjects are similar. Just as the British colonialist turned to the *Georgics* in the context of the dispossession of lands by an imperial power, and just as the pastoral poets and painters made their art for a gentry appropriating the countryside from peasants, so Americans built their green fields for rest and play on Igorots' land, remaking the political economy and lifeways of the Ibaloi in the process. On Ibaloi lands, colonialists built an oasis of America that reminded visitors of what the Philippines lacked and what could be had in the United States and through its aegis—good sanitation, smooth roads, sound nourishment, refreshing air, leisure, Modernity, Civilization, and Beauty—some of the preoccupations of the reform-minded professional-managerial class known as Progressives.[80] This enclave of America was defined by its difference: it was unrepresentative of the archipelago in its natural and built environments. In this sense, both American colonial labor and cultural differences instrumental to ruling were made and performed here. Far from being a retreat from politics, Baguio, though small in size, became a setting for colonial business where commissioners entertained Filipino elites and Igorots, too, making them guests on their own lands. If this setting was a reminder of home for weary and homesick colonialists, it offered Philippine visitors a form of colonial education much as formal political tutoring, schooling in English, and an industrial education were doing elsewhere across the archipelago. Like Louis XIV seeking a "location of power" to legitimize his state, the colonial government sought a site to remake into its domain, an oasis of American colonial power, something arguably far easier to do outside the Manila capital.

Yet even as the hill station stood as a sure sign of imperial power, Americans naturalized the imposition. In Baguio, the US occupation became associated with nature itself—as a greensward at the center of the town site, as landscape views, as the golf green peacefully tended by Igorots, as fresh strawberries and milk on the table. The United States undertook not only a small presence limited to the government reservation; it was in harmony with nature. This sanguine view, like the later pastorals described by Williams, "served to cover . . . contradictions of the time," namely the dispossession and enclosure of Philippine lands. Just as the sharp distinction between country and city proves to be a fiction, an ideological construction, so the notion of Baguio as an innocent retreat—"a landscape garden designed by Jehovah"—proved the same. This imperial pastoral smoothed over the contradiction of the imperial republic, a democracy crushing the first nationalist revolution in Asia.

## The Terrain Ahead

Given the impossibility of seeing the whole of Baguio—or any place—from on the ground, the chapters that follow offer a tour of sites and episodes in its transformation from Ibaloi pasture to American pastoral and in the hill station's design, construction, and use during early US colonialism. These chapters are neither perfectly chronological nor are they exhaustive of Baguio, which was comprised of many more spaces than those covered here. Rather, these chapters offer a set of overlapping vantages on the hill station and views of several of its most talked about and photographed sites—pasture, road, town center, local marketplace, country club, and an American's private home. Compared to the military reservation, the Philippine Constabulary training grounds, or the Baguio retreat for the colony's schoolteachers, which came to occupy significant stretches of the hill station, these places may seem quotidian or even superfluous to US colonialism. But to my mind, they are especially revealing zones for exploring the interplay of coercive and non-coercive methods of rule, forms of power that were formal and informal (in their broadest senses), the reach of US power and its limits, and, in all these ways, for pursuing the formation of the elusive, seemingly anticolonial empire.

We begin in the Cordillera Central mountains, which had long served Igorots as a "region of refuge."[81] There they enjoyed a measure of autonomy thanks to the topographical challenges presented to invaders by the Cordillera and due to the gold that Igorots managed to harvest from its rocks and streams and trade with lowlanders. Over time, the Ibaloi's mining and commerce allowed them to gain a significant wealth in cattle, and by the 1700s, this livestock, gold, and Spanish attempts to secure the latter, had shaped Ibaloi social relations and even the landscape, creating the grassy meadow that Americans admired.[82] The first chapter discusses how US colonialists would claim the pasture and a mountain refuge for themselves and follow European and British imperialists in building a hill station, a formation that emerged in the context of territorial conquest and attempts to consolidate colonial authority. Inasmuch as Baguio's transformation into a "government reservation" through enclosure demonstrated the power of an occupying force, the quest for the hill station also signaled the regime's vulnerability.

Once in the mountains, we make our way to Benguet Road, the American colonial road built to reach the hill station. It was one of the first major public works projects that the colonial government undertook, and it became the most celebrated road in the Philippines. Even today, a Tagalog alphabet book features the "zig-zag" road to represent that final letter of the alphabet.[83] For Americans, this twenty-six-mile highway was an opportunity to celebrate the marvels of US engineering and managerial power; for many Filipinos, the road was an extravagant waste—the colonial government spent millions on it, yet every time a strong storm hit, much of the highway was washed away. This second chapter explores how Americans recruited Philippine labor for the project in the face of disinterest if not outright resistance by Filipinos. Despite promises to "liberate labor" and teach colonial subjects "free agency," in recruiting workers Americans largely relied on the effects of counterrevolutionary war and, later, laws that circumscribed Philippine peoples' subsistence practices and movement. These forms of "primitive accumulation" freed labor in ways far removed from many Filipino workers' own vision of free agency and were a response to the contest Filipinos posed to US rule.[84]

Our third stop is the "government reservation," the Baguio town site. Much of this land had belonged to Mateo Cariño, an Ibaloi headman whose family had long occupied and used considerable areas of Baguio as pasture for grazing cattle. Cariño had granted Americans a portion of his land, but the US government claimed additional acres of Cariño's, leading to his protest. His case ultimately went to the US Supreme Court, where he won in a judgment that amounted to recognition of native title and articulated a distinction between the US occupation of the Philippines and the history of colonialism across the American West. But the Ibaloi headman had won the battle and lost the war. By the time of the Supreme Court decision, Daniel Burnham had come and gone and construction of the hill station had begun. The city planner, charged with creating an urban design for Baguio and Manila, brought the City Beautiful to the colony. This was a style and practice associated with Progressives, who aimed to temper the wild growth of the late-nineteenth-century city with planning and ameliorate social conflict through carefully designed public spaces, like parks. These shared spaces, City Beautiful advocates held, might help bring commercial imperatives and civic commitments into better balance and cultivate democratic culture. In Baguio, Burnham designed such a middle ground landscape, harmonizing nature and commerce and culture, for the enjoyment of American tutors in Filipinos' self-government. This pastoral was built on the dispossession of Ibaloi land, pastureland that had shaped the social and ritual life of women and men like the Cariños. Together, the Supreme Court's decision in the Cariño case and Burnham's designs manifested the ambitions and the contradictions of the imperial republic.

The most famous or infamous spot in the hill station was arguably its marketplace, the concern of chapter 4. Much as visitors to the Chicago World's Fair seemed compelled more by the amusements and commercial attractions of the Midway than the gentility of the White City, American visitors to Baguio were most taken with its marketplace. Its hustle and bustle were objects of consumption in photographs, postcards, and travel accounts. Some Americans came to see the infamous trade in dogs, sold not as pets but as consumable articles in Igorot ritual. Burnham had positioned the marketplace at one end of the central axis of the town, yet were one to follow the logic of US imperial ideology, the marketplace stood at the very center of the colony. To some administrators, the marketplace was a site of liberation: it was the place where Philippine peoples could enjoy self-rule as they traded the fruits of their labor. Here, they could become the independent agents who might then constitute citizens of a nascent Philippine republic. As this chapter shows, even as they celebrated the market as an incubator of individual liberty and nationhood, American imperialists judged Philippine people in part by what they traded and what they consumed. They used representations of Igorots' trade in dogs and Filipinos' supposed trade in persons to entrench perceived differences among Philippine peoples, and they made these into new justifications for continued imperial rule—for anything but liberty.

Our final stops, the Baguio Country Club and Topside, the private home of Baguio's biggest booster, William Cameron Forbes, take us beyond the center of town and into the social apex and inner sancta of Baguio. An heir to the Forbes family fortune, one amassed through trade with China, and a grandson of transcendentalist Ralph Waldo Emerson, Forbes made Baguio a famed American retreat. Inspired by the country club founded by a Forbes relative in Massachusetts—reputedly the first country club in the United States—Forbes brought the institution to the Philippine mountains. And Forbes also orchestrated the construction of Topside, which he hoped would serve as a "show-place" for all that was possible in Baguio and in the colony at large. Forbes invited friends and political opponents, American and Filipino, from former revolutionary general Emilio Aguinaldo to future Philippine president Manuel Quezon, to partake of the green grounds and gardens of Topside and the country club in the hopes of winning their approval for Baguio and US policy in the colony—to forge a Filipino "ruling class,"[85] partners in the process of the Philippines' benevolent assimilation and in building "the political conditions for capital accumulation."[86] In Baguio, Forbes became the new chief, and Cariño and Filipinos guests on their own lands.

## A Counter-Pastoral

As this sketch of the work ahead suggests, I treat the built environment as an expression of historical patterns of production and social and cultural reproduction, presenting Baguio as a "palimpsest of landscapes."[87] I show how rule and subjects of rule were generated in the face of Philippine resistance, the contradictions of imperial ideology, and in place.[88] By the 1920s, the end of our journey through Baguio, a class of Progressives had cut their teeth as colonial administrators. And the hill station had become a "show-place" of what the United States offered and performance space for setting in motion new relations of obligation and indebtedness. It helped to consolidate a Filipino elite and create a reserve of Filipino labor, who worked projects like the road to Baguio and who, like Carlos Bulosan, would travel across the Pacific for work in Hawaiian sugar plantations, Pacific northwest canneries, and California farms. And in part through Baguio, Americans made Igorot peoples instrumental to US rule. This story offers an account of one unique colonial place even as it telescopes the story of US imperialism in the Philippines—an account of dispossessions in Americans' quest for "free" trade.

Much literature since the cultural turn in US imperial studies has emphasized the gendered and racial dimensions of US imperial power—the process of creating colonial subjects by making and remaking gender and racial differences.[89] Recent imperial studies that focus on institutional and structural concerns, meanwhile, often take the imperial or colonial state as the primary object of inquiry, largely bracketing economic questions and capitalism.[90] The latter were problems that preoccupied New Left historians and revisionists like LaFeber, who focused on imperial ideology and political economy and pursued "the underlying causality of imperial expansion, the old Marxist question." By tabling that issue, the authors of a recent volume aim to "free [them]selves to examine the nature of the state that Washington developed, at the dawn of the twentieth century, to rule an empire of islands that reached halfway around the globe."[91]

A history of this government hill station and its built environment, by contrast, calls for attention to the state and also to capital and their interchange, and to the commerce between cultural and material aspects of rule. Inspired by work like Fernando Coronil's in developing a "cultural materialist" method "capable of apprehending the production of meaning and the reproduction of life as distinct moments of a unitary process," my hope is to connect the histories of revisionists and those since the cultural turn in imperial studies, so as to mark the discursive aspects and material conditions of colonial subject and capital formation.[92] A focus on Baguio, a compound of memories and symbols, and of stone and wood, grass and soil, seems to call for such a method, just as it provokes consideration of both imperial designs and colonial practice. Through an excavation of the colonial built environment we can see how Philippine peoples and specific conditions remade and challenged lofty imperial designs. Indeed, attention to the built environment demands attention to structure and history—the ideologies and architecture of US power and their contradictions, faults, and renovations once set on the ground and before their intended subjects. The built environment—colonial structures—also speak, if quietly, to the agency of non-elite Philippine peoples like many of the Ibaloi who dwelled in Baguio and the Filipinos who worked along Benguet Road and left behind few written records despite inspiring and pushing Americans into retreat.

Anthropologist Keith Basso writes that in contrast to the abstraction of space, places are forged through dwelling and engage "the forms of consciousness with which individuals perceive and apprehend geographical space."[93] "Place-making," he writes, "involves multiple acts of remembering and imagining," and "place-making is a way of constructing the past, a venerable means of doing human history."[94] I see this story as a history of imperial place making and itself a form of place making. It tells a history of capital—the making of the colonial state's summer capital, to be sure, and capital—a form of power sought after, appropriated, and used by colonialists—more broadly. Jeffrey Sklansky has recently described capital as "an elusive sovereign." "As well as a way of being, of seeing, and of believing," he writes, "capitalism is a way of ruling, of establishing and exercising social power."[95] This history attempts to locate the elusive and show its formations in the colonial context. In so doing, it pits a history of conflict, collaboration, and dispossession against the classic narrative trope of progress through ascent.[96]

ONE

# A Cure for Philippinitis

There's a malady terrific and it's very very sad
For you can't think of anything.
They call it Philippinitis and you have it very bad
When you can't think of anything.
You start to write a letter and you try your best to think,
You sit for half an hour and then overturn your ink,
Then drop your pen and paper and go out and take a drink
For you can't think of anything, can you?

Chorus:
It's so easy to forget a little thing like a thought,
When your mind is topsy-turvy and your memory is short.
I'd be a "savvy" hombre and I'd know a great lot
If I only could remember what I've quite forgot.[1]

"Philippinitis" inspired this American soldiers' song, a tune probably written around 1901 or 1902. Barmy as they may seem, the lyrics help to explain the appeal to American colonialists of a mountaintop retreat in the Philippines' Cordillera mountains. In this verse, an ailing soldier makes light of his forgetfulness and mental lassitude—the feeling that his mind is slipping away. These were the sure symptoms of a bout of Philippinitis. A 1908 Manila travel guide offered this description of the condition: "a word coined for the purpose of expressing facetiously a state of mental and physical torpor, with lack of interest in one's surroundings, ambition wanting, a general disinclination to mental or physical exertion, forget-fulness, and irritability." Besides taking precautions to avoid cholera and malaria, typhoid fever and dysentery, Americans who found themselves in the Philippines were to guard against overeating and remaining too sedentary. These were the vices that made one susceptible to that terrific malady.[2]

Philippinitis had equivalents across colonial South Asia, among them "Burmah head" and "Punjab head."[3] More generically, the ailment was known as "tropical neurasthenia," a south-of-the-equator version of the distressed condition that William James had popularized as "Americanitis" perhaps also somewhat facetiously.[4] American medical officer Dr. Louis Fales wrote on the nonfatal but troubling ailment in 1907. 50 percent of American women and 30 percent of men who visited the Philippines, he reported, "were struck by neurasthenia to such an extent that they are in a state of semi-invalidism." Women afflicted by tropical neurasthenia, Fales wrote, "become nervous, irritable, anemic, lose weight, suffer with neuralgia, spells of faintness, sleep poorly" and suffer from menstrual problems. Children were expected to fair poorly, too, with consequences for civilization itself. Dr. Fales cautioned that the offspring "of neurasthenic parents . . . will inherit an organism lacking in nerve force; being forced to live in an enervating climate, their small reserve will be still farther drawn upon, and in a generation or two there will result a race with little resemblance to the mother stock, small, puny, weak-minded in fact a degenerate race which would soon cease to exist if new stock did not continually come from the home land."[5] Such prospects may have thrown into doubt whether the acquisition of the Philippines and Americans' service there were truly to elect what Theodore Roosevelt had termed the "strenuous life."

Indeed, the specter of Philippinitis boded ill for a lengthy American presence on the archipelago and amplifies the disquiet of the soldiers' song. Statistics on the causes of death at the start of the US occupation were no more encouraging. At the height of armed conflict—between July 31, 1898 and 1900—600 Americans were killed or died from battle wounds, while 700 had succumbed to disease. The story for Filipinos was far bleaker. The Philippine-American War brought about the death of more than 700,000 between 1898 and 1902—some from combat and others from outbreaks "of cholera, typhoid, smallpox, tuberculosis, beriberi, and plague." The annual death rate on the archipelago had doubled in these years from thirty to sixty per thousand. Over the course of the US occupation, Americans would come to view Philippine peoples and unsanitary practices that made them vectors of those dreaded diseases as more threatening than the climate. But early on, tropical conditions were themselves most worrisome. Altogether, the threats posed by the climate, disease, mental distress, and combat in the tropics supplied the Philippine

Commission with motive to begin investigating possibilities for a health resort in the mountains of northern Luzon as early as 1899.[6]

Commissioners chose to explore the province of Benguet. Noted for its cool temperatures and its abundant pine trees, Benguet sits in the southern part of Luzon's Cordillera Central, a mountain range with peaks reaching as high as 8,000 feet.[7] Americans hailed the Cordillera as the Philippines' Adirondacks or likened them to the mountains of the US West. To the Spanish, the region and its climate recalled "the Peninsula."[8] For the Igorots native to Benguet, the landscape, largely open grassland, keyed up a distinct set of meanings. It was one they had forged as they reproduced their lives and responded to Spanish incursions over centuries.

The landscape gave physical shape to the Ibaloi's political economy and social organization, both fashioned in the context of Spanish colonialism. By the late nineteenth century, Benguet was composed of a couple of "*pueblos*," indicating Christian residents, and a larger number of Ibaloi "*rancherías*," groupings of six to twenty "scattered huts."[9] This social geography reflected the fact that through Spanish rule, Igorots across the Cordillera had made the work of *reducción*—the consolidation of Philippine peoples into tractable towns of 2,400 to 5,000, which the Spanish undertook in the lowlands—exceedingly difficult.[10] As we will see, some Igorots met attempts at colonial subjection with a willingness to negotiate, but others responded with warfare or with flight into "higher and lonelier mountains, swelling the hordes of head-hunters, into whose savage life they fall back," wrote one observer.[11] To the Spanish, for whom urbanism was a sign of civilization, this only heightened Igorots' backwardness. Igorots' dispersal and dwelling across the peaks and plateaus of the Cordillera signaled their "barbarism," as did their refusal of baptism and unwillingness to pay tribute.[12]

The highlands were not only home to those who had long resisted Spanish subjection; they also proved seductive to lowlanders seeking escape from Spanish authority. Lowlanders who set off for retreat in the Cordillera were known as *remontados* (those who return to the mountains, or runaways), and some anthropologists argue that at least some Cordillerans are the progeny of *remontados*.[13] "Igorot," in fact, first appears in Spanish accounts as "Ygolote," a term that referred to peoples on the mountainous edges of the Ilocos in the west and in Pangasinan, the lowland province just south of Benguet.[14] Some believe that the Igorots of Benguet, the Ibaloi, may be descendants of Pangasinan people who ascended the mountains to work gold mines even before the arrival of the Spanish; their customs changed as they adapted to a new environment.[15] Nabaloi, the language of the Ibaloi, reflects these migrations in its incorporation of elements of Pangasinan and Ilocano.[16]

When the Spanish formulated their understandings of these Philippine mountain peoples, they could look to precedent from the Peninsula. Indeed, conquistadors and administrators may have been reminded of "the mountains south of Granada," a region feared "a preserve of unconverted and politically dangerous Moors." For almost a century before they began expelling the Moors from the Peninsula, the Spanish worried "that the mountains harbored a permanent 'fifth column' of unbelievers."[17] Similar concerns plagued Spaniards and Americans in the Philippines. According to reports of the Philippine Commission, in the tumult of the early 1900s, "ignorant and superstitious people" chose to "withdraw to the mountains, under the leadership of leaders who profess to have divine attributes and to have the assistance of God in the protection of their followers." American commissioners recognized that these runaways followed the paths of insurgents during the Spanish period who had "had no refuge but the mountains, and being in the mountains conducted a free robber life, and about them gathered legions not unlike those of the Robin Hood days of England."[18]

Igorots and these *remontados*, then, were largely successful evaders of a tribute- and labor-seeking colonial state. "Far from being 'left behind' by the progress of civilization in the valleys," James Scott writes of hill peoples of Southeast Asia, "they have, over long periods of time, chosen to place themselves out of the reach of the state"—the colonial state. And this choice was largely enabled by the mountains; Scott describes how political authority "sweeps readily across a flat terrain," while "abrupt changes in altitude, ruggedness of terrain, and the political obstacle of population dispersion and mixed cultivation" tend to halt its progress.[19] In the province of Benguet, it was not only geography that presented challenges to imperial sovereignty; it was also the materials, like gold, that the indigenous people harvested from the earth that gained them a fragile and circumscribed measure of independence.

In this southernmost province of northern Luzon, gold would produce a wealthy elite whose degree of power over fellow Ibaloi was unique among Cordilleran ethnolinguistic groups. By the time of the US occupation, this gold had allowed the elite to amass herds of cattle through trading along the coast and in the lowlands. The extensive pastureland across Benguet hinted at the central place of cattle in Ibaloi social organization and economic and ritual life. Attempts to secure this wealth and the social status attached to it in the face of Spanish incursions would shape another Ibaloi response to the invaders: not flight and abandonment of lands but negotiation. These accommodations by the elite would create the circumstances under which Benguet became recognized as the first *comandancia militar*, or military district, in the mid-nineteenth century.[20] Some elites, it seems, traded degrees of sovereignty vis-à-vis the Spanish for economic and social power within their communities.

In setting their hill station on Ibaloi lands, Americans sought to capture some of the advantages offered by the Cordillera—its remote location and the amenability of at least some of its native dwellers. They followed the Spanish, who had first envisioned a mountain hill station in the later 1870s.[21] Not unlike the Spanish, who by the nineteenth century sought to consolidate colonial state power, Americans were undertaking colonial conquest and, in that context, sought a break from the heat of the tropics and the preservation and reproduction of their own culture and labor force. Americans were also seeking reprieve from the exigencies of colonial state and nation building and from Filipinos who themselves had beaten trails in the mountains to elude US forces. This insurgency is something that the tune "Philippinitis" evokes in later verses. The soldier sings of the "civil governor" and General Adna Chaffee, a veteran of the Indian Wars, the US occupation of Cuba, and a US relief expedition to China during the Boxer Rebellion, all before landing in Manila where he served as major general of the regular army. "They [the civil governor and Chaffee] do what they can," but as the song goes, "The Philippines are pacified, not tranquilized, said he; / He overlooked a little thing as anyone mag see— / He forgot the Filipinos, and it's plain as plain can be / That he can't think of everything—can he?"[22] In building their hill station in Benguet far from the Manila capital, Americans would seek an antidote to "Philippinitis," that hazard of colonizing the archipelago. On Baguio pasture they would advance Spanish work of turning what had for centuries been a space of limited colonial state authority into a domain of colonial power. Americans would claim Baguio for a "government reservation"—a distinctive zone of US authority within the nation-state Americans were building. In the process, they would reorient the landscape again, ultimately turning Ibaloi pasture into American pastoral and making ascent into the mountains appear not a sign of barbarism but a privilege and perk of the civilized.[23]

## "A Sort of Thermopylae"

In early 1900 Robert Rudd mustered out of the Presidio in San Francisco, a one-time Spanish military base, and set out for what had until recently been another important nexus of Spanish imperial power, Manila. Rudd was a member of the Ninth Battalion of Ohio Infantry Volunteers and the Forty-Eighth Regiment, an organization of African American men, some of whom had already served in Cuba.[24] Volunteer troops came to compose a sizeable portion of servicemen in the Philippines as Congress had capped the regular army to 28,000 soldiers.[25] Rudd was among the 125,000 volunteers that President McKinley had called up in April 1898.[26] When he arrived in the Philippines, the American flag was already flying over the capital of Manila, and McKinley had issued his "Benevolent Assimilation" proclamation, defining daunting objectives for the US military occupation: "to win the confidence, respect, and affection of the inhabitants of the Philippines by assuring them in every possible way that full measure of individual rights and liberties which is the heritage of free peoples, and by proving to them that the mission of the United States is one of substituting the mild sway of justice and right for arbitrary rule."[27] McKinley instructed American forces to execute the United States' commitment to individual rights and the right of property, to preside over the opening of ports and the reestablishment of trade, and to perform all this while exercising absolute authority over the archipelago and its peoples.[28] Many Filipinos met this seemingly contradictory mission—political liberation under military rule—with protest and, eventually, a protracted, second war for independence.

Rudd, together with other members of the Forty-Eighth Regiment, was sent to help squash this rebellion through service in territory designated the First District, an 8,000-square-mile expanse of northern Luzon that included the interior mountain provinces of Benguet, Bontoc, and Lepanto and also the western, coastal provinces of La Union, Ilocos Norte, Ilocos Sur, and Abra.[29] Like many African American troops, he was not assigned to combat so much as to scouting and guarding duties,[30]and in that capacity, he operated out of La Trinidad, the one-time Spanish capital of Benguet and the home of a Spanish agricultural experimentation station.[31] La Trinidad stood as the only American garrison in the Cordillera mountains; the other upland bases occupied positions on the western coast.[32]

At first glance this post in northern Luzon may have seemed a backwater of the war. Commanders initially assumed that the First District would be easily pacified given what they understood as a long-standing antagonism between the Ilocanos and the Tagalogs, the ethnolinguistic group from the central plains of Luzon that had produced many leaders of the Filipino revolution. Experience proved otherwise; the district was a hot bed of insurgency, and by the time he was promoted to major general of US forces in 1900, Arthur MacArthur had come to regard the district as the worst in Luzon.[33] Owing to the insurrectionary activity of leaders like Manuel Tinio and Juan and Blas Villamor, the district's American commander found that his "patrols were ambushed, his supply lines raided, his communications destroyed, and his contacts among the population kidnapped and killed."[34] In fact, not long after his arrival in northern Luzon, in May 1900 Rudd received a cable from his commander, Colonel of the Forty-Eighth William Penn Duvall, emboldening Rudd with the promise that there in Benguet he might well seize the opportunity for "'landing' the biggest fish of all'"—revolutionary General Emilio Aguinaldo.[35] Aguinaldo would eventually surrender to US troops in March 1901 in the province of Isabela, northeast of Benguet. By this point, Americans had indeed realized Spanish fears of the mountains as a home to rebels. During the fall of 1899, Aguinaldo had led units of his Army of Liberation across Pangasinan and seemed to be moving further north through Ilocos and into the tobacco-rich Cagayan Valley.[36] A rumor circulated that Aguinaldo had taken shelter in a Benguet residence disguised as an Igorot servant woman.[37] The costume cloaked a general with supposed plans "to make of the Benguet mountains a sort of Thermopylae, where a final stand would be made."[38]

Perhaps in a sign of Americans' utmost confidence about victory over Aguinaldo and his men, the First District simultaneously figured into US plans beyond counterinsurgency and military-led pacification. Even before Rudd mustered out of northern Luzon, Americans had caught wind of Benguet's suitability for a health resort. In July 1899, the Philippine Commission interviewed residents of the Philippines who had experienced the Cordillera, querying them about the province and about Baguio in particular. Horace Higgins, an engineer and Manila Railroad builder, who had lived in the capital since 1887, likened the temperatures of Benguet to those in "the south of Europe." Members of the commission would include him in an exploratory trip to Benguet the following year.[39] Commissioners also interviewed a tea and coffee planter named F. H. Donaldson-Sim, who had resided in the province for three years before revolutionary general Antonio Luna forced his evacuation. Donaldson-Sim enticed Commissioners with descriptions of Benguet's balmy temperatures, its "pure and fresh" air and "splendid" waters, and the Igorots' impressive stock of cattle—"as large as the cattle in Europe."[40] He testified that the Igorots of Benguet were "half-civilized and . . . awfully quiet, very peaceful"; they disliked the Spanish, and he believed they would be "favorably disposed toward" Americans.[41] These Igorots were likely the Ibaloi, who numbered between 12,000 and 15,000 in 1905 and lived over the greater portion of the province; the Kankanaey dwelled on the remaining land area of Benguet.[42] In his interview with commissioners, Jose Camps singled out the Ibaloi as "the most advanced" and "the rightest" mountaineers of the Cordillera.[43] His assessment had been based on a nearly three-year-long sojourn in La Trinidad where he had developed a hotel patronized by the Spanish and by foreigners and merchants from Manila and Iloilo.[44] He confirmed what Donaldson-Sim had observed—that besides the Spanish, "rich natives" had begun making pilgrimages to Benguet seeking cures for "dysentery, diarrhea, and fever."[45] Camps also reported on commerce in La Trinidad, one mainly in gold and coffee and in cattle and carabaos, and "a great business" was conducted with Igorots, he said, who had a taste and stomach for "strong drinks."[46]

Its interest in Benguet sufficiently piqued, the following year, the Philippine Commission sent a group including Higgins to explore the prospects for a "health resort" in the Cordillera.[47] An army medical inspector, the United States' former chief health officer in Manila, and a Filipino weather observer accompanied Higgins, Dean C. Worcester, and Luke Wright, two commissioners, to assess the province's offerings for themselves. Worcester, though a zoologist by training, would become the self-appointed advocate of "Non-Christian Tribes" in the Philippines as the controversial secretary of the Interior, the department housing the colony's "Bureau of Non-Christian Tribes." Luke Wright would serve as governor general of the colony between 1904 and 1906. They approached Benguet from the coast on a twenty-four-hour trip by gunboat, then mounted horses and set off over land with a couple dozen troops from the Third Cavalry. Along the way, the group came upon the remains of insurgents' matériel—a canon foundry, a furnace, and a collection of church bells that were to be repurposed into guns.[48] Aguinaldo reportedly had had the bells torn from churches on the coast and hauled to the site by Igorots.[49] By the eleventh hour of their expedition over land, the group reached La Trinidad, where members of Rudd's Forty-Eighth greeted them. Worcester learned that only three troops in La Trinidad were then suffering from illness. To him, the good health of most soldiers gave evidence in favor of Benguet as an American resort; "colored troops do not always take the best care of themselves," he reasoned.[50]

His group's visit to La Trinidad was short, and the next morning, they decamped for Baguio to meet Otto Scheerer, a German expatriate and student of Igorot culture and language. US military authorities like Duvall were wary of Scheerer. They believed he fraternized with revolutionaries, and Duvall advised Rudd to "watch [him] like a hawk."[51] But to civil au-

1.1 “Panoramic view of Baguio: Benguet—1900; 1823; 1900.” Dean C. Worcester Photographs of the Philippine Islands, University of Michigan (Special Collections Library).

thorities, Scheerer was a friend, and American civilian personnel looked to him to understand Benguet. He must have been an effective tour guide, as Worcester and Wright's visit proved a tremendous success. By the end of their stay, they had concluded that Baguio might serve not only as a "health resort" but even as the political and military capital of the colonial Philippines, ostensibly supplanting Manila altogether.[52] Benguet's natives appeared just as the commissioners had heard, "illiterate pagans" but "pacific, industrious, and relatively honest and truthful people who have never taken any part in the insurrection." They had supported the Americans as informants and porters and in that way "rendered our forces invaluable service."[53] The fact that General Luna had made plans to seek refuge in the region was seen as a sign of the security of the province; once revolutionaries were flushed out of the mountains, "any serious attack from insurgents or ladrones," the commissioners believed, "is so remote as to be hardly worth taking into consideration." That Baguio's climate brought to mind "northern New England in the late spring or early summer" recommended it even further, and Worcester and Wright saw the province as the perfect spot for a reserve camp of soldiers and for "the acclimatization of newly arrived men." What is more, the natural beauty of Baguio stunned them: the commissioners described the Baguio plateau that raised the town high into the sky, setting it some 4,700 feet above sea level. They remarked on the pines that had also charmed the Spanish and the lush tree ferns and fresh spring waters. In Scheerer's garden, commissioners discovered "coffee, oranges, lemons, cacao, and bananas" along with potatoes, peas, and raspberries, offerings that could surpass New England's in the early summer.[54]

Baguio's promise as pasture did not go unobserved either. Scheerer had noted the open fields of Benguet in describing the province as it existed around 1893. The future center of the American hill station appeared then as "a lonesome *Ranchería* showing hardly a score of dwellings scattered here and there over the surrounding hills and sentineled, in a central location, by a desolate-looking tribunal or council-house."[55] The pastureland that greeted Americans had also met Spanish soldiers earlier in the century. In 1829, Lieutenant Colonel Guillermo Galvey, who had commenced his destructive raids on northern Luzon in Benguet,[56] described significant herds of cattle, carabaos, and horses across the province. [57] Later in the century, Manuel Scheidnagel noted that the area offered "ample field for horse breeding," and its colder climes would support the cultivation of hearty livestock. Chickens reared in the mountains, he wrote, yield meat "reminiscent of the countryside of Spain."[58]

## The Limits of *Reducción*

While the pastoral landscape may not have conveyed it to newcomers like Worcester and Wright, it was one that hinted at the degrees of autonomy that Igorots had managed to eke out and, in the case of Benguet's elite, the measure of profitable accommodation they had made to colonial authority. Across the lowlands, the Spanish had undertaken a process of *reducción*—an effort at drawing Philippine peoples from scattered villages, or *barangays*, into more tractable towns centered around a church. With Spanish manpower limited, the missionary operated at such settlements "as the most potent agent of colonial rule." He collected tribute, appointed natives to local posts, and attempted to keep Philippine people pacified, becoming in these ways essential "to the actualization of Spanish power in everyday life."[59] Daniel Nemser has written of this policy of *reducción*, also known as *congregración* (congregation) in colonial Mexico, as a "pastoral technology," invoking Foucault's concept of pastoral power.[60] This "power over men," a spiritual flock, exercised by the Christian pastorate, Foucault argues, became a "model and matrix of procedures for the government of men," the business of the modern state emergent in the sixteenth and seventeenth centuries.[61] Nemser contends that in Mexico, *congregración* served as a "racializing procedure" that worked together with "language, discourse, and representation" to produce "a new form of racialized subjectivity," the "Indian," and make possible forms of "economic extraction" by the Spanish.[62]

The scattered huts of Benguet pointed to the limits of that Spanish pastoral technology in the mountains. While the Spanish, through missionary and military personnel, had certainly exerted considerable influence and even force, they had not managed to reduce the Igorots into that social geography so favorable to rule, at least not until the nineteenth century. In the Cordillera province of Benguet, the Ibaloi had built several distinct kinds of settlements through the end of the Spanish colonial period, each with an attendant social order. The province contained rice-based agricultural settlements. It had communities that subsisted on root crops, especially *camotes* (sweet potatoes) and *gabi* (taro), grown in swiddens.[63] These root crops, which leave behind few signs of their cultivators, have long been a choice of people who have wished to preserve "physical mobility" and "escape detection."[64]

Besides settlements oriented around agricultural production, the province contained two other kinds of communities, ones devoted to gold mining, which "clustered around privately-controlled tunnels," and others built on gold trading, which rose along routes to lowland trade hubs.[65] The Baguio area, in fact, was a key center of this gold trade. By 1878, Benguet Igorots were selling about "four to eight thousand ounces" of the metal.[66] Igorots mined the gold and also collected it from riverbeds, making the region especially attractive to the Spanish,[67] who did manage to build garrisons in the area in 1620, 1623, and 1625.[68] Dr. Antonio de Morga, a lieutenant governor of the Philippines, described this gold mining in the early seventeenth century,[69] noting that Igorots mined only the quantity of gold they needed, and at designated places along the coast and in the lowlands, they traded the treasure for necessities like carabaos, cattle, rice, salt, iron, and woven goods.[70] Such trade shaped oral traditions of southern Benguet,[71] and the commerce also enabled Igorots to maintain a measure of autonomy. It complicated the proselytizing, the pastoral efforts, of Augustinians and Dominicans who sought a spiritual flock in the mountains after working across Ilocos and Pangasinan. According to Padre Vivar, the first Augustinian missionary assigned to minister to the Igorots in the mid-eighteenth century, attempts at religious conversion were stymied as agricultural work preoccupied women, and with the men, he wrote, "constantly and indefatigably occupied in trafficking."[72]

This persistent trade with lowlanders (not much trade seems to have transpired among the Cordillera's ethnolinguistic groups)[73] and Igorots' resistance to conversion posed a serious problem to *reducción* and, more generally, Spanish sovereignty. Indeed, in their unceasing trading and in tax- and tribute dodging, Igorots had managed to carve out a relationship to the Spanish colonial state that differed from lowlanders'. The latter's was defined by "economic extraction" like in colonial Mexico—the payment of tribute and forced labor, or *polo*—and of course, Christianization, all enabled by that process of *reducción*. Tribute was a tax paid by male Indios between the ages of eighteen and sixty, eventually succeeded in the 1880s by the *cédula personal*, a head tax.[74] The *polo*, meanwhile, referred to labor—up to forty days of work—expected of male Indios, a practice that emerged with the end of the *encomienda* system in the mid-seventeenth century.[75] The work, which could entail constructing government buildings, cutting trees, or military service, was often punishing and came at a cost: inattention to agricultural work at home and failure to amass tribute payment. These colonial measures served as the Crown's main means for accruing what it lacked, namely revenue and manpower. Through these taxes and their labor, Philippine peoples financed Spanish colonialism and underwrote the Crown's military ventures and also its trade with China.[76]

In the mountains, by contrast, the Spanish met particular challenges. The Benguet *comandante* Manuel Scheidnagel claimed in the 1870s that "The Igorrote . . . has been able to enjoy the widest freedom for his own aims, business and trade, without seeing any obligation more than paying annually his meager *reconocimiento de vasallaje*," a poll tax.[77] He voiced a view communicated by Antonio de Morga, who had noted over two centuries earlier that "as long as [Igorots] could exchange their gold, wax, etc. against articles which they lacked, and which were necessary for their subsistence, they would never allow themselves to be converted."[78] Only an end to that trade and a condition of economic dependence, de Morga believed, could drive "the infidels" to conversion. Scheidnagel seemed to agree, and he believed that Spanish policy had given Igorots the perfect reason to continue rejecting *reducción* and conversion, a transformation of status that seemed to bring only the burden of taxes.[79] In these ways, ineffectual Spanish policy together with mountain gold, the mountain terrain, and Ibaloi subsistence practices helped to make the project of Cordillerans' political subjectification through *reducción* a difficult one.[80] Still, the Spanish colonial presence and, more narrowly, colonialists' attempts to intervene in the gold trade through the centuries did come to reshape Igorots' economic practices and their social organization.

## Making Marlboro Country

In the years following the conquest of the Philippines, a number of factors led Igorot gold production to decline. For one, Mexican silver and its exchange for Chinese products through the galleon trade came to consume the Spanish, and interest in pursuing a Philippine El Dorado diminished correspondingly. Beyond this, it seems that Igorots made their own assessments of the costs and benefits entailed in the gold trade and purposely limited their mining of and commerce in the metal.[81] De Morga had noted that although agents of the Crown wished to learn the whereabouts of their mines, Igorots shared little, wary of what might befall their land and people should the Spanish learn too much. Recognizing that their power lay not in conspicuous displays of wealth (except during ritual feasts among themselves) but in what remained hidden, these Igorots of Benguet told visitors "they have more in safe-keeping in the earth, than in their homes."[82] Over time, these strategies for keeping gold out of Spanish hands and reducing the risk of Spanish invasions of Benguet conditioned changes to the Ibaloi's social structure. They advanced its stratification, heightening a characteristic of Ibaloi communities since at least the 1500s and reformulating the basis of social power from one rooted in "achievement" and bravery to inherited status and wealth.[83]

Limiting trade with lowlanders in the last decades of the sixteenth and the first decades of the seventeenth centuries created circumstances that some members of wealthy Ibaloi families, or the *kadangyan*, effectively exploited. They took charge of the gold trade that did persist and crafted for themselves the crucial and lucrative role of trade broker.[84] The capital amassed through this control of the gold trade afforded the *kadangyan* the opportunity "to dominate other aspects of the local production sphere"[85] and acquire cattle, which by the late nineteenth century constituted a "chief fountain of wealth" for Benguet's Igorots.[86] Even by Padre Vivar's time, a "petty plutocracy" rooted in cattle ownership and premised on gold had been "consolidated and solidified" in Benguet.[87] "Hither is brought all the gold that comes dirty from the mines," Vivar wrote. "[T]hey refine it somewhat and take it down to Ilocos for sale, bringing up in return so many buffaloes and cattle that to this day I do not know where they are consumed."[88] This cattle and gold wealth enabled leading Ibaloi to further bolster the status of their families by claiming land rights. And the Ibaloi property system, allowing for these assertions of ownership, was far from simple. While Ibaloi land and its resources were communal in the sense that they laid "within community territorial boundaries," individuals and groups could and did assert exclusive rights to land based on "the principle of *primi occupanis*."[89] The more labor an individual or group invested in the land and its material features, the more extensive the rights the party could claim to it. For instance, while the Ibaloi viewed streams that contained surface gold as open for use by all community members, they saw mines as individual property, "their yield privately controlled."[90] Likewise, if Ibaloi made substantial improvements to land by planting trees or constructing fences, their rights extended forward in time and could be inherited by descendants.[91] This is just what some leading Ibaloi managed to do, advancing that new social system of inherited status and wealth.

Gold trading and cattle ownership, then, as they developed under Spanish colonial rule, effected change in the very landscape of Benguet. They shaped the pastureland that would greet Americans at the turn of the twentieth century. Cattle rearing and the gold trade under the Spanish also shaped the Ibaloi's social geography. As Otto Scheerer observed, the ownership of livestock together with "the barrenness of the soil," especially in higher elevations of the province, led the Igorots "to live dispersed all over the mountains, be it in real individual isolation" or in small groups of huts, "*rancherías*."[92] Cattle ownership and gold conditioned the Ibaloi's highly stratified social structure as well. This social hierarchy made the Benguet Igorots distinctive across the Cordillera, for while forms of social stratification existed among other Cordilleran groups, "the degree of differentiation between elite and commoner" was unique to the Ibaloi.[93] Leading *kadangyan*, the *baknang*, displayed and distributed their extraordinary wealth at ritual feasts, or *cañaos*, among their dependents—children, "fictive kin" acquired through political partnerships, and also tenants and slaves, who could undertake swidden farming, rice cultivation, and home building on a headman's lands.[94] The most "lavish" of these ritual feasts was called the *pachit*, which could go on for days, even weeks; the more livestock a host slaughtered, the higher the *baknang* elevated "his prestige and power."[95] As much as these rituals grew that prestige and legitimized the *baknang*'s ascribed status, they also served "practical" functions.[96] The feasts supplied the *abiteg*, or common and poor Ibaloi, with most of their "animal protein,"[97] supplementing those root crop staples like *camotes*. In this way, ritual feasts turned cattle ("alienable objects") into meat ("inalienable objects"); through the *cañao*, the elite "create[d] commodities out of the normally inalienable and gifts out of the normally alienable," drawing to wealthy Ibaloi labor and laborers—that factor and those figures crucial for claiming more significant rights to land and additional productive resources.[98] The wealth in gold and cattle so dramatically displayed at these feasts had been shaped in part by Spanish colonial interests; in turn, it would condition Ibaloi negotiations with colonial authorities, who, after centuries of pacification efforts, finally managed to make Benguet northern Luzon's first *comandancia militar* (military district) in the mid-nineteenth century.

## Benguet as *Comandancia Politico-Militar*

The establishment of *comandancias militares* in the Cordillera followed Spanish efforts through the mid-eighteenth and nineteenth centuries to undertake serious and more effective approaches to Igorots' trade practices, refusing to countenance the ongoing, unregulated commerce between mountaineers and lowlanders. Their efforts coincided with a broader change in colonial policy: colonial administrators' choice to double down on attempts to keep the Pacific colony in line and in the red. They were pushed into action for several reasons. The British invasion and occupation of parts of the archipelago during the Seven Years War in the mid-eighteenth century led the Spanish to rethink their logic for maintaining the Pacific colony. "[A] divinely sanctioned mission or sovereign right of conquest" would no longer suffice in the aftermath of the British assault. The Spanish needed something more. The resolution that emerged was an effort to make the Philippines a "modern colony" similar to the possessions of the British and Dutch.[99] In the context of imperial competition and contemporary Bourbon Reforms and, eventually, the end of the galleon trade in 1815, the Spanish made attempts to develop their Pacific transshipment port into a profitable agricultural colony like others in Southeast Asia. This process was attended by "military centralization" and "the subordination of both provincial governors and religious missionary orders to authoritarian control."[100] In effect, the Spanish were recalibrating the balance between state and church power in the colony in favor of the state and also attempting "to make state power coterminous with state boundaries"[101] and state power coterminous with the reach of colonial trade.

The government's creation of a tobacco monopoly in the 1760s exemplified the priorities and methods of the new regime.[102] It was also a state project that some Igorots, who grew tobacco, had complicated. Seizing an opportunity afforded by the monopoly, some who grew the plant cultivated an eager market: lowlanders seeking an untaxed share of the crop.[103] Over time, the trade between the two sharply reduced the taxes that the colonial state could collect in places like Ilocos. Just as the Spanish sought to control economic and in that way political power in the colony at large through tax and tariff policies and by instituting the monopoly itself, so they would crack down on the illicit Igorot commerce.[104] These attempts at seizing control gave way to particularly dramatic action in 1829, when Lieutenant Colonel Guillermo Galvey undertook his notorious raids in northern Luzon beginning in Benguet.[105] Igorots' initial resistance to Galvey's campaign provoked the burning of 180 of 500 Igorot houses; ten years later, only 100 houses remained in Benguet's Trinidad Valley.[106] Galvey had made Benguet's "the first 'tribe' of Igorots to be listed as Spanish subjects."[107] From there, Galvey, like "a determined conquistador," would move further into the mountains.[108]

In the years following Galvey's blaze through Benguet, the Spanish would divide northern Luzon into four *comandancias militares* (military districts), the first of which was Benguet.[109] In 1854, the province, foreshadowing a future as the US colonial summer capital, became a *comandancia politico-militar*.[110] This administrative zone was led by a military man "not merely to protect Spanish subjects against attack by non-Spanish subjects, or prevent invasion of Spanish boundaries by the introduction of contraband, but to enforce law and order throughout actual Spanish territory."[111] *Comandantes* of these units chose the leaders or headmen for *rancherías* across the province, who were expected to collect tribute and enforce corvée obligations.[112] Igorots were exempt from the *cédula personal*, but they paid the *reconocimiento de vasallaje de infieles* ("recognition of vassalage of non-Christians"), a poll tax. "Newly converted Igorrotes (*nuevos cristianos*) continued to pay this tax for a term of ten years, when they had to take out a '*cédula personal*.'" Igorots were also required to labor a certain number of days on roads and public works, and according to Otto Scheerer, their labor was certainly exploited.[113]

By 1880, after centuries of effort and at a moment when the Spanish empire had been reduced to a small collection of island colonies, Benguet was pacified so far as colonialists were concerned.[114] The province was composed of nineteen townships, two of which were "*pueblos*," referencing the presence of Christians, in this case Ilocanos. The rest of the province, with non-Christian Cordillerans, contained "*rancherías*" like Baguio.[115] In 1896, some 12,092 of these Igorots "were paying an annual head-tax of 50 centavos and making no overt opposition to colonial status."[116] And further, commerce that Igorots of Benguet had been carrying on for centuries advanced, only now with the sanction of the Spanish colonial state: potatoes from Benguet could be found in Manila while horses, carabaos, coffee, and beans made their way to La Union in Ilocos by 1880. Lowlanders, meanwhile, were making their way into northern Luzon to peddle their goods.[117]

This military occupation and diminished autonomy did not mean the same thing for all the Ibaloi. Through the period during which the Spanish intensified their efforts to bring Benguet under colonial authority, Ibaloi elite like the Cariño family chose paths of negotiation with the Spanish.[118] Galvey reported that following his 1829 raid on Benguet, "the Igorots of Benguet shortly afterwards asked me for peace and have since been my friends."[119] Biguñg, the grandfather of Mateo Cariño, the Ibaloi man who would challenge the US government's appropriation of Baguio land in the first decade of American rule, "assembled the leading men at his house where a head of cattle was slaughtered in honor" of General Galvey.[120] And Biguñg later served as a companion to the general on his tour through the province. Biguñg's son Pablo, father of Mateo, meanwhile, "was baptized under the sponsorship of the Spanish Comandante Oraá"; Mateo's older brother Juan was also given the name Oraá at baptism.[121] These attempts at accommodation were not new. As far back as the mid-eighteenth century, Kidit, a headman whom Scheerer called the "reputed founder of Benguet's nobility" and who may have converted to Christianity, tried to convince other Ibaloi to accept *reducción* and accede to permanent settlement.[122] Kidit had introduced Padre Vivar to other Igorots of Benguet,[123] winning the regard of the Spanish priest, who complimented the Ibaloi leader for possessing "the genius of a mandarin."[124]

As Galvey's report may suggest, the *baknang*'s willingness to negotiate with the Spanish could reflect simply the dire circumstances and imbalance of power they faced at the hands of Spanish forces. It was a survival strategy, a means of defending Ibaloi from further military attacks. And its pragmatism may have been revealed when Kidit, though friendly with Padre Vivar, chose to defend his fellow Ibaloi during the Spanish invasion of Benguet back in 1759.[125] Over a century later, when the revolution against Spain arrived in the Cordillera, Juan Oraá Cariño, Mateo's brother, although named after a Spanish *comandante*, allied with insurgents and was appointed the governor of Benguet during the brief First Philippine Republic.[126] Sergio Osmeña, the first speaker of the Philippine Assembly and a former revolutionary, recalled his journey with General Aguinaldo through Baguio and Trinidad where he met the Ibaloi leader (and also Otto Scheerer).[127]

But the inclination to compromise with the Spanish earlier in the century likely also signaled the degree to which gold and cattle ownership had made negotiation with the Spanish the preferable choice for the elite. Kidit's negotiations with the Spanish and his descendants' willingness to slaughter cattle and play hospitable hosts to the likes of Galvey, in this view, may reflect the interests of a land-owning class. This elite preserved its wealth, and further, "[a] more sedentary life of the *kailians* [dependents of the *baknang*[128]] could only be in the interest of the propertied class of Ibáloys," Scheerer wrote based on interviews with members of the Cariño family.[129] If gold and cattle ownership afforded the Ibaloi degrees of autonomy, they also conditioned elites' collaborations with the Spanish. In these respects, as desolate as the Benguet landscape may have seemed to Americans, so that it could appear like New England's Adirondacks, the landscape was, in fact, a "palimpsest," constituted by changing patterns of subsistence, social organization, and negotiations of power.[130]

## The South of Europe in the Cordillera

By the time Americans arrived in Benguet, the southern part of the province is said to have been "one huge 'Marlboro Country,'" where by 1908, something on the order of 25,000 cattle grazed the land.[131] Baguio became the most famous piece of Benguet prairie.[132] To the Ibaloi, it was known as Kafagway, an Ilocano term meaning "grassy clearing."[133] On one section of the prairie now near the city hall, horses, cows, and carabaos were known to converge from across Kafagway to feed.[134] And, in the late nineteenth century, some twenty-seven families, including Mateo Cariño's, were using this land as pasturage for their livestock.[135] Land had become Mateo's in the 1870s through his marriage to Bayosa, the daughter of another prominent Benguet family, marriage serving as a means by which the elite could further consolidate their power.[136] When Mateo arrived at Kafagway, he added further value to its lands: he promoted the planting of rice, a practice in Tublay, the part of Benguet from which he had hailed. And he repaired and raised fences around his and Bayosa's Kafagway pasturelands, improvements that advanced their claims to the land.[137]

Americans saw a future for Kafagway different from the one Cariño seems to have imagined. Their view had already been envisioned by the Spanish. Indeed, when Americans began formulating their plans for a health resort, they drew on an older Spanish study for a similar project. The Spanish report was not dated, but references within it suggest that its authors, a Spanish colonel, an engineer, and the "first surgeon of the military health department," prepared it in the late 1880s or early 1890s.[138] Their interest in Baguio was much as Americans' would be; for them, a retreat in the lower Cordillera would be a boon for sick and injured soldiers, who could enjoy respite in a climate similar to Spain's. The province was also known to contain "medicinal mineral waters."[139] The doctors noted that former Spanish Governor General Valeriano Weyler (1888–1891) had already chosen Benguet as the choice district "for the acclimatization of Europeans," and established a "military-agricultural colony" in La Trinidad, Benguet's capital.[140] Manuel Scheidnagel had described La Trinidad in 1878, noting its *casa real* (central municipal building) surrounded by gardens, barracks for the Guardia Civil, jail, church and convent, tribunal, and promenade and small botanical gardens featuring "rare plants" that Scheidnagel had collected on his journeys through northern Luzon.[141] These were hallmarks of a hill station in formation.[142] Under Weyler, who would become notorious for his reconcentration program in Cuba later in the decade and earn the moniker "Butcher Weyler," there had been plans to expand the station and build an infirmary. However the Spanish ultimately ruled out building on La Trinidad just as the Americans would.[143] To US colonialists, the Spanish capital, was lovely—"green with growing crops, dotted over with Igorrote houses . . . a panorama of striking beauty." But the investigating party was less enthralled by the clouds, which arose from the surrounding canyons and cast the town in fog.[144] Some time after Weyler's tenure as governor general, a military surgeon was sent to study

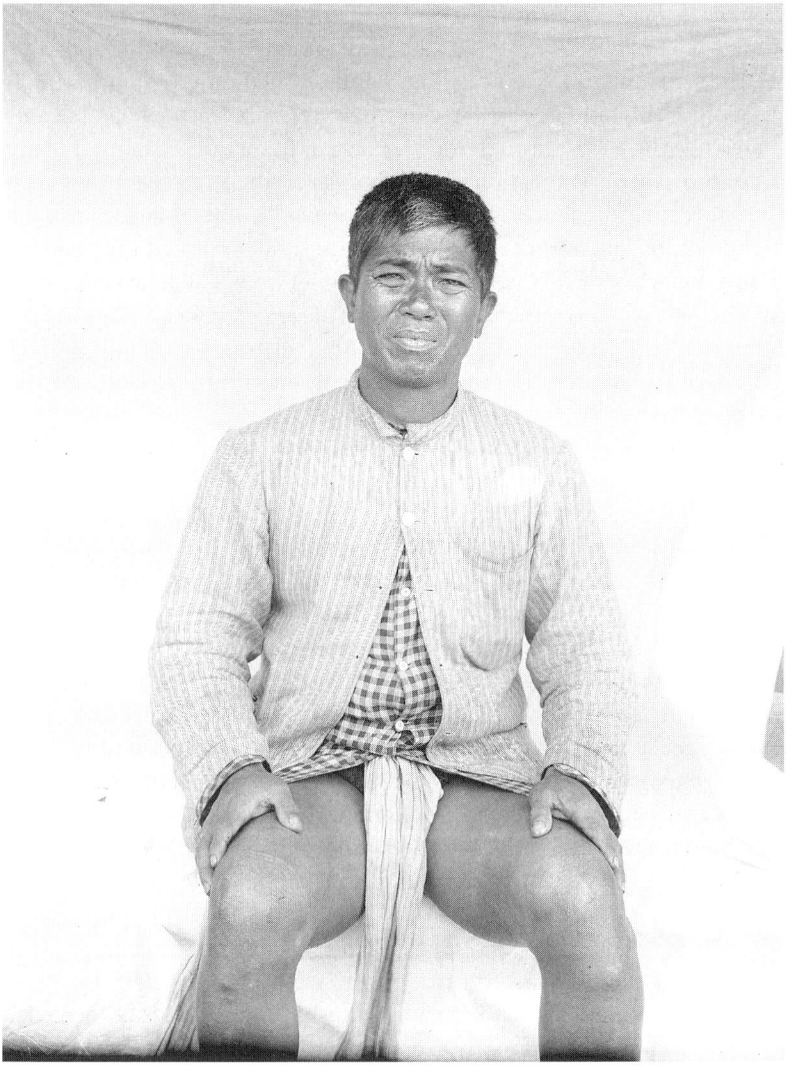

1.2 "Benguet Igorot, Mateo Cariño, Type 1. 3/4 length front view." 1904, Baguio, Benguet, Photo 10A003, Dean C. Worcester Photograph Collection. Reprinted with the permission of the University of Michigan Museum of Anthropological Archaeology.

Benguet's climate. He recommended Baguio as preferable to La Trinidad.[145] To the Augustinian priest Angel Pérez, it had been poised to become one of the best sanitariums in the Far East.[146]

These Spanish efforts to develop a mountain retreat gained ground just as the Crown sought to make the colony economically productive and

1.3 "Benguet Igorote woman, named Bayasa [sic], wife of Mateo Cariño: shows typical dress of wealthy Benguet Igorote woman; Baguio, Benguet—1901; 1739; 1901." Dean C. Worcester Photographs of the Philippine Islands, University of Michigan (Special Collections Library).

square state power with colonial borders following the end of the galleon trade and the loss of most of the Atlantic empire. While the colonial state was trying to reassert its authority in these ways, capital entering the Philippines from outside Spain was pulling the colony in other directions. This wealth, more so than the Spanish state's interventions like the tobacco monopoly, was "moving the colony decisively toward an export economy and linking its fortunes to the booms and busts of the world capitalist economy."[147] These dynamics, which helped to put people and ideas in radical motion, laid groundwork for Filipinos' revolution. By 1892 José Rizal had founded La Liga Filipina to call for reforms in the colony's relationship to Spain. That same year, Andres Bonifacio organized the Katipunan, a secret society of largely working-class Filipinos committed not to reform but to revolution and independence. Even for decades before that, the colony had hardly been at ease. Between 1850 and 1890, some twenty-four different governors had come through the colony.[148] This gives some indication of the political tumult on the archipelago, which would erupt in war in 1896. This context also suggests that the hill station was a spatial form with particular appeal as the Spanish turned to territorial reconquest and colonial development in the face of threats to its sovereignty. The endeavor required a greater number of civilian and military personnel, and such population changes arguably put a premium on maintaining distinctions between ruler and ruled—something that a hill station, a colonial enclave, could enable. These circumstances were in certain respects similar to those in British India in the late nineteenth century. Especially after the 1857 Indian revolt, which heightened British security concerns, "the hills" grew ever more attractive.[149] The rebellion also made the presence and health of British troops critical concerns to the regime. One British army surgeon believed the hill stations might serve as both health resort and military depot for new arrivals to "become acclimatized" and for a worn-out soldier to "'find a resting-place that would soon restore his constitution to its former vigor.'"[150]

The hill station's appeal was also informed by what the British called "medical topography," the science upon which proponents of these retreats drew. Medical topography "was concerned with those aspects of place—airs, waters, customs, etc.—which had a bearing on the health of a region."[151] Spanish thinking would follow along similar lines. Indeed, shaping late-nineteenth-century Spanish plans for Baguio was an assumption about how climate affected human health and how the climate, if one were suited to it, could prove therapeutic to the mind and body. In the Spanish report, this application of the climate to human health was called "climatoterapia"—"to convert the action of the climate into a medicine."

To the pines, for example, the Spanish doctors attributed emollient powers: "besides purifying the air, [the pine trees] contribute to modify the local currents of the winds, engender at the time pleasant sensations and agreeable recollections of distant native lands, and for this reason stimulated the spirits, producing happiness of mind and joyful hopes." Nature would serve as salve as patients "would find mitigation of their moral suffering in the life that can be led in the mountains of Baguio watching the growth of plants like those of the south of Europe, enjoying the temperature of moderate climes, and indulging in beneficial physical exercise." The doctors wrote of "improvement coming perhaps with the distractions that nature offers those who are in such nervous states."[152] Angel Pérez meanwhile, referenced "Oda á la vida del campo" ("Ode to the Country Life"), written by his fellow Augustinian Luis Ponce de Leon, and described the province as a place delightfully apart from the "maddening crowd."[153] Benguet's nature might nurture weary Spanish colonialists back to health.

If the mountain climate held such therapeutic potential for white colonizers, then what about for the Igorots? The Spanish were silent but to note a potential wrinkle in their theory: "It is certainly to be regretted that with such climatic modifying influences these Igorrotes should lack those conditions of physical energy and power of resistance which would make their customs better."[154] The benefits of "climatoterapia" fell short for the Igorots, they believed. But really, the powers that the Spanish assigned to the environment were not wholly foreign to Ibaloi "pagans'" views of nonhuman nature in Benguet. The mountains were central to Igorots' ritual life.[155] For the Ibaloi, Mt. Pulag, the highest peak on Luzon, was a creation site of their ancestors.[156] And these animists' "spirit-relatives"[157] do not inhabit the mountains alone; they dwell in timber, in the water, and in rice fields.[158] In his 1920 ethnography of the Ibaloi, C. R. Moss wrote that "The old men say that that are seven Kabigats," an Ibaloi god: "Kabigat of where the water rises in the north; Kabigat of where the water empties in the south; Kabigat of where the sun rises; Kabigat of where the sun sets; Kabigat of the sky-world; Kabigat of the underworld; and Kabigat of the space between earth and sky."[159] To the Ibaloi, these spirits are immanent in nature. Kabigat and other Ibaloi gods are understood to protect or aid the Ibaloi through life—in times of war, in the quest for gold, in agricultural pursuits, and in the healing of people and animals. Spirits of deceased ancestors exist in a hierarchy below these gods, and Ibaloi view them as "intermediaries" between the living and the very highest spirits.[160] They intercede on behalf of the living in prayers for future gains in livestock, rice, or money or on the occasion of rites of passage like birth, marriage, and death.[161] William Henry Scott summarized the role of worship in the life of Igorots in 1962: "pagan worship sanctifies the material world, forms the basis of the composition of society, and is the sanctifier of moral conduct." These, he warned, would be domains "left a vacuum by the decimation of Igorot worship."[162] Given the great importance of place in Igorot ritual life—Moss noted that "[e]ach ceremony must, in order to be effective, be celebrated in its appropriate place"—dispossession of Igorot land or its transformation would have serious consequences.[163]

The same could not be said for the Spanish who transplanted their spiritual life to the colony, raising the cross and erecting churches. However, colonialists' descriptions of the restorative influence of the Benguet environment did contain an assumption about the power of nonhuman nature. For them, the climate, especially, had a profound effect on human well-being. They believed that the tropics, if not fatal, were at minimum insalubrious to their health. Time in the tropics, it seemed, rendered Spaniards weak. The doctors, whose report Americans drew on, described the humidity and heat of Manila as "sedative, debilitating, and depressing in its effects on the human organism."[164] Benguet and the foothills of Mt. Pulag might offer deliverance from such conditions. The doctors quoted "Levy," possibly Albert Lévy, a French scientist who worked at the Observatory of Mount Souris in Paris. To him, they attributed the belief that "To elevate one's self in the tropical regions is to again place one's self in the salutary conditions of Europe"; the doctors crafted an association between ascent and western civilization itself.[165] The Spanish doctors cited several other French scientists in support of their assertions of Baguio's curative potential. Like the British in the nineteenth century, the French were engaging in debates on the capacity of European "acclimatization" to tropical conditions. By the later nineteenth century, racial thinking had entrenched notions of "climatic determinism" and made matters of heat and humidity serious concerns for empires.[166] The nineteenth-century science of *climatisme*, like Spanish climatoterapia, "involved 'reimmersing' patients into 'clement' climes over the course of a standardized 'cure.' . . . [T]he objective was *ressourcement*, an untranslatable term connoting a return to one's place of origin, climate, milieu, or spring."[167] The French pursued *ressourcement* through hydrotherapy in spas and springs across their empire, while the British situated military cantonments on the hills. Their imperial theories would change, such that the environment was largely acquitted and native peoples would come to be seen as the primary threats in "the growing racialization of germ carriage and distribution."[168] But initially it was this form of environmentalism that shaped Spanish interest in Benguet and that informed Americans' too. For Americans suffering from

Philippinitis, Benguet meant *ressourcement*—not to a place like the Vichy springs or the Spanish Peninsula but to New England.

## US Government Reservation: New England in the Cordillera

Worcester and Wright's exploratory party's encounter with Benguet—a vision of New England in the Cordillera—ultimately reinforced earlier reports that it was a choice site for a health resort, and the Philippine Commission wasted no time in passing measures to advance the province's development beyond what the Spanish had accomplished in La Trinidad. In its second act as an administrative body, the commission appropriated monies for survey work on a railroad route to the province.[169] About two months later, in November 1900, it called for the installation of civil government through Benguet. Just as Benguet was the first province of northern Luzon that the Spanish made into a *comandancia politico-militar*, a seat of political and military authority, under US rule it became the first province to be granted local civil governments. This measure would replace the military's pacification program conceived of by General Elwell Otis following McKinley's "benevolent assimilation" orders. It had entailed the rebuilding of municipal governments, roads, schools, and more across the archipelago.[170] Commissioners justified the move to civilian control by identifying the population of Benguet as predominantly Igorot and claiming that conditions in the province demanded greater "control and direction" by the insular government.[171] They claimed that these Igorots—growers of rice, cultivators of cattle, and miners and traders of gold—required both lessons in "the ways of civilized people" and protection from Ilocanos of the western coast of northern Luzon, who "display[ed] a tendency to take undue advantage" of them.[172] And if Spanish rule and Ilocano abuse were not sufficient to justify Americans' intercession, commissioners pointed to the caciquism of the Igorots themselves. Worcester, like the Spanish, commented on "Malayan society" as being "oppressively aristocratic" and, in that way, unfavorable to the man of little wealth and status.[173] The Igorots, commissioners held, had yet to "develop greater capacity for self-government."[174]

In describing Benguet's Igorots thus—as unthreatening, even as welcoming of Americans, and depicting the majority as victims in need—the commission attempted to justify the installation of civil government and Americans' very presence. Commissioners may have been eager to wrest control of Benguet from military authorities, who had been combing it and other parts of northern Luzon for insurgents.[175] And, in fact, when the commission sent General MacArthur a copy of its bill to install civil government in Benguet, he replied, "'the time was not propitious for the proposed legislation.'"[176] Whether or not the commission was challenging the military for authority in the region, it is clear that commissioners took care to caution that the privileges accorded Benguet were unique and not necessarily to be extended elsewhere; they were "special measures designed to meet the needs of a primitive and illiterate people. They are in no sense indicative of what is intended to give the civilized tribes."[177] Benguet and its Igorots were to be treated as a distinct population and place.

In calling for the establishment of civil governments through Benguet, the commission largely followed the Spanish, organizing Igorot places into "townships," like Baguio and Trinidad, which were further divided into "barrios." These barrios would send representatives to a council that joined an elected president and vice-president, in governing the township; the Cariños—Mateo and his brother Juan—would each take a turn holding the position of Baguio president between 1900 and 1903.[178] The Philippine Commission placed a powerful check on the power of these elected representatives by giving the provincial governor, appointed by the governor general, oversight over all ordinances and rules passed until the council showed it had "gained sufficient knowledge and experience properly to exercise, without intervention, the powers" accorded it.

Commissioners also sought to organize the tax base of the province. Over two days in early 1901, residents over eighteen who claimed ownership to property were expected to appear before the township's president. Those residents of the province whose property "real or personal" did not reach $200 (Mexican) were exempt from property taxes but would owe an annual tax of $1.00 (Mexican).[179] Those with property surpassing $200 would be expected to pay "an annual property tax an amount equal to one-half of one per cent" of the property's value. The law did away with the *polo* but not entirely. Those who did not pay their taxes were subject to labor for the province.[180]

The Philippine Commission would revise the terms of civil government when it passed a special provinces act in 1905. The measure, which initially applied to the provinces of Benguet, Nueva Vizcaya, and Lepanto-Bontoc in northern Luzon and Mindoro and Palawan, two islands southwest of Manila, defined the colonial government's relationship to the Philippines' "non-Christian" provinces, excluding what Americans called the Moro Province.[181] The logic of the law was consistent with that providing for civil government in Benguet, and it had been implicit in the 1902 Organic Act that created the insular government; the latter held those areas inhabited by Moros and "non-Christians" under the direct authority of the Philippine Commission. These peoples needed protection from predatory Filipinos, or as Dean Worcester put it, "the non-Christians are not to be subject to the municipal officials of Christianized towns, at whose hands they have in the past suffered so much oppression and whom they cordially hate, but are to be taught to govern their own settlements." Through the Special Provinces Act, the insular government reserved the right to intervene in the lives of "non-Christians" through the agency of provincial authorities who were appointed by the governor general.[182] Through the measure it also claimed authority over about 20 percent of the archipelago's population and regions of the country with valuable forest and mineral resources.[183]

The irony of American administrators' paternalist policy in the Cordillera is that for centuries, as we have seen, the Igorots had been carrying on a trade with many of the Filipino peoples who were now deemed only their abusers. Filipino lowlanders, meanwhile, had long been known to make the Cordillera their home as *remontados*. And, of course, some Igorots had been skillfully maintaining their autonomy from the Spanish or engaging in accommodations with them for generations, a past unknown or willfully ignored by those colonialists asserting the need for Cordillerans' special protection. The construction of difference and division implicit in the special provinces act and expedient to establishing US rule and to affording it "*ressourcement*" certainly constitutes an exercise in imperial power. Americans' laws of civil government, the Organic Act, and the special provinces act advanced Spanish work of bringing the region and its people under the authority of the colonial state, paradoxically by insisting on their separateness from the rest of the archipelago. In this respect, the Cordillera's colonial history accords with Lauren Benton's observation that imperial powers generated and maintained "highly variegated legal geographies." "Even in the most paradigmatic cases," she writes, "an empire's spaces were politically fragmented; legally differentiated; and encased in irregular, porous, and sometimes undefined borders."[184] These irregularities in sovereignty manifest an imperial regime's power to carve out exceptions essential to rule.

This "legal complex[ity]" within colonial spaces, the creation of "special provinces," even *comandancias militares*, indeed indicates that "political authority" was "a work in progress."[185] And in this regard, it also signals limits to imperial power, the need to adapt imperial designs to conditions on the ground. Americans' response to Philippinitis—their therapeutic interest in the Cordillera—did, like Spanish anxieties revealed in *climatoterapia* or French *ressourcement*, also hint at latent weaknesses of the regime.

All these colonialists believed they needed an antidote for what they variously called "forgetfulness" and "nostalgia."[186] They needed escape from the tropical heat and even from their colonial charges—perhaps members of the "maddening crowd" Angel Pérez referred to. Worcester and Wright even thought to relocate the colony's capital in the Cordillera, isolating it in the mountain fastness. These concerns suggest the vulnerability of colonial power; colonial state builders sought a "region of refuge" from the exigencies of imperial rule. And so in Baguio, colonialists would proceed with plans to transform the mountain valley, adding a new script to the palimpsest. They would transform Ibaloi pasture, itself a colonial landscape, into pastoral: a health resort with golf and polo greens, gardens and a country club, and camps for soldiers and teachers. They would build a "government reservation," an American domain to reproduce an American state of mind. And they would begin constructing this imperial pastoral by building a road to get there.

TWO

# Liberating Labor: The Road to Baguio

Begun in 1900 and called complete in 1905, the twenty-six-mile road to Baguio put the cool, pine-scented air of Benguet within closer reach of Manila. It shortened a trip that had previously taken twenty-four hours by sea followed by several days on horseback; with the new Benguet Road, visitors could take a train northwest from Manila to Dagupan in the province of Pangasinan. To complete the fifty miles left to Baguio, one mounted a *carromata* (horse-drawn carriage) to reach one of two towns further inland before ascending to Baguio along the Benguet Road.[1] In no time, this road became one of the most famous in the US colonial Philippines. Visitors marveled at its "zig-zag" stretch that snakes along the Cordillera, and the highway was photographed many times over. In 1909, the *Washington Post* called the road "one of the most remarkable highways in the world," a foretaste of the kudos the United States would win five years later with the completion of the Panama Canal, another node of the United States' emergent market empire.[2] By 1913, the *New York Times* was comparing the United States' road building to that of imperial rivals: "American automobilists who have been raving over French roads might find fresher fuel for their enthusiasm in the Philippines," the writer claimed. "Here are just as good thoroughfares as any in France."[3] The protagonist of Filipino novelist Carlos Bulosan's *America Is in the Heart* (1946), Allos, wrote of Baguio roads, too. Made of asphalt, they ranked as "the most modern and beautiful in the Philippines."[4]

2.1 "The famous zig-zag on the Benguet Road," in Dean C. Worcester's *The Philippines Past and Present, Vol. 1* (New York: Macmillan, 1914), 224–25.

Given the importance of Benguet Province to American officials, it is no surprise that this road was one of the first major, colonial public works projects and, given its scope and cost, that it was administered directly by the Philippine Commission. It also emerged as a touchstone in Americans' and Filipinos' disputes over US rule. To Americans, the Benguet Road signaled the colonial regime's modernity. It represented Americans' engineering prowess—their ability to defy the limits of a merciless geography—and it also spoke to the managerial talents of American road chiefs who gathered some 4,000 men of forty-six nationalities for the job.[5] Today some natives of Baguio celebrate the road for having assembled that diverse workforce and laid the groundwork for Baguio as a "multiethnic city." Japanese and Chinese workers would stay on and gain footholds in and around the hill station as farmers and merchants.[6] To some Filipinos earlier in the twentieth century, however, the road to the hill station exposed mainly imperial extravagance and wastefulness. In the year of the road's completion, the newspaper *El Renacimiento* declared that to "Anyone who has observed the tendency to needless expenditures for measures and men displayed by the government . . . ," "we say merely: Benguet Road."[7] Bulosan's protagonist Allos described yet another experience of the road. He had labored on a later extension of it and nearly drowned in floodwaters spawned by the torrential rains that beat down on the highway and its builders.

As the conditions detailed by Allos may suggest, the fate of Benguet Road hardly seemed certain. The geology and topography of the Cordillera mountains rose as towering obstacles. Even worse was the trouble road chiefs faced in recruiting workers like Allos and maintaining a labor force. As one American visitor writing on the "conditions in the Philippines" recalled, "the Benguet road was repeatedly cited to me as proof" of the colonial labor problem.[8] All too often Filipino workers disappointed if not infuriated their bosses and other American observers of colonial labor. Some attributed their inadequacy to constitution, others to culture. Blessed by abundant nature, one theory went, Filipinos could abstain from work and still secure sustenance and, for that reason, tended toward idleness and indolence. Others attributed native labor culture to Spanish colonialism: the former regime had cultivated not a sense of free agency and initiative among subjects but instead a bent toward passivity and dependence. Given these circumstances and the greater aspiration of making the Philippines not only an entrepôt to China but also a productive colony in its own right, Americans came to watch the Benguet Road project with great interest. Road chiefs joined administrators in hoping to liberate labor and inspire in their colonial charges a sense of free agency and initiative.

The history of Benguet Road shows the limits of this ambition. As it became understood as a test of the reliability and efficiency of native workers, the road also became a moving workshop in how to acquire, motivate, and retain them. And, ultimately, Americans amassed labor power not by employing market mechanisms and incentives and giving lessons on free agency. Instead, Americans secured native labor largely through various "extra-economic" means—through charismatic authority, the counterrevolutionary war, and successive laws of dispossession that forced "free" labor by separating people from means of subsistence, producers from the means of production.[9] In doing so, the makers of new empire drew on methods with a long history in the colony and even on the US mainland.

The story of Benguet Road thus leads us to the summer capital and through a chapter on the formation of capital in the colony. It reveals an important context for the transfiguration of Benguet pasture into Baguio imperial pastoral. Leo Marx has noted that the pastoral form, an idealization of the country as a place of simplicity and innocence, finds particular appeal amidst change in the ownership and control of land—in the context of dispossession. This is the process that underwrote the making of colonial Baguio, as we saw in the last chapter, and the construction of Benguet Road, as we will observe here. The history of this public works project also points more broadly to the challenges entailed in building the way for free trade and growing capital. While in many respects Benguet Road was unique, it was also just one of the many road building projects undertaken early in the US occupation. American administrators saw roads as essential to resolving two central and related colonial problems. The first was pacification. The Philippine Commission's 1900 report had stated expressly: "The most serious hindrance to rapid and effective movements by our troops has been the inaccessibility of the country in which the insurgents have hidden themselves. The difficulty has not been to overcome, but to get at them."[10] Roads were crucial to subduing the colony and the "brigands" and thieves, or *ladrones*, who eluded US forces. Roads and pacification itself were also essential if effective market building was to follow conquest. Writing in 1904, then-Secretary of Commerce and Police William Cameron Forbes drew the connection between commerce and pacification: "With the advent of good roads comes the advent of peace, for in the presence of quick transportation the ladrone moves out, and until the ladrone has gone industry can not begin. The practical extinction of the ladrone brings the commercial use of the road into first place. It is necessary that the main avenues of commerce should be made passable."[11] In other words, roads served as instruments of conquest and of commercial development, for the movement of troops and then goods, for building a nation and a market in the Philippines. The history of Benguet Road's construction demonstrates the uphill task of making all Filipinos amenable and the Philippines tractable toward these related ends and the ways in which "extra-economic" means were consequently enlisted in making a way for freer trade.[12]

## Roads: An "Educator of the People"

As the *New York Times*'s boast about US colonial roads in the Philippines suggests, the Benguet highway was just one of the many the US regime called for. American colonialists understood roads as key to the economic development of the archipelago, as did members of the class of educated Filipino professionals and merchants, writers and readers of newspapers like *El Renacimiento*, who saw in poor roads a reflection of the colony's dismal commercial life.[13] Americans pointed to these public works projects as a signal difference between their rule and Spanish forerunners' on the archipelago. Americans, in defiance of their predecessor's practices, refused to maintain their wards in a state of "savage" isolation.[14] For a couple of centuries, the Spanish had prioritized overseas connections between Manila, China, and Acapulco in the galleon trade. Not until the turn into the nineteenth century did they seek to develop provinces beyond Manila and grow an export economy. Domestically, the Spanish pursued *reducción* of Philippine peoples into towns, units more easily managed by the limited number of colonial personnel, primarily priests. US colonialists, by contrast, emphasized connections between places and sought to productively link one Philippine region, its people, and resources, to another. They did so while making English the language of instruction across the archipelago. Road building, like language instruction, was a key means by which the United States aimed to build a nation and market in the Philippines.[15]

American colonialists also viewed roads as crucial in the immediate term—for the pacification of the colony where ordinary Filipinos maintained resistance to the US occupation.[16] Although civil government under William Howard Taft succeeded detested military authority under General Arthur MacArthur in 1901, the colonial government did not enjoy authority over all parts of the archipelago even as late as 1905. Some Filipino elites had acquiesced to US rule and had begun a long and complicated history of collaboration through the Federalista Party, one that avowed support for Philippine independence publically while "privately hop[ing] for the Philippines' annexation to the United States."[17] The majority of these political elites enjoyed landed wealth in the largely agrarian colony; under US rule, Federalistas believed, their political and economic interests might best be secured.[18] But not all Filipinos had made their peace with "benevolent assimilation." Playwrights, tailors, farmers, cooks, and also so-called brigands and highway robbers showed the continued resistance of "the poorer and less-educated classes" to US authority.[19] Their unrest suggests that at least some Filipinos sought something more than nationhood status and political autonomy; their vision of liberty, expressed in Tagalog as *kalayaan*, called for rebuilding the social order altogether, a revolution feared by the likes of Federalistas.[20]

Through the 1900s, banditry and highway robbery, some undertaken by these revolutionaries, some wrongly attributed to them, beset the country. In August 1900, a distressed colonel of the Twenty-Fifth Infantry Division, stationed in the northern Luzon province of Zambales, described these highway robbers. He warned that though the insurgency might appear over, peace did not yet prevail: "In this province certain small bands of robbers come out of their mountain jaunts every few nights and run off stock and sometimes abduct the laborers in the fields in day time." "My troops have been chasing these robbers incessantly for the past six months but with small results," he admitted. "The trouble being that when being closely pressed they hide their guns and go about dressed like any other natives, which is a complete disguise to us."[21] These robbers, also referred to as "outlaws" and "marauders," were a thorn in the side of Americans much as they had been to the Spanish, who referred to them as *ladrones* (thieves) and their deeds as *bandolerismo* (banditry), ways of figuring their aggression that Americans readily adopted (Taft called *ladrones* the "pest of centuries").[22] By the end of August 1902, the constabulary had killed 663 *ladrones*, captured 2,802, and brought about the surrender of another 707.[23] In March 1903, the *Manila Times* reported that "outlaws" were crowding the main prison of Bilibid: "Ladrones with Long Sentences Are Trying Capacity of Prison. Convicts Come in Bunches," ran one headline.[24] By 1908, about 4,800 Filipinos had been accused of *bandolerismo*.[25]

The Philippine Commission's belief in the prudence of road building grew out of these circumstances. Commissioners held that roads would not only aid the development of the country but also that "[t]heir value from a military standpoint can hardly be overestimated, and indirectly they would only be second to primary schools as an educator of the people."[26] Or, as one observer wrote, "It is clear that the roadmaker fully as much as the school-teacher must be the evangelist of the Philippines."[27] This was the view of David Doherty, a Chicago physician and American Anthropological Association member, after three months of travel through the Philippines around 1903. The administration, although keen to emphasize its benevolent aims and disavow colonial intentions, echoed Edward Gibbon on the roads of imperial Rome. "They united the subjects of the most distant provinces by an easy and familiar intercourse," Gibbon wrote. "[T]heir primary object had been to facilitate the marches of the legions; nor was any country considered as completely subdued, till it had been rendered, in all its parts, pervious to the arms and authority of the conqueror."[28] US authorities might well have declared the same. To make frictionless, maritime highways to China as Albert Beveridge imagined them, it turned out that Americans also had to build colonial roads, an infrastructure long recognized as a crucial method of state building. Fixed capital was required to get capital in motion, and the road is a good example of how the demands of a market empire seemed to multiply, exceed initial estimates, and incur concrete costs to the state. To make Manila the Pearl of the Orient, Americans needed the Baguio retreat, and to access Baguio, they needed a reliable road to get there, and so on.

Given the dreams of making the Philippines instrumental to reaching Chinese markets and the ongoing insurgency across the archipelago, then, the Philippine Commission wasted no time in allotting funds for raising roads. The very first act of the Philippine Commission in 1900 and its first major appropriation directed by corporate lawyer turned Secretary of War Elihu Root provided \$2,000,000 (Mexican) for roads construction.[29] By August, Americans had laid down or bettered some 1,000 miles of colonial roads.[30] The commission's second act, passed in September of 1900, called for a survey for a railroad line from Baguio to Dagupan, the town closest to Baguio from Manila by rail, a measure that offers some indication of the momentum behind plans for the mountain resort.[31] In December, the commission allotted another $75,000 for a thirty-five-mile road from the town of Pozorrubio (northeast of Dagupan) to Baguio.[32]

Commissioners expected the road to be completed by July 1901.[33] This proved too tall an order, and the grant proved to be the first of many appropriations for a project that would frustrate and embarrass the colonial regime. By 1905, the road had run a tab of nearly $1.5 million (US currency)—three-quarters of the total the United States compensated Spain for the entire archipelago.[34] When completed, salaries and wages had consumed about $980,000, or nearly two-thirds of the sum spent on the road. Construction materials came in as the next closest expense at about 15 percent of the monies used.[35] These costs made almost comical that initial $75,000 appropriation. *El Renacimiento* questioned further appropriations for Benguet Road upkeep in 1905, noting that the road did not benefit the general public even as it placed "a heavy burden on taxpayers."[36] As late as 1912, seven years after the road was declared complete, *El Renacimiento*'s successor, the Philippine newspaper *La Vanguardia*, ridiculed it as "the highway of millions," where "more money [had] been buried than in the Isthmus of Suez."[37] This was a view echoed in the American press the year before, when a writer assailed the Baguio hill station as "the costliest disaster, from a monetary standpoint, from which these Islands have suffered since American occupation."[38] Even the pro-American *Manila Times* had earlier taken a swipe at the project, noting that since the initial survey, "work has been going on, dollars have been piling up, two appropriations having been exhausted and another one is in sight."[39]

## Land and Labor

Despite the clear priority given road building (and the Baguio hill station) and the substantial appropriations accorded it, the Benguet Road project troubled the colonial regime before it was eventually touted as one of the greatest highways in the world. Three American engineers headed up the project before it reached completion. The road's first engineer was Captain Charles W. Mead, a civil engineer. His survey had served as the basis for the $75,000 appropriation. When the road did not materialize by his target of July 1901, one Colonel N. M. Holmes took the reins in August of that year. In June 1903, Major Lyman Kennon, an officer who had served in Cuba during the Spanish-American War, succeeded Holmes and directed the project to its finish.

The road proved to be a far greater challenge than Captain Mead had anticipated. He had chosen a path through the Bued River Valley as the most efficient and viable route to Baguio from the lowlands in part because a horse trail already ran over a portion of the land.[40] But the conditions through the valley overwhelmed workers. Mountain peaks soared to three to four thousand feet, some even 7,000 feet above the road's path;[41] at one point, road chiefs considered building a nine-mile stretch with a 6-percent grade to cover a distance of a mere mile and a half.[42] Eventually, engineers shifted the course from Mead's proposed route, which had proven a failure.[43] The natural environment still posed terrific hurdles, as Igorots and *remontados* and their Spanish pursuers would have known well. This topography and also the instability of the land and its kinetic energy imperiled builders, threatening their physical safety. As Holmes described the land, the steep slopes and their geological structure allowed for a "slight disturbance of their outer surface" to "cause a small slide to form which would grow larger and larger so that finally we would be fighting a huge slide hundreds of feet long on a slope that had been previously smooth, and apparently easy character."[44] The menace of calamity lurked beneath a seemingly smooth surface. Rains and typhoons, meanwhile, seemed to will its dissolution and add further to the expense of the road. A writer for *La Vanguardia* agreed, observing in 1912 that "The Benguet Road is one of the most unstable. More than four times it has been the object of the attention of the government and the engineers because every time a storm comes, the bridges are dashed to pieces and the road disappears."[45]

The work itself was backbreaking as laborers sawed lumber for eleven bridges by hand, blasted huge masses of rock and earth to stabilize the road, and carted supplies up from Dagupan.[46] Workers tumbled off cliffs; others plunged to their deaths when a footbridge gave way; cholera and dysentery ravaged other men. Allos, the protagonist of Bulosan's *America Is in the Heart* (1946), described these hazards along a later extension of the highway. "The work went on for three months," Allos narrates. "[S]ometimes it rained torrentially and the water washed away the soft shoulders of the road." "We worked on toward the river that separates Binalonan and Puzzorobio [sic], until one day the water came rushing upon us. I was swept away into a deep bend of the river and was pasted there against the bank, struggling."[47] What Allos described seemed par for the course on the first portion of Benguet Road. "Few days pass without casualty," stated William Cameron Forbes. "[U]p and down the huge canyon," he wrote in 1904, "can be heard the sound of the pick, the rattle of rocks bounding down the mountain, then a load splash in the river, the booming of dynamite and powder as the great blasts go, and strange cries of men."[48]

Among the earliest of these road workers were Igorots. Beginning in January 1901, the government employed several hundred from Benguet, paying them up to forty cents a day in the insular currency.[49] They were managed by US foremen under an army lieutenant.[50] In his initial assessments of the road, Captain Mead described Igorot workers as "fairly industrious, and, as laborers . . . superior to any other natives" whom he had supervised.[51] After he had taken charge, Holmes generally agreed. Though "disinclined to present himself for labor," he judged the Igorot's value at "about three Filipinos." Holmes also contrasted their "cheerfulness and intelligence" to the "stupidity and apathy of the Filipino."[52] Kennon would employ Igorots largely to haul materials including rations, which amounted to about sixty tons on ration days.[53] At least some Igorots embraced waged employment on the road.[54] But roadwork as an exercise in contractual freedom seems to have been relatively short lived for Igorots. When the commission passed the Special Provincial Government Act in 1905, the year of Benguet Road's completion, it required that every man between the ages of eighteen and sixty pay an annual tax of two pesos for the upkeep of roads or commit ten days of work on these public works projects.[55]

In any case, two hundred Igorots were an insufficient labor force for the job. In 1902, a year after visitors should have been ascending the Cordillera along the highway, Holmes, then the road chief, struggled on with labor problems. A cholera epidemic had struck the workforce, claiming the lives of thirty natives and five Americans. This stalled work for a couple months before it resumed, only to be arrested again upon the Christmas season. Holmes groused just as Mead had about "Filipino holidays"—Catholic feast days—as the cause of delays.[56] By the beginning of 1903, a monthly average of only 270 men worked on the road.[57]

To improve these numbers Holmes sent out labor agents to scour for more workers north in Ilocos, and south in Pangasinan, those provinces known for centuries to Igorot traders. His representatives approached provincial governors and local headmen, and they made pitches to "the men themselves." But their "persuasive powers" were to little avail. Holmes did not disguise his irritation. "It was found in some of the northern provinces that the food supply was short, the villages were overflowing with able-bodied men, no work to do, and no money," Holmes wrote in one glum report. What might seem dire circumstances to the colonel could not move Filipino men to work: "[T]hese men have preferred to lie idle in a state of semi-starvation."[58] Holmes could find few Filipinos willing to abandon their provinces and likely their own subsistence obligations, growing rice, *camotes*, corn, or tobacco as Allos's family did, for the grueling and dangerous work of carving out a path along the steep Cordillera. According to 1903 census statistics, the majority of residents interviewed across Pangasinan, Ilocos Norte, and Ilocos Sur either had no known occupation or no "gainful" employment or were involved in agricultural production.[59] Agricultural production in these provinces largely reflected the general conclusion reached by census takers—that just over 80 percent of farms in the Philippines were cultivated by owners and that most individually held farms, whether owned, rented or shared, were on small parcels of land. The greatest majority were less than five hectares, and in Ilocos Norte and Ilocos Sur no less than 64 percent of farms were under one hectare in size.[60] If Filipinos were not farming primarily for subsistence on these small parcels, they were growing crops for sale.[61] These conditions may explain the limited enthusiasm for work on the road.

Even those whom Holmes had found to toil seemed less than thrilled at the task. He wanted his crews to display a willingness to work, as if such an appearance could affirm the rightness, even the benevolence, of his demands. But they held out "with a perfect stoicism the termination of the day." Or, they worked, but withheld even a jot of pep: "In handling a pick the native will raise it in the air, allow it to drop by its own weight, striking a glancing, infinitesimal blow," until an American foreman intervened and corrected him.[62] If his commentary was testimony to his incompetence in motivating labor, Holmes's assessment also hints at Filipinos' "camouflaged resistance" to foremen, Syed Hussein Alatas's description of the "silent protest" of those Malays, Filipinos, and Javanese whom colonial authorities so often took to be "indolent."[63] Holmes's men soldiered, performing the barest minimum necessary for staying on the job. And workers knew how to throw other wrenches into the imperial gears. Holmes reported that "There is . . . a spirit of maliciousness prevailing among these people"; they stole, scattered, or destroyed tools; they "slic[ed] and nick[ed] . . . freshly painted bridges." They even destroyed a large "Benguet Road" sign set at the beginning of the highway.[64] The colonel concluded that "the native laborer" was "idle, shiftless, and stolidly indifferent, approaching his work with no degree of intelligence or judgment, of a deceptive and treacherous character, wantonly careless and frequently maliciously destructive, uninterested in and indifferent to his work."[65] So desperate was Holmes that at one point in 1903, with only 173 workers on the road, he began hiring Americans at $2 gold per day; this was about double the rate that Kennon would later offer them.[66]

Holmes was not alone in some of his views of Filipino labor. In mid-June of 1902, the commission reported that "The laborers under the Spanish regime were accustomed to do their work at the bidding of some superior, not from a motive of earning wages, but in obedience to the order of one entitled to command."[67] Perhaps commissioners had in mind the supposed legacies of the *polo*, the forty days of work required of natives by the Spanish colonial state. This work could involve construction of government buildings, enlistment in military ventures, and the felling of trees and preparation of lumber for the Spanish galleons.[68] In light of this past association with labor, commissioners saw the United States' duty as "teaching the Filipino laborers the independence and dignity of labor under a free government."[69] As late as 1911, William Cameron Forbes, then governor general of the colony, sounded the same concern about workers' lack of free agency, and consequently, Americans' duty in the colony: "[W]e have not yet succeeded in liberating labor. By reason of long custom the native still lives in a state of peonage and we have not yet secured that freedom of investment in the individual to buy and sell that makes the poor man a free agent."[70]

As these comments from colonial administrators suggest, the labor challenges along Benguet Road set in relief a problem that some Americans perceived as endemic to the colony. And this made the road to Baguio a test of Filipino labor, one studied by those with a stake in the political and economic future of the colony. Kennon even submitted his final report on the road as a contribution to the issue of labor in the Philippines, a question "of general interest," he noted.[71] He recognized that the lessons of Benguet Road had great consequence. And if the road proved a trial for workers, it had also become a workshop in labor management and a challenge to road chiefs, who experimented with further means to secure reliable and tractable labor while keeping costs down.

Besides using labor agents to secure workers and drawing on the local Igorot population, chiefs tested the efficacy of employing white Americans as foremen. They generally came from military ranks or had worked as miners and were thought key to extracting consistent work from Filipinos.[72] While temperatures were cooler along the road, Holmes did observe that the Philippine climate compromised "white m[e]n's efficiency"; it was assessed at two-thirds the efficiency of white men in the United States. These workers also bore a greater susceptibility to falling sick; "a large American force," Holmes wrote, "means a large sick report."[73] But for those able to perform the work, labor on the road could be transformative, at least by one account. It could redeem crews of ex-soldiers sinking into moral decay in the aftermath of their military service. Grace Helen Bailey, a visitor to the Philippines, who wrote of her travels in *The Overland Monthly*, claimed that soldiers who had slid into "a dissolute existence peculiar to the tropics" following the war "reached up to the level of men once more, moved to the virile impulse, to new ambitions by the bracing air and play of muscles as pick rang out to pick in swing and strengthening of brawn." Their "redemption" owed to "the God-sent labor on the Benguet road."[74]

If the road project gave former soldiers an alternative to postwar dissipation, it also gave chiefs an opportunity to study, more generally, the strengths of American labor relative to that of other workers on the road. Holmes presented assessments of workers' productivity based on nativity in that way, serving not just as a manager of labor but also of race and an implicit "race competition."[75] While "Americans," a category that included workers from Europe and Africa,[76] shoveled twenty yards of dirt in nine hours, he noted, Igorots moved nearly eleven yards, Filipinos a measly three.[77] Kennon's wage scale, meanwhile, showed that American laborers earned from $1 to $1.50 a day; Japanese and Chinese workers, $1; "native laborers," twenty-five cents, although a note does appear that that wage could rise based on time on the job and "individual capacity."[78] In these reports, race and nationality were grafted onto the value of labor power so that pay reproduced and monetized racial and national difference much as it had on railroad projects in the US West, on the Panama Canal, and in US domestic factories, too.[79] They also illustrate Lisa Lowe's argument that in US history, "capital has maximized its profits," or in this case, minimized costs, "not through rendering labor 'abstract' but precisely through the social productions of 'difference.'"[80]

The most controversial attempt at imposing discipline and managing costs through labor competition was Holmes's call for the recruitment of Chinese workers. The suggestion raised the ire of Filipinos. As an article in *El Renacimiento* protesting the introduction of "*amarillos y negros*"—Chinese and black workers—insisted in February 1903, "Our laborers are abundant, our laborers work, our laborers can challenge any other laborer in whom work is entrusted."[81] The year before this, the United States had extended its 1882 ban on Chinese immigration to include the archipelago. Philippine Commissioners had worried then that the introduction of Chinese "coolie" labor "would furnish the strongest and most taking argument" to those who opposed a reduction of the tariff imposed on Philippine imports into the United States. This was a measure that commissioners supported to ease strains on the colonial economy, one dramatically weakened by war. Introducing Chinese labor could lead opponents to argue that free trade between the metropole and colony "would bring American labor and its products into direct competition with cheap Chinese labor and its products in these islands."[82] But in 1903, as colonialists' attention began to shift, if fitfully, from war to recovery and development, some Americans like Holmes and foreign commercial interests again issued calls for the recruitment of Chinese workers.

President Theodore Roosevelt pursued other possibilities, too. In 1902, he appointed T. Thomas Fortune, an African American journalist with the *New York Age*, a "special commissioner" and sent him to investigate the possibilities for black workers and African American colonization in the Philippines and in Hawaii.[83] Fortune found the Philippines rife with "white supremacist thought," but he departed the archipelago not wholly down on the prospects for African Americans.[84] He held that naming a well-known African American like Booker T. Washington to the position of governor general might persuade both Filipinos and African Americans of the opportunity that lay in migration.[85] These hopes amounted to little. *El Renacimiento* largely rejected the idea, maintaining, as many African Americans also did, that the intelligent and enterprising "are not exactly those who would come."[86] Calls for the recruitment of Chinese workers enjoyed more traction.

Bernard Moses, a member of the first Philippine Commission, for example, made the case to the *San Francisco Chronicle* in 1903. Moses, a professor of history and political economy at the University of California, wrote that most Filipinos worked small landholdings or were retained "in a traditional relation not greatly unlike that of feudal dependents." Rapidly transforming this social order and the culture it bred, he believed, would be difficult, so he promoted the recruitment of Chinese workers. Regarding immigration from India or Java as "undesirable" even if possible, he envisioned "a superior product as a result of [Chinese and Filipinos'] union."[87] Moses, like Holmes, shared the view of the *Manila Times* early in the US occupation. In 1899, an editorial promoting Chinese immigration disparaged Filipino workers as "a worthless lot" and warned that closing the doors to Chinese labor would render capital "shy of coming here to invest in any enterprise, or to build railroads, docks, or any improvement."[88]

*El Grito del Pueblo*, a pro-labor, anti-imperialist newspaper of radical labor union Union Obrera Democratica, struck back at these representations of Filipino workers just as many pieces in *El Renacimiento* did. Founded by Isabelo de los Reyes, an Ilocano who had once circulated among anarchist and socialist intellectuals in Spain and gained recognition for writing on what he called Filipino folklore, the Union Obrera constituted the Philippines' first federation of workers. The "Barcelona-style free-wheeling central" led strikes across Manila in 1902 and brought together printers, tailors, carpenters, tobacco workers, dock workers, barbers, and clerks—

urban workers who emerged as a distinct class as the Philippines moved toward an export economy through the nineteenth century.[89] In 1903, *El Grito del Pueblo* contrasted Filipino and Chinese laborers on Benguet Road. The Chinese performed sound work until waylaid by their opium habit; Filipino workers, meanwhile, although known to register regular complaints with supervisors, proved to be reliable workers once their demands for better food rations were satisfied.[90] The paper also published a Spanish translation of an article written by Edward Rosenberg of the American Federation of Labor in 1903. Rosenberg had been brought to the colony at the behest of Governor General Taft, who aimed to disarm the radical Union Obrera.[91] Rosenberg also concluded that poor pay and working conditions plagued Philippine workers, not laziness.[92] The Philippine Commission's 1903 annual report, meanwhile, as if to blunt Holmes's unflattering report on Philippine labor, or "Exhibit F," and calls for the recruitment of more Chinese workers, published "Exhibit $F^1$" and an "Exhibit $F^2$," a series of ringing, even gushing endorsements of Filipino labor relative to Chinese from army quartermasters.[93] "One hears a great deal of the necessity of introducing Chinese labor into these islands to meet the demands," wrote one commentator. "It has been my experience that any labor which can be performed by the Chinese can be performed equally well by the Filipinos."[94] Given the constraints posed by these politics, then, and the trials and errors of his predecessors, and, of course, the imperative to get the Benguet Road done, on taking the helm, Major Kennon would pursue other methods to raise a labor force. These paired with colonial laws that were simultaneously compelling work would reveal the limits and contradictions in Americans' commitment to liberating labor by teaching Filipinos, whether figured as idle and indolent or as poor serfs, the virtues of free agency.

## Poblete's Workers

In 1903, after Kennon had taken charge, the *Manila Times*'s editors deemed the ongoing road project "An Opportunity to Test Native Labor," and an editorial dared workers to make good on the claim that they have "only been waiting an opportunity to work at reasonable wages."[95] Pascual Poblete, a Filipino journalist and labor leader, responded. Together with Isabelo de los Reyes and Dominador Gomez, a physician, he had founded the pro-independence Partido Nacionalista. The political party drew from Manila's "urban middle sector" over the capital's elite,[96] and US authorities associated it with the continued political agitation and even banditry disturbing areas in and around Manila.[97] In 1903 Poblete took to *El Grito del Pueblo* and advertised his readiness to mobilize 200,000 workers "to whatever business that requests them of me so long as it would offer a salary of fifty cents gold at least." As he reflected on his offer later, Poblete explained that he had intended to show that no need existed for importing foreign workers into the Philippines; he also hoped to "liberate" Filipinos in the provinces of Luzon "from the cruel indigence that overwhelms them."[98] So far as he was concerned, work along Benguet Road might do just what Holmes had believed it could: alleviate conditions of hunger precipitated by drought and blight and prolonged war that plagued the lowlands especially around 1903. This was a view of road work that *El Renacimiento* had shared earlier that year, believing that "Nuevas Carreteras" could extend a sort of new deal to Filipinos. Road building had the "noble purpose of providing occupation" for those left unemployed since the war.[99] In showing the eagerness of large numbers of Filipinos to take up work, Poblete also believed he might intervene in the question of Chinese labor. Kennon, then recently appointed to the position of chief of Benguet Road, responded to Poblete's irresistible offer in early July.

What followed became a source of some dispute and even an official investigation.[100] Poblete managed to amass some 1,000 workers.[101] The first 250, some with wives and children, left Manila and began the journey to the road on July 17. Before the end of the month, however, most had returned to the capital. Poblete wrote articles in *El Renacimiento* explaining the affair and reported that American workers had maltreated his men. Some were deprived of a decent place to sleep, shelter from driving rain, and even adequate food.[102] Beyond that, they were denied the wages that they had come to anticipate. Poblete reported that Kennon had initially agreed to compensate recruits not with the fifty cents gold that Poblete had advertised as their going rate, but twenty-five. Invoking that racial hierarchy of labor, Kennon claimed that Filipinos did not work as hard as Americans did, but he also pledged that if it turned out that Filipinos possessed the same capacities as American workers, "they would have the same pay." Poblete, invested with faith in his workers' ability and, apparently, in Kennon, accepted these terms, believing Filipinos surely ought to be paid as well as Americans.[103] The later investigation showed that the workers were misled by the message of *capataces* (foremen) recruited by Poblete. The workers had come to expect as much as $1–2 gold a day—the going rate for American workers—but wound up with a forty-cent wage.[104] Worse, as workers abandoned the project in northern Luzon and began the long, exhausting journey south to Manila, some men reportedly fell victim to hunger, and reports emerged that several men even perished along the trip. Most of Poblete's other recruits, having heard these accounts of Benguet Road, never followed the first group to Dagupan.[105] And Poblete, who had been ridiculed in the *Manila Times* for his initial offer (the paper referred to his advertisement as "Poblete's Filipino Labor Bluff"), became the object of derision again. The paper published a report claiming that Poblete would have enjoyed a tidy commission had he successfully supplied the mass of workers for Benguet Road.[106]

Historian Greg Bankoff has suggested that Poblete's supposed "labor bluff" was more like a ruse: Poblete may have devised an opportunity to embarrass the American colonial government or at least test its promises of justice and benevolence.[107] This could well have been the case. Poblete certainly bore the credentials of an insurgent nationalist committed to overthrowing US rule. In 1902, he had been questioned by the constabulary over rumors that his political organization aimed to reignite the Katipunan, the secret society composed largely of rural peasants and urban workers, like members of the Union Obrera.[108] The Katipunan had pursued not reform but revolution in 1896. And further, as Bankoff points out, Poblete's highly public offer of labor and, later, these stories of death and starvation on Benguet Road came shortly after the colonial government's offensive on the Union Obrera.[109] Earlier that spring, the colonial government had seized the accounting books of the union and brought up its leader Dominador Gomez, then also at the helm of the Nacionalista Party, on charges of embezzlement.[110] They tarred one who had claimed to stand for the people ("*el pueblo*") with the charge of corruption. Perhaps Poblete did indeed see an opportunity to strike back.

Still, Poblete was careful, at least in his initial public statements, to avoid incriminating Kennon specifically or issuing a blanket condemnation of the project; in fact, he made a point of noting that ultimately sixteen of his workers remained and continued to labor on Benguet Road and reported fair treatment.[111] And perhaps with good reason; with the installation of Kennon as chief, the road and labor management on it had begun to turn a corner. As Bankoff has suggested, Kennon brought both methods of "systematic management" and "racial paternalism" to the road project.[112] Kennon critiqued his predecessor, observing that workers had been "little considered or consulted" in the process of their recruitment.[113] Kennon, who had effectively gathered large groups of Ilocanos to serve the US military as guides, as seamen, and as transporters of military equipment around 1901,[114] claimed that under his predecessor, Americans on the road had used coercion but not what he called "special inducements" to maintain steady labor.[115] This was Kennon's best practice for recruiting workers—setting a music band on the road, hosting dances, permitting men to bring along their wives, and honoring holidays.[116] Filipinos, notably a large number of Ilocanos, worked for Kennon, stories went, because they liked him and because they wanted to.[117] "The life of it all is the major," Forbes wrote of the highway. "Everyone loves the major. He is the great moving spirit. They come to him with their woes and with their joys."[118] In effect, Kennon performed like a headman, or *baknang*, distributing surplus capital as a sign of his status and as a way to gain his charges' consent and their labor power. Other reports on colonial labor affirmed the efficacy of this management strategy; a vice president of the Atlantic, Gulf, and Pacific Company tasked with building the Manila port works asserted in 1903 that the best way to maintain a Philippine labor force was to supply workers with homes, amusements, Sunday fiestas, and schools—needs, services, and pleasures that successive wars had compromised.[119]

## Liberating Labor

Kennon's upbeat report on Benguet Road and his success in recruiting workers, however, at times betrayed itself and made the voluntary labor that Kennon was eager to promote look anything but freely given. Kennon's charisma and the fanfare on the road distracted from quiet and more coercive means at work for securing and maintaining labor on the highway. For instance, Kennon experimented with prison labor from Bilibid, the main prison in Manila that had filled with *ladrones*; this was a practice used throughout the archipelago and eventually in Baguio. Forbes would employ prisoners to keep Baguio beautiful: In 1909, in exchange for pardons, convicts were required to work on the hill station for several months; "we get laborers there without the necessity of guarding or of paying them," Forbes wrote approvingly.[120] Kennon was not so successful with this brand of worker along Benguet Road. Prisoners grew "homesick" for the prison, he admitted; disease took its toll on others; "some died, other escaped, and the remainder were demoralized and useless as laborers."[121] Unwittingly, Kennon's dismissal of penal labor, a practice that would become the object of Progressive reform in the United States in succeeding years, seemed to undermine his upbeat account of Benguet Road. The fact that prisoners would become demoralized and come to long for their cells speaks volumes.

In the end, according to Kennon, only the most desperate men took to working on the highway to Baguio—tractable men, perhaps, unlike Poblete's who had claimed the free agency and the power, if severely circumscribed by dire postwar conditions, to negotiate the terms of their pay and work. Kennon's team sent Filipino agents—not unlike Poblete's—to towns and districts "in which starvation conditions were reported to exist."[122] While these conditions had not moved men while Holmes was at the helm, it seems that by Kennon's time, they did. The secret, in short, was not educating Filipinos on the liberating properties of free labor or hiring those well acquainted with contractual relations as Poblete's men seemed to be. It was not inviting Filipinos to make rational choices to maximize their incomes. It was targeting the hungry—those who had to work—then plying them with cigars and dances. "The Benguet road," Kennon wrote, "was a harbor of refuge for all of the unemployed of the Philippines."[123] He noted that rations along the road presented "a greater variety and abundance than [workers] could obtain ordinarily at home. The mere fact that meat was issued regularly to them," he wrote, "undoubtedly brought many laborers to the road."[124] Kennon's findings echoed those of political scientist Paul Reinsch, who wrote books on imperial politics and colonial government between 1900 and 1905. In the first of these, Reinsch commented on labor conditions in the Philippines. Reprising that enduring, imperial understanding of colonial subjects, Reinsch claimed, "The wants of the natives are few. They soon earn what they consider a competency and with it retire to their native villages to live their accustomed peaceful life." Employers, he heard, "have rather preferred men who drank, gambled, and played the *gallera*, because such men, having more wants to satisfy, would work with greater energy and persistence."[125] Kennon had help in gathering hungry, dependent workers like these, and it came not only from labor recruiters seeking out the most unfortunate. It also came from wider colonial conditions: from the cumulative effects of successive wars and legal responses to the losses and social dislocations they precipitated. These extra-economic means stood to turn the "idle, shiftless, and stolidly indifferent" Filipino described by Holmes into a hungry worker and generate at least the impression that his work was freely done on the road and across the archipelago.

The first of these means was a measure passed by the Philippine Commission in the fall of 1902, about a year after military authority under Arthur MacArthur gave way to civil government under William Howard Taft. While the law itself did not generate a ready supply of labor, it points to the way in which war and subsequent counterinsurgency efforts could generate conditions that compelled work. The law defined what constituted "highway robbery," the theft of personal property, especially carabaos or water buffalo, by armed groups traveling the country. Conviction carried a punishment of imprisonment for at least twenty years if not death.[126] The law addressed itself to the preservation of private property and to those *ladrones* and brigands who had made roads into the far reaches of the archipelago crucial to the pacification of the colony. As late as 1905, William Cameron Forbes, as secretary of commerce and police, noted the trouble posed by these *ladrones*. But as the Spanish had done, Forbes, like Taft before him, chose to attribute no political significance to their acts. To Forbes, the direct English translation of *ladrones* made the most sense: these shadowy figures were no more than thieves.[127] *El Renacimiento* ran a series of articles in 1905 with contributors offering their assessments of this "*bandolerismo*." Writers debated whether the causes were political, the result of caciquism, or the consequence of desperation and hunger. It was difficult to parse "bandits'" motives, but it was clear that their offenses mattered. Some like Isabelo de los Reyes believed that Americans might seize on banditry as cause for delaying the formation of a popularly elected Philippine Assembly, a promise of the 1902 Organic Act.[128] And indeed, Henry Bandholtz, an army officer and head of the colonial police force, the Philippine Constabulary, threatened that the formation of the Assembly was contingent on the elimination of banditry.[129]

The measure and the punishment meted out to those convicted—at least twenty years, if not death—suggests just how seriously officials took the subversions of *ladrones*. The perceived gravity of their acts implies that more was at stake than herds of carabaos. The brigandage law can be interpreted as fundamentally "anti-nationalist," much like the 1901 law against sedition and the "seditious plays" written by Filipino literati committed to Philippine independence and much like action taken against Dominador Gomez and the Union Obrera.[130] In fact, the Philippine Commission linked the seditious plays, Gomez, the Partido Nacionalista, and ladronism through Luzon in its 1904 report on "disturbances" through the colony.[131] The following year, Isabelo de los Reyes, in one of his contributions to the "*bandolerismo*" series in *El Renacimiento*, implied a similar relationship between ladronism and the politics of anticolonialism. He argued that only independence would "bring permanent peace to our country" and eliminate the "pretext"—the excuse—claimed by the makers of "insurrection, or banditry, as you like to call it."[132]

As early as 1900, the American colonel stationed on Zambales had also recognized the implications of this insurgency, only he had in mind different ends than de los Reyes. The colonel had warned then that "The early suppression of these *Ladrones* who are making war under the name of *Insurrectos*, is a matter of vital interest . . . to our Republican Administration. If these marauders are not stamped out, they can keep up this appearance of war indefinitely, thus giving campaign ammunition to our enemies at home."[133] Perhaps, then, it is no surprise that Taft and Forbes would declaim the political valence of brigands' deeds and refuse to recognize them as anything more than robbers just as they sought to limit the purview and, in that way, the radical potential of the Union Obrera. When Edward Rosenberg had made his pitch for the AFL following the Gomez scandal, he cautioned Filipino workers that Taft would not tolerate a union engaging in political agitation.[134] In these ways, Americans followed their colonial predecessors who "treat[ed] any form of dissent and rebellion as banditry and therefore without political significance."[135] Taft and Forbes criminalized and in that way attempted to depoliticize the actions of the dispossessed through language and through the law, appropriating for the imperial regime the power to define the public order and justice itself.

The brigandage law targeted not only those dispossessed of political sovereignty but also Filipinos who had lost means to reproduce their own lives. Forbes and those who argued that the object of the banditry was merely "booty" may have been right in some cases. Across Luzon, carabaos were crucial for cultivating rice. But war had dramatically reduced the stock of carabaos to a supply further diminished by disease. In 1902, the colony's stock had declined to 10 to 15 percent of its size in 1896.[136] One consequence was that the price of carabaos increased precipitously from $20 (in Mexican silver) to $200 since war had visited the archipelago.[137] Another outcome of the sudden loss of carabaos was that land went uncultivated and "intense hunger, verging at times on wholesale starvation" plagued Filipinos. "In many districts, food became a far more precious commodity than human life," with rice at 25 percent of pre-war production.[138] Mid-1902 witnessed a food shortage that forced the Philippine Commission to import rice from India; rice even became payment for construction work on provincial schoolhouses, those other key sites of colonial nation building.[139] The next year, *El Renacimiento* described receiving reports of inexorable hunger and loss of crops due to "intense drought or . . . the voracious fangs of the insatiable legions of locusts."[140] These were conditions that Poblete aimed to ameliorate in supplying Filipinos with work—both the scourge of hunger and the violations of property spurred by sheer deprivation. As *El Renacimiento* put it in one article, hunger inspires "the savage instincts of the beast [to] dominate."[141]

The colonial law against banditry, then, aimed to preserve property that the US war for the Philippines had helped to undermine and halt action that the imperial war had incited. A war that devastated lands and traditional means of subsistence created those conditions of need that Kennon concluded were vital to accruing tractable labor, even as it inspired continued insurgency or "brigandage." The US counterrevolutionary war in the

Philippines thus at once complicated and simplified the task of recruiting colonial labor for such projects as road building.

The commission's law against vagrancy can be interpreted as another measure that could serve counterinsurgency functions while also creating a labor pool. The commission enacted it on the same day as that outlawing highway robbery. The law defined the vagrant as "Every person having no apparent means of subsistence, who has the physical ability to work, and who neglects to apply himself or herself to some lawful calling."[142] The law went on to enumerate seemingly endless forms of doing nothing, or what *appeared* to administrators as the absence of productive labor. In this way, the law was tantamount to mandating work of every body. And not working, or not working in ways discernable to colonial authorities, became no mere moral lapse. It was provocation to an apparatus with recourse to punishment.[143] The commission intended to target men like those ex-soldiers described by Grace Helen Bailey—"dissolute, drunken, and lawless Americans who are willing to associate with low Filipino women and live upon the proceeds of their labor," Taft wrote. But he also noted that the vagrancy law was useful in prosecuting "Filipinos suspected of complicity with *ladrones*, who have no visible means of support and who are probably spies of the *ladrones* for the purposes of enabling them to make lucrative raids."[144] Some one thousand Filipinos were accused of vagrancy, a "crime against public order," between 1903 and 1908.[145] If the law against highway brigands aimed to still bandits, or latter-day "Robin Hoods," the vagrancy law would put them to productive labor.[146]

The third body of laws, forestry conservation laws, which Gifford Pinchot, the first chief of the US Forest Service, helped to put into practice, was similarly aimed at creating public order by regulating how people self-provisioned and their very mobility.[147] Adopted in 1904, the chief goal of the US forestry program—apart from the classification and identification of species and forest districts—was devising a licensing and surveillance system to control loggers' and local peoples' use of forests.[148] Americans, like the Spanish, recognized "communal" forests, which were granted to any town that sought the designation. Within these forests, Americans extended "free-use privileges."[149] According to George Ahern, Forestry Bureau director from 1900 to 1914 and one-time secretary to Sitting Bull, communal forest rights would allow Filipinos to "exploit their own forest resources," and it would curtail "useless destruction" of woods.[150] In this way, the colonial government could sell "surplus timber," generate revenue, and ultimately reduce tax burdens.[151] Forestry law thus looked to the fiscal welfare of the colony and to its material development—Manila was in dire need of lumber for construction early in the US occupation.

But it also regulated and remade Philippine peoples' relationship to their own lands. Forestry law defined which species could be harvested based on market values, and it limited the geographic bounds within which an individual could secure firewood and timber for house building. In this regard, "free-use" policy within communal forests could function much like the rule against *ladrones* and vagrancy—as a way of fixing the population spatially and minimizing the potential for the social and political challenges that highly mobile individuals and groups could pose. And much like the vagrancy law, in creating these rules, the bureau set up new categories of transgression. If Philippine peoples proved unwilling or unable to abide by the bureau's rule of communal forests, if they secured materials from other areas or used timber for non-domestic purposes, they were to be identified as criminals. Stipulating "free use" had simultaneously defined what constituted criminal theft.

Violations of "free use" alone did not aggravate American foresters. The continued practice of *caiñgineros*, or swidden (or "slash and burn") agriculturalists like some Ibaloi, also vexed them.[152] Benguet Province Governor William Pack noted this practice among Benguet Igorots who utilized fires to produce grazing land and fertile ground for cultivating *camotes*.[153] To Dean Worcester *caiñgin* demonstrated Philippine peoples' heedlessness and even indolence: "A large majority of the inhabitants of the Philippines," he wrote, "will not fight, for any length of time, the tropical weeds and grasses which invade their cultivated fields, and rather than attempt to do so [they] prefer to clear forest lands, slaughtering the trees indiscriminately and burning them where they fall," he wrote.[154] Ahern's 1906 report named the making of *caiñgins* "by far the most destructive agency in the Philippine forests";[155] a later report put timber losses due to *caiñgins* in the millions.[156] Even as administrators were proscribing native forest practices, however, they were granting American entrepreneurs the right to fell trees for profit. As early as 1902, the Benguet Commercial Company, founded by H. Phelps Whitmarsh, the first civil governor of Benguet, secured a license following Whitmarsh's departure from office to cut up to 3,000 cubic meters of Baguio pines. In three years, he reported sales of 1,895 pesos of a resource that the Ibaloi had viewed as communal.[157]

Ahern believed that without sufficient bureau manpower and strict surveillance, it would be difficult to stop *caiñgineros*.[158] And so the bureau tried to convert *caiñgineros* into owners and operators of small farms in a policy akin to allotment in the US West. There, the US government aspired to civilizing and assimilating Native Americans by converting them into owners of discrete parcels of tribal lands, a process that ultimately freed portions of the West for sale and settler colonialism.[159] In the Philippines, the 1904

Public Land Act allowed "for the homesteading, sale, and lease of public lands that are more valuable for agricultural than for forest purposes."[160] "People desiring to make *caiñgins* should be required to make out homestead applications for the land," Ahern wrote.[161] He assumed that private ownership of land created a disincentive to *caiñgin* making—the former itinerant becomes vested in a single plot by mixing his labor with the soil. Assigning people to plots of land and making them property owners may have granted a measure of independence to those under the tight fist of local and provincial leaders. But this policy against swidden agriculture did more: it contained native peoples' movements and limited the spaces across which they could carry on subsistence- and even profit-generating practices.[162] In this way, the spirit of this law can be likened to that against *ladrones* and brigands. Besides creating new responsibilities and limiting an agriculturalist's options when land became exhausted or infertile, it shaped a social geography better suited to control. When assigned to specific plots, individuals could be more easily found, their homes located by an agent of the state, searched, and made subject to taxation, another objective of the measure and one that Igorots had for so long managed to subvert.

Finally, in altering long-practiced means of subsistence if not profit making, forestry laws helped to make wage labor—like work on the roads—not a choice but a necessity. People would have to work not for food and shelter but for means to afford and access them, a process already underway as money and production for the market rose in importance in the nineteenth century. In this light, the homesteading provision and imposition of Progressive forestry principles appears no mere means of conserving nonhuman nature and saving virgin forest from the slash and burn of indolent Philippine peoples, or of putting the colonial administration on sounder financial footing. Like early forest conservation policy in the United States, forestry laws in the Philippines could serve as an instrument for mastering nature together with native peoples and for forcing "free" labor by enclosing common lands and resources.[163] Through a war of dispossession that undermined political sovereignty and devastated Philippine lands and conditions for self-provisioning and means of economic sovereignty, through the charismatic authority of figures like Kennon, and in the passage of laws that continued the process of conquest through legal means, Filipinos were put to work. This, it would seem, was colonial schooling in free agency and in the dignity of labor.

Altogether, then, we have an instance of what Karl Marx called "primitive accumulation." Marx used this phrase to identify a process of "separation between labourers and conditions of labour"—between producers and the means of production—that takes place, he argued, at the transition from one mode of production to another.[164] The account of primitive accumulation with which we may be most familiar is the enclosure of "the commons," a process that marks the transition from feudalism to capitalist agricultural production. This enclosure as a form of primitive accumulation entailed "social forces" beyond "impersonal 'pure' economic laws";[165] Marx wrote that "methods" of primitive accumulation "employ the power of the State, the concentrated and organised force of society, to hasten, hothouse fashion, the process of transformation of the feudal mode of production into the capitalist mode."[166] Scholars have debated whether to understand primitive accumulation as a historical event—one that is the "premise" of the capitalist mode of production[167]—or as "an ongoing" process, one that persists within capitalism.[168] Several have recently made the case for the latter, noting that Marx himself identified colonialism, "the national debt, the modern mode of taxation, and the protectionist system" as forces of primitive accumulation.[169] The story of Benguet Road's construction suggests the possibility of enclosure experienced as both event, or primitive accumulation, and ongoing process, or "accumulation by dispossession," geographer David Harvey's term. For the Ibaloi, the separation of producers and means of production may have had a "primitive" or "*ex novo*" character given the limited reach of the capitalist market in Benguet. Many Filipinos, however, already well acquainted with capitalist market relations, experienced enclosure as ongoing: Spanish colonialism ultimately gave way to a counterrevolutionary war—an extra-economic means of dispossessing Filipinos of economic and political resources to reproduce life and even create society anew. This, in turn, yielded to a system of laws that coerced "free" labor.[170]

## Coxey Poblete

If these stories of Benguet Road remind us of the extra-economic forces enlisted in capital formation, they also prompt review of the notion of "new," market empire. They suggest that making the way for free trade entailed practices that looked anything but new. Americans' methods of accruing labor for the road and for other colonial projects, shaped in part by forestry laws and rules against highway robbery and vagrancy, replicated that of the United States' predecessor. The Spanish had also impressed Filipinos to build ships for the galleon trade and required work by natives through the *polo*. US colonial laws—generated in the context of insurgency and counterinsurgency—also had precedent in the post-Civil War American South. Black Codes, state laws enacted in the South by the end of 1865, were intended to enable newly freed men to be property owners, to marry, and to make labor contracts, among other privileges. But vagrancy laws and the theft of horses and mules as capital crimes were also tools for "stabiliz[ing] the black work force and limit[ing] its economic options."[171] White, southern lawmakers' efforts to limit freedmen's rights to hunt, fish, and pasture livestock had a similar purpose.[172] As Amy Dru Stanley has argued, "slave emancipation furnished a native model of forced labor."[173] War and its exigencies did the same in the Philippines, and in these respects, new empire can appear less than novel from the vantage of the colony.

This history of Benguet Road recovers, then, some of the political and extra-economic work of market making and bears out Jane Burbank and Frederick Cooper's point that "[t]he imperialism of free trade was always on the verge of becoming something else—that was why it was imperialism and not just trade."[174] Through charismatic authority, war, and laws of dispossession, colonialists appropriated labor power to improve an infrastructure for "free" trade. These methods entailed naming and classifying Filipino subjects as robbers, brigands, and criminals and setting workers within an instrumental hierarchy of ethnic and racial difference.[175] Through these material and discursive means, American administrators liberated labor and generated subjects of rule like Carlos Bulosan and his analogue in fiction, Allos.

In succeeding years, Americans tested additional ways to build and repair colonial roads. Drafted by William Cameron Forbes, a first measure proposed a corvée system, wherein men were expected to submit to five days of labor or pay a tax. When provinces refused to adopt the measure, Forbes proposed a road toll, a measure succeeded by one that permitted provinces to double the *cédula*, or registration, tax to help subsidize provincial road-building.[176] This policy, paired with the special provinces act, which mandated labor on roads, emerged in the wake of Benguet Road's construction. Colonialists had learned how to distribute the cost of road building and maintenance to all Philippine peoples who continued to experience acts of dispossession. The protagonist of Bulosan's 1946 novel, in fact, sets off for work in Baguio and eventually in the United States to redeem his family who had been gradually dispossessed of land in Pangasinan, the province south of Baguio where early road engineers repeatedly tried to recruit workers.

Back in 1903, the *Manila Times* referred to Pascual Poblete as "'Coxey' Poblete," referencing Jacob Coxey, the Ohio businessman who led a march of his "Industrial Army" to Washington in 1894 during the economic depression.[177] Demanding recognition of the poor and unemployed of the nation, he declared at the Capitol, "We stand here to-day in behalf of millions of toilers whose petitions have been buried in committee rooms, whose prayers have been unresponded to, and whose opportunities for honest, remunerative, productive labor have been taken from them by unjust legislation, which protects idlers, speculators, and gamblers."[178] The *Manila Times* had crafted an apt allusion. Both Coxey and Poblete were insurgents. One questioned appropriations of wealth in the United States, wealth so extravagantly displayed the year before at the 1893 World's Fair and the year before the Benguet Road's completion at the 1904 World's Fair. The other challenged the United States' appropriation of wealth and political sovereignty in the colony. Their insurgencies suggested that imperial expansion would be no simple solution to what ailed the American economy and society, contrary to Albert Beveridge's great expectations. Indeed, its terms would be thoroughly challenged in the Philippines by the likes of Poblete, Isabelo de los Reyes, and Dominador Gomez. Bulosan would himself go on to organize workers on the west coast of the United States. At least for a time these men, together with some so-called *ladrones*, presented a challenge to both the US occupation and to the collaborationist Federalistas. They offered a competing nationalism that aspired to more than recognition of Philippine nationhood. Printers, tailors, carpenters, and common workers sought *kalayaan*, a Tagalog term commonly translated as "independence" but conveying meanings apart from Americans' and even some elite Filipinos.'[179] As Reynaldo Ileto explains, members of the revolutionary Katipunan believed that the realization of *kalayaan* would "bring about a condition of brotherhood, equality, contentment (*kaginhawaan*) and material abundance (*kasaganaan*)."[180] They sought a new social order through the decolonization of political *and* economic life.

In light of these ongoing contests, perhaps it is no wonder US colonialists sought a road to retreat, a pursuit that recalls José Rizal's 1890 reply to Spanish notions of the indolent Filipino. Rizal argued that indolence was not "the cause of backwardness and trouble" in the Philippines, as Spaniards held, but the "effect of the trouble and the backwardness" of the colonial regime itself. Rizal described how Europeans lived in the tropics: "Surrounded by a numerous train of servants, never going afoot but riding in a carriage, needing servants not only to take off their shoes for them but even to fan them!"[181] This was hardly all American colonialists' vision of life in the Philippines, or, for that matter, in the Baguio hill station, but it underscores the irony of the notion of "indolent" Filipinos. They were put to work building a road to a health and leisure resort for the new occupiers of their land.

THREE

# "A Hope of Something Unusual among Cities"

Following his visit to the colonial Philippines, Daniel Burnham mused to an architect friend: "The dive into the Orient has been like a dream. The lands, the people, and their customs are all very strange and of absorbing interest. It surprises me to find how much this trip has modified my views, not only regarding the extreme East, but regarding ourselves and all our European precedents."[1] In many respects, the US colonial government could have done no better than in choosing Burnham to design plans for Manila and Baguio. By the time he had reached the Philippines, Burnham was a renowned skyscraper builder and urban planner.[2] With his partner John Root, he had raised the Rookery and Monadnock office buildings—structures that towered over Chicago's streets and remade the urban skyline. In the 1890s, Burnham set his sights on even more immense spaces: He took on whole cities and urban life themselves, seeking to build not just upward but outward. Prior to his work in the Philippines, Burnham collaborated on a redesign of the Washington, DC Mall and a plan for Cleveland, Ohio. He departed for the Philippines from San Francisco, breaking from work on a scheme for "the American gateway to the Orient and the South Seas," as Herbert Croly described the West Coast city.[3] Construed that way, San Francisco was the perfect entrée to work in the Philippines. So, too, was the project that launched Burnham's career in planning: the Chicago grounds of the 1893 World's Fair.

Hosted by a city he proudly identified as "the first territorial acquisition of the new republic," the exposition commemorated Columbus's discovery of the Americas.[4] Burnham served as the exposition's director of works, and in that capacity, the "impresario" managed a team of American architecture's luminaries, who designed structures to showcase the United States' agriculture, industrial manufactures and machinery, and arts across nearly 1,000 swampy acres of southern Chicago.[5] Burnham and John Root worked out the positions of these buildings and their scale. Near the center of the exposition complex, one prepared with Central Park architect Frederick Law Olmsted, they situated what became known as the White City, an ensemble of neoclassical buildings—steel skeletons encased in white plaster. Evoking the majesty of imperial Rome or Athens, the White City left little doubt that Burnham was well acquainted with European precedents. The project gained him fame and popularized the "City Beautiful," a practice and a style that emerged as an effort to ameliorate through space and the built environment social problems produced in the headlong advance of industrial capitalism. City Beautiful's advocates, whom we call Progressives, aimed to check commercial interests that, like the skyscrapers housing them, seemed to dwarf civic commitments and even supplant them.[6] They insisted on the value of comprehensive urban planning to counter haphazard urban development conditioned primarily by market imperatives.[7] In aesthetic terms, the City Beautiful, in borrowing from the Beaux Arts, privileged the order of the neoclassical—a style signaled by the clustering of civic buildings, the provision of open space around them, and the axial design of city streets. This aesthetic for public buildings and urban spaces communicated City Beautiful advocates' reverence for and commitment to civic life and their aspiration for using public spaces as antidotes to immanent class conflict and catalysts of social harmony.[8] Careful planning and architectural design might help urban dwellers find a new balance between the civic and the commercial dimensions of urban life.[9]

Architect Louis Sullivan, among others, would criticize Burnham for resorting to the neoclassical, a nonindigenous architectural form, in the White City. "Architecture died in the land of the free and the home of the brave,—in a land declaring its fervid democracy, its inventiveness, its resourcefulness, its unique daring, enterprise and progress," he wrote of the White City's effect.[10] Burnham seems to have been unmoved by the criticism when he took up work in the Philippines. Much as he drew on European architectural precedent to mark the status of the United States and its "first territorial acquisition" four hundred years after Columbus's arrival, he would look to Old World capitals when fashioning civic structures for the new US territory in the Pacific.

Burnham's work embodied such ironies, and they also lay in the very task he had assumed—to craft a style and physical presence for an imperial republic. The seeming incongruity of an imperial republic was tacitly recognized by those charged with administering the United States' new overseas possessions. In its 1900 report, the Philippine Commission issued a "protest against the suggestion of calling the Philippines a colony." Commissioners explained that to Filipinos, "No other word in their whole political vocabulary is so ill omened, so terrible, so surcharged with wrongs, disasters, and sufferings." To call the US occupation "colonial," they held, "would insure the emphatic and universal condemnation of the Filipinos for the most perfect system of free government which the mind and heart of man could devise for the inhabitants of that old Spanish colony of the Orient."[11] The desire to disassociate the United States with colonialism was also one reason that administrators sought to replace military with civil authority as swiftly as possible. To them, the distinction between the rule of William Howard Taft and that of General Arthur MacArthur appeared fundamental. Secretary of War Elihu Root believed that the arrival of civil government with "advancing armies" demonstrated "that our sovereignty meant justice and not oppression, liberty and not slavery, protection of law and not the license of arbitrary power."[12] Others were more explicit about the instrumentality of the distinction between military and civil authority. A friend wrote Taft in 1901 that the onset of civil government "would give the people to believe that they themselves had the responsibility for maintaining order."[13]

Americans understood that, along with Filipinos and compatriots at home, the world was watching their immersion in extra-continental empire building. Writing on the need for a well-prepared colonial service in 1899, Yale historian Edward Gaylord Bourne observed that "these islands are not in a remote corner of the earth like Alaska, where failure would be hidden or unnoticed, but they lie at the very meeting place of nations, and all that we do there will be under a white light of publicity. The most energetic and ambitious powers of Europe will be our neighbors and critics."[14] The Philippines' position at that "meeting place of nations" was precisely why they were so desirable to Americans and perhaps why Burnham was eager for the commission. But as Bourne's comment makes clear, this position of power also intensified the stakes of the US project. These constraints and the need for legitimacy came to shape the literal and also the more figurative architecture of US colonialism.

If only to ward off other aggressors like Japan, Britain, and Germany, the United States had to flex its muscles as had other imperial powers. Yet to gain hegemony and abide by its founding myths, the United States also defined its imperial character as anticolonial. Burnham's plan for the government reservation at Baguio and the US Supreme Court's resolution to a critical land dispute involving Baguio native Mateo Cariño expressed Americans' efforts to walk the ideological tightrope and served as interpretations of US imperial sovereignty. In trying to give "adequate expression of the destiny of the Filipino people" and "witness to the efficient services of America," Burnham employed Beaux Arts neoclassicism, a style drawn into celebrations of civic life by proponents of the City Beautiful and also a style of Napoleon Bonaparte's Paris—one with a military and imperial history.[15] In keeping with the City Beautiful aesthetic, Burnham set Baguio's colonial offices amid grass and trees, surrounding them with green, creating a pastoral, middle-ground landscape and associating imposing structures and martial allusions—and institutions of colonial power—with the virtues of nature itself. The US Supreme Court's decision in the Cariño case, meanwhile, set an important precedent for native title while also legitimizing the imperial regime and distinguishing it from the colonial history of the US West. One was a physical architecture, the other part of an imperial legal infrastructure, but both can be interpreted as sanguine avowals of good intentions, even utopian visions, that rationalized the imperial republic's dispossessions of Philippine peoples' land and sovereignty.

## From the Capital to the Hinterlands

Burnham arrived in the Philippines in December 1904, and as a guest of Governor General Luke Wright, he occupied a room in the former residence of Wright's Spanish predecessors. Burnham described its interior as "royal" in a letter to his wife. "There are old Spanish paintings and furniture," he observed, "and many bronzes and Chinese vases."[16] Apparently the new occupant of Malacañang Palace had yet to clear out the relics of Latin rule. Beyond the mansion, Burnham found much promise in the capital. Burnham admired the city's inland waterways, or *esteros*, and compared the capital to some of the finest cities in Europe. Manila, he believed, had "an opportunity unique in the history of modern times, the opportunity to create a unified city equal to the greatest of the Western world with unparalleled and priceless addition of a tropical setting."[17]

Burnham embraced this opportunity to create a "unified" city, a place that was anything but. Manila presented challenges that faced many American cities, like population density and inadequate sanitation, and it also was a setting disturbed by war. Edith Moses, an early visitor to the capital, described the scene in 1900 this way: "handsome houses were jostled by nipa shacks; canals filled with laundry women and naked babies invaded aristocratic quarters. There was but one street of uniformly good houses."[18] This may have been an improvement over the conditions that met American soldiers in the summer of 1898. They found a capital with failing public services and, for that reason, a city beset by disease, dirty water, and choked sewers[19]—circumstances suffered by Filipino residents and American colonialists alike. While the Spanish had created Intramuros, "a segregated and militarized zone" within the capital for themselves, Americans, despite the segregated society from which they had come, chose to dwell among Philippine peoples in the capital city.[20] In an effort to improve the morale of Americans in the city and to make good on McKinley's "Benevolent Assimilation" Proclamation, the army had gone some distance in relieving the capital of its sanitation troubles.[21]

But to Burnham, some of the military's fixes had created an aesthetic problem. He lamented the army's crude building style. "At the present time," he warned, "corrugated galvanized iron roofs are taking the place of beautiful Spanish tile, to the serious detriment of the city's appearance."[22] This landscape may have signaled US sovereignty but in a most infelicitous way. Its utilitarian look could convey a lack of refinement or a sense of impermanence, or, perhaps, stand as a reminder of war. A future First Lady, Mrs. Howard Taft, noted the distinct martial feel of the capital in the early 1900s. Introduced to the Manila house that her husband, then governor general, had selected for their family, she was jolted by the military presence at the site: Once behind a "high stone wall," she was driven "past a row of soldiers who stood at attention, with their guns held stiffly in front of them. I knew our house had to be guarded," she wrote, "but it was something of a shock for a moment, just the same, to see the guardhouse and the trim soldiers with their business-like equipment."[23]

There was good reason for such a military presence in Manila. One year earlier, the capital had seen a standoff between American and revolutionary forces, which circled the US occupied city, erupt into outright conflict. February's Battle of Manila took place largely on the outskirts of the capital and had led to the decimation of poorer districts beyond the commercial and political centers.[24] But even when it had appeared that Americans had gained the upper hand in the conflict, Manila seemed on the brink of invasion. "Sentries and individual soldiers were attacked by knife-wielding assailants, snipers fired into troops moving about the city, buildings were torched, and standing orders confined soldiers to their barracks."[25] General Antonio Luna supposedly instructed Filipinos in Manila to "rise and wage 'war without quarter.'"[26] Such threats to colonial authority and rumors of revolt traveled through the capital as Filipinos maintained resistance to the US occupation throughout the archipelago.[27] *Ladrones* were known to descend from mountain areas just outside Manila and infiltrate the capital "whenever pursuit was hot," "remaining in concealment until the danger had passed."[28] Mrs. Taft recalled that "[t]here were occasional rumors about uprisings . . . when the guard at our gate would be doubled—but Mr. Taft assured us that Manila was as safe as New York or Chicago."[29] As long as corrugated iron roofs and trim soldiers occupied Manila, however, the urban landscape would represent the US occupation in an unflattering, even nakedly imperial, form.

If addressing himself to a landscape and urban life shaped by the effects and exigencies of war was a task facing Burnham in Manila, in Baguio, he encountered a distinct set of challenges. He spent no more than two weeks in the mountains, assessing the "government reservation,"[30] a creation of a February 1903 resolution by the Philippine Commission. It had called for some five square miles of Baguio to remain "exempt from settlement and claim."[31] This enclosure followed measures that had placed Benguet Province under civil government and made Baguio the provincial capital. Across these acres little level land could be found for the commission's wish list of buildings, transportation, and a reliable water supply—all to make Baguio habitable by the insular government's employees for up to four months each year.[32] Besides the challenging topography of his site, Burnham had little by way of American precedent in devising plans for the resort. A writer for *The Overland Monthly* observed in 1912, "Summer capitals have long been the rage everywhere, except in the United States." "The crowned heads of Europe" have any number of resorts from which to choose. By contrast, he wrote, "democracies are happily not much given to these luxuries, any more than perspiring wage earners are prone to encumber themselves with summer homes and gardens."[33] And so if the absence of American precedent gave Burnham free rein, it also hinted at a potential ideological problem, as *The Overland Monthly* writer intimated.

This did not deter Burnham, however. The Baguio that he saw was likely similar to the one described in a January 1903 article in the *Manila Times*. The former governor of Benguet Province told a reporter that Baguio "greatly resembles a mining town out west. At night the prospectors gather round the 'village store' and swap yarns and tobacco and generally play the role of miners the world over."[34] These prospectors may have looked out to the "lonesome *Ranchería*," described by Otto Scheerer.[35] The dwellings, few though they were, likely featured nipa roofs and pine boards that Igorots prepared using their axes and bolos, tools secured from Chinese traders on the coast. Sitting about five feet off the ground,[36] these structures included "a space . . . between the earth and the floor of the house . . . which serves as shelter for the pigs and chickens."[37] In 1920, the anthropologist C. R. Moss described the process by which these homes were built. Large groups of Igorots, generally men, secured the timber and constructed the dwelling and did so expecting no wages. Compensation came by way of *cañaos*, or feasts given by the homeowner in which "the value of the animals killed always exceeds what would have been paid the workers had they received regular wages."[38]

The Spanish doctors who wrote on Baguio's prospects as a sanitarium took a dim view of the most modest of these Igorot dwellings. "Through the great filth that prevails in it and the large number of beings that at times inhabit it," they wrote, "it becomes transformed into a veritable beasts' den."[39] Edith Moses, who found her way up to Baguio, likewise saw Igorots' dwellings as "poor huts, mere sleeping holes."[40] But in C. R. Moss's rendering, Igorots' homes revealed native ingenuity: "No nails are used, but the timbers and boards are fastened by means of mortises and grooves, while the rafters are tied to the joists with rattan."[41] And these structures were suited to life on the move; he noted that the houses could "be taken apart and moved without injury to the lumber."[42] To the Spanish, such mobility had been a source of frustration, for together with their ability to uproot and replant their *camotes* and treasures of gold, it made Igorots "so intractable" and even elusive to Spanish colonizers.[43]

In not bearing the obvious impress of Spanish colonialism as tile-roofed houses in Manila did or the scars of the recent counterrevolutionary war or a substantial built environment, it could seem that Baguio offered Burnham *tabula rasa*. But this was not the case. The abandoned tribunal and Igorot dwellings and their arrangement over Baguio's hills hinted at the colonial past. As discussed in chapter 1, Spanish incursions, together with Igorots' agricultural practices, cattle rearing, and gold mining, had shaped the dispersed settlement pattern in the region. Burnham gazed over land with as complicated a past and present as the city of Manila, and some years after his visit, ownership of these lands would become an especially fraught question. Some land, over which Baguio native Mateo Cariño claimed ownership, was seized by the insular government for the future Camp John Hay military reservation. Advancing his claim on this land during Spanish rule and then unsuccessfully during the American occupation, Cariño would take his case to the US Supreme Court. His challenge raised the issue of what US sovereignty would mean for the Cordillera's indigenous peoples. And its resolution, much like the plans that Burnham would draw up for Manila and Baguio, sketched a vision for a benevolent and even ennobled form of colonialism.

## Mateo Cariño v. the Insular Government

Cariño was known, if not to Burnham, then certainly to some of the earliest American visitors to the region.[44] Edith Moses visited his house in 1902 and recognized his as one of the "better houses" in Baguio. "Mateo's wooden house was low," she wrote. "[T]he roof was thatched with grass bleached by time to a soft dark gray. The main building and a small ell were raised some ten feet above the ground. The upper floor of the house contained one large irregularly shaped room, where we were entertained, and a small space screened off as a bedroom, for the parents of the family. The ell contained one room, which served as a storeroom and kitchen. The furniture consisted of bamboo beds, a few willow chairs, a rough table, and one wardrobe. The space under the house was used as a storeroom for rice and *camotes*, a stable for horses, and sleeping place for the retainers of the family."[45] In featuring several rooms, Moses could not dismiss Cariño's house with those of other Igorots. Mateo's house indicated his elevated status.

As discussed in chapter 1, the Cariños had been a powerful Benguet family for generations, one that had amassed a substantial wealth in gold and cattle. Angel Pérez described the impressive fortunes of Mateo's brother Juan in the early 1900s, listing the contents of one of Juan's Benguet houses: "golden lamps," walls decorated with landscape scenes, "magnificent table center-pieces," table "consoles of the finest wood," full-length mirrors, and solid silver tableware.[46] Mateo himself claimed some 370 acres of land, which he had used to pasture cattle just as his ancestors had and also to raise *camotes*.[47] Perhaps following the example of his forebears, who had at times chosen paths of accommodation with the Spanish, Mateo gave some of his land to the US colonial government. In fact, his home became the center of the Americans' hill station in the passage of the February 1903 act forming the "government reservation,"[48] and between 1902 and 1903, Mateo served as the president of the Baguio township.[49] In acquiring the land, the commission had added to the government's existing holdings; in the fall of 1901, the Philippine Commission had issued 11,000 pesos for the acquisition of two houses and eighty acres of land from Otto Scheerer.[50] In October 1903, the government claimed yet another piece of Baguio land for a military reservation, the future Camp John Hay. Hay, author of the 1899–1900 "Open Door Notes," sought to secure for the United States access to the China market equal to that enjoyed by European powers. The secretary of state's work would be commemorated by this military camp a good distance from the Manila entrepôt. Cariño

3.1 "Former House of Mateo Cariño." (no date), Baguio, Benguet, Photo 10A196, Dean C. Worcester Photograph Collection. Reprinted with the permission of the University of Michigan Museum of Anthropological Archaeology.

asserted ownership over part of this land, through which the US military had already cut a public road.[51]

If Baguio called to mind the Adirondacks for some Americans, for Mateo it stood, perhaps, as a reminder of his family and the relatives from whom he had inherited much land. He testified, "When my father was about to die he delivered this land to me and I called the old people—the 'principales' of the place—to witness this delivery; and I gave a fiesta."[52] By this ritual, a *cañao,* wherein Ibaloi communities gathered for at least a day of food and drink, rather than through a will (which seems to have been first used only in 1917), the land passed down to Mateo.[53] He maintained the old fences, some sod, and some wire that he himself had promoted to limit the spread of disease among Ibaloi peoples' livestock.[54] As this act of enclosure reminds us, the Ibaloi possessed a concept of individual and group land rights, rights whose spatial and temporal scopes were determined by the amount of labor applied and improvements made to the land. Mateo testified that his father, who had also made such investments in the land, counseled him "never to let this land go but to hold it always."[55]

Not long after the US occupation of the Philippines and after at least forty years in the Cariño family, however, the land appeared destined for new ownership. The case revealed that in 1901, Mateo had agreed to option some of the land in question to an American merchant in Manila named Metcalfe A. Clarke.[56] To complete this transaction Mateo had to acquire proper title.[57] That year, he successfully obtained "possessory information title" from the Philippine Court of Land Registration, an effort that followed attempts to "secure a written title to the property" in the 1890s.[58] The US colonial government, already set on Baguio as home to its summer capital and military reservation, contested Cariño's application to the colonial Court of Land Registration. And when overruled in 1904, the insular government made a successful appeal to the Court of First Instance in Benguet. Mateo would challenge this decision in the Philippine Supreme Court but to no avail.

In the course of their proceedings, both colonial courts denied Cariño's assertion of land ownership by the doctrine of prescriptive title. While they conceded that he and his ancestors had occupied the land for decades—according to Cariño's attorneys their tenure may have preceded the arrival of Magellan—both held that Mateo had never enjoyed title to it.[59] Indeed, despite his two efforts to comply with Spanish royal decrees in the 1890s, he had never been accorded written title. Mateo and his brother Juan testified that Spanish officials assured them that they need not secure such documentation—their ownership was recognized and would never be denied.[60] As *baknang*, "[t]hey were people guaranteed their property rights (by traditional law) if they toed the line" drawn by the Spanish.[61] But in those last years of Spanish rule, such a promise may have seemed insufficient. An 1893 mortgage law had been promulgated to afford, finally, "a systematic and comprehensive [land] registration scheme" in the colony.[62] The Maura Law, the Spanish regime's last measure on land rights, followed in 1894. Ostensibly, it was a means of providing Philippine peoples with access to land for agricultural cultivation, but its details amounted to a reversal of Spain's historic "respect for customary land rights," a commitment made by Felipe II in the sixteenth century.[63] The Maura Law allowed the regime to claim as state property all lands except those held by owners possessing legal titles. In other words and at least in theory, undocumented customary land rights would no longer be recognized by the colonial state as they formerly had been.[64] From early in their occupation, Americans followed this policy, and it was reflected in the 1902 Public Land Act; it allowed "native settlers" to secure free patent to lands in the public domain, which Americans claimed to characterize nine-tenths of the Philippine landmass.[65] For those without the resources and wherewithal to pursue and secure patents or documentary titles, this meant the loss of a measure of security through customary "group rights in land."[66] Indeed, the land law's effect was to eliminate "the primary or territorial right of political sovereignty of indigenous peoples, which in turn supports the secondary or usufruct right of individuals in property."[67]

Like the other "special provinces," Benguet had initially been excluded from the measure. But in 1905, Benguet Provincial Governor William Pack reported that inhabitants sought the extension of the Public Land Law. They regularly asked him "when their lands will be given them," Pack wrote, a statement that implies acknowledgment of the colonial appropriation of Igorot land. Pack encouraged the commission to bring Benguet, beginning with Baguio or La Trinidad, at last within the land law's reach. He believed his charges possessed "moral rights" that "demand legal rights." "[A]nd," he went on, "there are no legal rights among Americans that do not recognize property rights."[68] Pack offered another reason to apply the Public Land Law to Benguet. "The sooner the nomadic style of cultivation of land is stopped," he wrote, "the sooner we will know positively which may be regarded as state lands without question or counterclaims. The sooner these people are anchored to one place by vested property rights the sooner we may expect to be able to lay upon them at least a part of the burden of the provincial expenses." Not only would anchoring indigenous peoples to one place render taxation a simpler matter and more profitable practice for the colonial state, but it would also put unclaimed lands squarely within the public domain and potentially available for auction. "We would then be in a position to exploit this district to the homesteader or land leaser," Pack wrote, noting that the province was particularly hospitable to "the American settler on account of both climate and crops."[69] In January 1906, the portion of the Public Land Law that gave "native settlers" a means to secure free titles was extended to Benguet.[70] By July of the following year, Pack believed that natives of Benguet had had sufficient time to secure patents. That year, the governor reported that native settlers had submitted claims on over 1,000 parcels of land.[71] It was now time that the province come under the whole of the land law, which would make homestead leasing opportunities available widely.[72]

It was in this context of developing land laws and their extension to Benguet that the US insular government considered Cariño's assertions of ownership. The government, challenging the findings of the colonial Court of Land Registration, contended that without official documentation, Cariño's land remained part of the public domain when the Spanish and American governments signed the Treaty of Paris, and the land continued in that state upon the inception of US sovereignty. In December 1906, the Philippine Supreme Court offered this analogy to make sense of

Cariño's claim: "We do not understand that a person in possession of unsurveyed public land in the State of Minnesota, for example, whose ancestors had occupied that land for forty years," judges wrote, "could maintain in court a claim that he was the legal owner of the lands by reason of the presumption that the United States had granted the land to his ancestors, a presumption founded not upon any proceedings taken in the General Land Office to acquire a patent thereto, but upon the mere possession for that length of time."[73]

The justices likened Cariño to a white squatter in the US West. And yet in the ruling, they also marked Cariño's difference as an Igorot: they rejected the notion that Cariño could ever have pursued title to the land. Noting the Spanish failure to convert Igorots to Christianity, the court deemed incredible the possibility that "these uncivilized people" might have "taken advantage of" "the provisions of laws relating to the grant, adjustment, and sale of public lands."[74] They had remained outside the colonial state's authority and hence its land laws. An uncertainty about Cariño runs through the court records: Was he to be understood as a liberal subject, capable of and interested in owning and selling property, or more properly recognized as a cacique, a traditional leader who held land only by virtue of his social position? Was he an innocent and ignorant savage, vulnerable to the predations of Americans eyeing newly desirable lands in Benguet, or had he been among those insurgents conspiring against the new colonial power? Officials clashed over representations of him.

Drawing on testimony from Benguet Governor William Pack, the government put forward the view that Cariño possessed the land only through his status as "chieftain." While conceding that Cariño's reputation was impeccable—"none better in the world"—Pack affirmed the government view that "Igorrote chieftains like Mateo, simply by force of character, maintained their dominant sway over large tracts of land in the province of Benguet."[75] This representation of Cariño was difficult to square with Mateo's family history and Ibaloi notions of land ownership. To imply that Cariño was not well acquainted with individual property rights was to overlook his efforts to maintain enclosures surrounding his land and register the land in the 1890s. Justice Oliver Wendell Holmes seemed to acknowledge this past when he described Mateo as "a man of more than ordinary intelligence and thrift."[76]

Still, if one were to follow the assertions of the insular government and assume that property ownership of a kind meaningful to westerners like Pack was foreign to Cariño, what, according to the government, led him to claim Baguio lands as his own and seek title? Attorneys for the insular government insinuated that Cariño's efforts shadowed others' with a stake in the development of Baguio, namely Metcalfe Clarke and Otto Scheerer. They attempted to show that Scheerer, who had already sold land to the government, had prompted Cariño to secure title to the land. "Didn't Mr. Schearer[sic] tell you that the land in question would acquire a high value after a time?" the insular government's attorney pressed Cariño.[77] Cariño conceded that Scheerer had urged him to make the assertion and that then-Benguet Provincial Governor Whitmarsh also had counseled him to pursue "possessory information."[78] Scheerer prepared the notices that Cariño had posted on his land to deter Americans who had begun entering the province, even squatting on his land.[79] Whitmarsh in his capacity as provincial governor had reported on these circumstances to Taft in January 1901. Hardly opposed to their arrival, Whitmarsh noted that soldiers and "speculators, merchants, miners, and the adventurous float which precedes a wave, are coming in to spy out the new land of Canaan." They might help "Americanize the pagans of this Province."[80] Whitmarsh was among these prospectors and Baguio entrepreneurs. Following his service as governor, he founded the Benguet Commercial Company, which by 1903 was supplying visitors with hotel accommodations, butchered meats, various commodities, and even services for the construction of buildings.[81]

While he and Scheerer enjoyed favorable relations with civilian leadership, they had run afoul of the US military and supplied civil and military leaders with another opportunity to tangle over the reach of their respective authority. The two had antagonized military leaders by questioning the army and its practices in the Cordillera. Scheerer, for instance, wrote General MacArthur that Igorots were "abjectly frightened of the military authorities, especially by the negro officers in Trinidad"—this was Robert Rudd's unit.[82] Scheerer's loyalty to the new colonial government was in particular dispute, and for much of 1900 and into 1901, military and civilian officials exchanged correspondence on the German expat. Colonel Duvall, commander of troops in La Union and Benguet, deemed Scheerer "a confirmed *insurrecto*, a restless agitator, a pernicious intriguer, and exploiter, a busy informer and spy."[83] Military authorities like Duvall viewed Scheerer as in league with Filipino revolutionaries who had sought refuge in the mountains, and they took umbrage at the Philippine Commission's appointment of him to the post of provincial secretary under Whitmarsh.[84]

In response to the military's charges, in the fall of 1900, Scheerer approached Captain Rudd, who had been instructed by Colonel Duvall to watch Scheerer "like a hawk."[85] He asked to "take the oath of 'amnesty'" and clarified that while he sympathized with the Filipinos, he had never supported the insurgency.[86] Taft himself would come to Scheerer's and Whitmarsh's defense in a letter to General MacArthur, writing of the former's service to the commission and the cordial relations that he and Igorots in the region enjoyed and of Whitmarsh's invaluable experience in Benguet.[87] Duvall held firm; but for the Igorots of Baguio and Tublay (from where Cariño had hailed), there were few in the province who supported Scheerer's appointment, he insisted. As if Scheerer's alleged fraternization with revolutionaries was not enough, Duvall added peculiar cause for suspicion. "[A]nother indication of [Scheerer's] reactionary rather than edifying and elevating tendencies," he remarked, was his "habit of dressing in the Igorrote garb, the 'gee-string.'" Duvall interpreted Scheerer's Igorot costume as consistent with the German's objectionable politics—"his well known socialistic, if not, indeed, anarchistic, tenets."[88] In these ways, the animus between military and civilian authorities in Benguet and the challenges to US sovereignty that it recalled churned below Cariño's case, one that would determine the fate of a military reservation. Military officials drew connections between Cariño's older brother Juan and insurgent Pedro A. Paterno. Duvall described Juan, who had served as the president of Baguio during the brief tenure of the revolutionary Philippine Republic, as "the opulent, powerful and tyrannical *insurrecto* Governor of Benguet."[89] Mateo himself had been identified as an abettor to his brother, and Mateo's son had been jailed for at least five months in connection with these charges.[90]

Beyond the insurrectionary activity believed by some to discredit Cariño and Scheerer, government attorneys suggested that unscrupulous business lay at the heart of the Cariño land suit. They suggested that Metcalfe Clarke, an ice cream merchant in Manila to whom Cariño had optioned his property, was orchestrating a grab of Baguio land—land that had dramatically increased in value since the Benguet Road had been built and since Baguio had become the official summer capital of the colony. Whitmarsh had testified that in one year, the value of land had increased from $60 to $100 gold per acre, and the ruling of the Court of First Instance in 1905 determined that Cariño's land was worth $25,000.[91] Clarke rejected the charges and denied prior knowledge of the military's plans for a compound. He was familiar, however, with other ways to turn a profit in Baguio: Clarke knew something about gold prospecting.

By the time Cariño's case had reached the US Supreme Court, Clarke had founded what is now recognized as the oldest mining company in the Philippines, Benguet Consolidated, located about five miles from Baguio. By 1903 he was serving as its president.[92] Clarke was not the only party to Cariño's case with a financial stake in the hill station. One of Cariño's attorneys before the Court of First Instance and the Philippine Supreme Court, John W. Haussermann, would serve as a director of Benguet Consolidated beginning in 1907. Years later, a 1939 *Life Magazine* profile described how Haussermann "indulged superstition of 5,000 Igorot miners that feasts must be held to placate the gold god"—an Ibaloi Kabigat, perhaps.[93] In the 1930s, as the "Gold King," he would call on fellow Republicans to reconsider relinquishing the Philippines to independent status in light of the archipelago's potentially lucrative mineral resources.[94] Daniel R. Williams, another of Cariño's lawyers and a secretary to the Philippine Commission, served as vice president of this mining company beside Clarke and Haussermann.[95] While Clarke's and Haussermann's dealings with Cariño may have been aboveboard and while the US Supreme Court dismissed the insinuation of profiteering as an attempt at "influencing the Court to believe [Cariño's] claim to be merely speculative," it is clear that they were parties with special interest in the disposition of Igorot lands.[96] If the insular government had cause for denying this particular claim, Cariño's supporters had reason to equip him with legal means for disposing of land, much as early English settlers in North America developed an interest in arguing that Native Americans enjoyed property rights. Besides Indians' demographic advantage, over time, to deny Native American property rights became tantamount to discrediting the land claims white settlers had acquired.[97]

In its brief for the Supreme Court, counsel for the insular government sidestepped these issues and made a simpler case. Cariño, "although an occupant of the lands in question by 'ancient possession,'" had "failed to perfect said claim under the laws of Spain in force at the date of . . . [the] change of sovereignty" and, for that reason, became a "trespasser" on public lands.[98] In short, Cariño had failed to comply with the rules established under the Spanish occupation and in that way forfeited his right to the land. Cariño had powerful counsel crafting a response. His attorneys before the US Supreme Court included Frederic Coudert, who had engineered several of the first "Insular Cases." On behalf of clients seeking free trade between the new colonies and the mainland, Coudert had argued for recognition of the United States' new possessions as "domestic"—as part of the United States—for tariff purposes.[99] When they took on Cariño's case, Coudert and his partners, perhaps seeing the implications of the decision for foreign investors, argued for recognition of the Ibaloi's property rights. To them, the implications of the government's claim were absurd: the requirement that Cariño show "paper title" from the Spanish regime meant that "the whole Igorrote nation may be driven as 'lawless squatters.'"[100] They also advanced anthropological knowledge, citing a study of the Bontoc Igorots—an ethnolinguistic group living north of Benguet Province—to explain Ibaloi understandings of property and argue that "Cariño and his predecessors held the land not as 'squatters' on Crown lands but according to the laws and customs of their people."[101] Recognition of Igorot law and custom—its difference and likeness to western notions of property—became the premise for accepting Cariño's land as alienable and transferrable in the colonial market.

Cariño's attorneys also drew on Spanish colonial law, jurisprudence with which they would have been familiar given their work on the Insular Cases. "Whatever may have been the brutality of the fierce 'Conquistadores,'" they wrote, their law "recognized and protected the rights of the native occupants of its Indian possessions even more fully and scrupulously than our own Government has done in the case of its Indian wards."[102] Still, the lawyers identified American precedent for recognizing Cariño's claims. "American Courts have uniformly held the individual possession of the American Indian to his land as sacred," they wrote. "While the United States Government has claimed a general right of paramount control over the Indian tribes as such, the doctrine that the right of the individual to the land which was his by tribal law could be taken away at will and that he was a mere lawless 'squatter' has never been an official American doctrine."[103] While the US government might enjoy sovereignty across the continent, they argued, it did not enjoy property rights to Indian lands.[104] It followed for the Philippine colony. All of the attorneys made their cases, then, by analyzing their clients' circumstances in light of legal pluralism and imperial precedent, a necessity recognized by William Franklin Willoughby, an analyst of colonial government administration. In 1905, Willoughby, later the first director of the Brookings Institution, wrote of the challenges presented by a long colonial history in the Philippines. These were not "open field[s]" permitting the easy "building-up of American communities with American institutions and laws." The islands "are in every way more complex and difficult than those that have been presented in the case of prior annexations."[105] They were a legal palimpsest.

Justice Oliver Wendell Holmes wrote the US Supreme Court's opinion, and he, too, made comparisons to the United States' former territorial annexations in addressing what he took as the central question facing the court: "whether the plaintiff owns the land."[106] Theodore Roosevelt had appointed Holmes to the Supreme Court in 1902 at the recommendation of Holmes's fellow Massachusite Henry Cabot Lodge, that stalwart champion of American imperialism. Given the justices' close votes deciding early Insular Cases, these men saw Holmes's support for the United States' imperial undertaking as crucial to his appointment.[107] Holmes delivered the court's opinion in 1909, the year that a city charter was drafted for Baguio and by which time Burnham had come and gone from the archipelago. In large measure, Holmes's response resonated with the case made by Cariño's attorneys. Upon review of Spanish land law and in setting these policies in relation to US congressional legislation on the colony, he disputed the decision of the Court of First Instance. Holmes concluded that "Spain did not assume to convert all the native inhabitants of the Philippines into trespassers or even into tenants at will." The former colonial power had indeed accorded indigenous Philippine peoples the means with which to secure title to land over which they asserted ownership. The Maura Law, a reversal of Spain's longstanding recognition of customary property rights, he held, "should not be construed as confiscation, but as the withdrawal of a privilege." For these reasons, to maintain that in the transition of sovereignty, Cariño became a "mere trespasser," as the insular government would have it, would "amount to a denial of native titles throughout an important part of the island of Luzon," Holmes wrote.[108]

Holmes's opinion may be taken as an articulation of the meaning of US sovereignty in the Philippines. For one thing, he expressed its significance for the Cordillera's indigenous people. Holmes held that the Organic Act of 1902, which pledged that "all the property and rights acquired . . . by the United States are to be administered 'for the benefit of the inhabitants thereof,'" surely included those "savage tribe[s]" of Benguet and was not exclusive to the Hispanicized and Christianized Filipinos of the lowlands. That initial pledge contained in the Organic Act would be violated and rendered void were the colonial government "to treat as public land what they [the Igorots], by native custom and by long association, one of the profoundest factors in human thought, regarded as their own." Holmes, setting precedent for the recognition of "native title" in the Philippines with legal power today, maintained that "when, as far back as testimony or memory goes, the land has been held by individuals under a claim of private ownership, it will be presumed to have been held in the same way from before the Spanish conquest, and never to have been public land." Holmes recognized difference—Cariño represented one of the "savage tribe"—but maintained the applicability of property rights; these were to be protected even if the custom of "non-Christians" rather than written documentation evidenced ownership.[109]

Second, Holmes made clear that the new sovereign's prerogatives were paramount even as the court drew on Spanish precedent. "How far a new sovereign shall insist upon the theoretical relation of the subjects to the head in the past and how far it shall recognize actual facts are matters for it to decide," he wrote.[110] His ruling, granting a means by which all Philippine peoples could claim land rights, also asserted US sovereignty over that part of Luzon that had long existed as a space beyond the operable authority of the Spanish colonial state. In that respect, even as Holmes's decision recognized indigenous land rights, it also advanced state-building efforts in the Cordillera begun under the Spanish.

Finally, in framing his opinion, Holmes articulated the raison d'être of the United States' overseas empire. "The acquisition of the Philippines," he wrote, "was not like the settlement of the white race in the United States. Whatever consideration may have been shown to the North American Indians, the dominant purpose of the whites in America was to occupy the land." "[T]he reason for our taking over the Philippines," he asserted, "was different." Here, "our first object in the internal administration of the islands is to do justice to the natives, not to exploit their country for private gain."[111] Cariño's attorneys had likened Mateo to a Native American and called on the United States' past dealings with indigenous peoples of North America to justify a finding for the Ibaloi, but Holmes made the case for difference at least in regard to imperial intentions and their colonial consequences. The 1902 Organic Act, which set the groundwork for US colonial government in the Philippines, had already reflected this disposition, as Holmes noted. If only to placate American agricultural interests fearing competition from Philippine sugar and tobacco, the measure limited the size of agricultural land, forest, and mineral claims and, in that way, the "private gains" that could be made in the archipelago.[112] To Holmes, settler colonialism and practices in the United States' extracontinental empire should diverge. This contrast could make the latter, which sat under the "white light of publicity," look more innocuous, even benevolent.

The insular government had also made an argument for distinctiveness but on different grounds and to different ends. Its attorneys asserted that Supreme Court decisions on "the tribal rights of Indians have no possible application" in this case. "[T]hose cases," they insisted "were decided with relation to the rights of Indians conferred by treaty, or relating to tribal lands or reservations made by the Government of the United States."[113] Ironically, the government made this argument just as the Supreme Court was helping to attenuate the significance of these treaties and, in that way, native peoples' sovereignty. The court's 1903 *Lone Wolf v. Hitchcock* decision reflected this change. Referred to as the "Indian laws' Dred Scott decision," the court's ruling advanced the process of allotment "without tribal consent" and "confirmed the power of Congress to unilaterally abrogate treaties."[114] With good reason, allotment has been understood as a "colonial land policy," one that can be interpreted as a form of enclosure. "Creating and dispossessing visible private property, in this case land, served to expropriate intangible political sovereignty," David Chang argues. "Visibly, the federal government took millions of acres of land from Native nations, called it private property of Indian people, and then permitted the capitalist market . . . to transform them into non-Indians' farms, ranches, homes, and summer cabins."[115]

The effect of *Lone Wolf* was to weaken the status of Native Americans. The ruling made their condition resemble the indeterminate position of men and women in the United States' new possessions, which in the Supreme Court's obfuscating formulation in the Insular Cases were "foreign in a 'domestic sense.'"[116] And, if we follow Chang, the effect of allotment, an enclosure of sources of economic and political sovereignty, was to create private property *and* advance state power. So even as Holmes was marking a distinction between the settlement of the US West and the administration of the new insular possessions, the Supreme Court was also creating new and similar categories of subject peoples by making possible the alienation and commodification of land across the US West and concluding that the Constitution did not follow the American flag into the Philippines. Discursively, however, the court seemed inclined to emphasize distinctions. While the dispossessions that attended expansion across the Old West were inevitable and necessary, in the new frontier it would be different. The court emphasized the temporal and spatial distance between these projects, denying the dispossessions underlying both.

Though the court decided in his favor and called on the army to award Cariño a $10,000 settlement, Mateo never enjoyed the fruit of his legal victory.[117] He died before the ruling came down.[118] And, further, going forward in time, "the insular US regime systematically and successfully blocked any wider application of the Cariño precedent" in the colony.[119] In practice, it blunted the potential power of Holmes's decision by continuing to embrace Spain's Regalian Doctrine, which had held that all colonial lands but those for which Philippine peoples had secured documented titles belonged to the state.[120]

By the time the court reached its verdict, Burnham's plans for Baguio were already underway. Like the US Supreme Court, Burnham was operating in the interstices of indigenous, European, and American spatial practices and architectural traditions. Holmes's opinion defined and legitimized US imperial rule, associating it with benevolence and justice. Burnham's designs for Baguio land operated similarly, shaping an architecture and a model town for agents of a democratic republic undertaking imperial conquest.

## Colonial Cities Beautiful

At the time Burnham and his assistant Pierce Anderson drew sketches for Baguio, Cariño's case had not been resolved and the government had not yet completed surveys of the reservation. Burnham regarded his designs as "tentative and incomplete," but they illustrate well how the Chicago skyscraper builder and city planner envisioned transforming Ibaloi pastureland to serve an anticipated future population of 25,000.[121] Burnham chose to develop the most level land available within the government reservation, the parcel on which Cariño's home sat at the center and one sitting slightly north and west of the future Camp John Hay (see fig. 3.2). The elliptical-shaped tract extended "about one-half mile wide by three-fourths of a mile long," and at the southern end it connected to the new Benguet Road. As Burnham described its topography, the land appeared as "an enclosed hollow dominated by low hills and connecting ridges."[122] His plan for this stretch, submitted to the Philippine Commission in 1905, shows an arrangement of "public, semi-public, and private institutions," a street system, and recreational areas.

When he reflected on his Baguio design in his 1909 *Plan of Chicago*, Burnham compared it to L'Enfant's scheme for Washington, DC.[123] This was a plan that he, Frederic Law Olmsted, Jr., Charles McKim, and Augustus St. Gaudens, the team commissioned to redesign the Washington, DC Mall, had revisited in 1901. Like L'Enfant's late eighteenth-century plan for the United States' capital, the Baguio design "provides for such public buildings as may be needed for government offices, for the service of the city itself, and for the healthfulness, convenience, and recreation of the people," Burnham wrote. "[A]ll of these functions are so arranged as to make a unified and orderly city."[124] The likeness to L'Enfant's plan was not only on the level of function. Apart from his selection of the town site, Burnham's key maneuver was to draw a line running northwest to southeast through the Baguio meadow and place government offices atop the low hills at either end of the axis. This was similar to L'Enfant's recommendation for placing the Capitol and the US president's residence at either end of a long avenue, the "Mall," which Burnham and his team doubled in width and length.[125] In Baguio, town office buildings were to sit at the northwest terminus, closest to the business district. The cluster of "National buildings," for commissioners and colonial administrators, was to

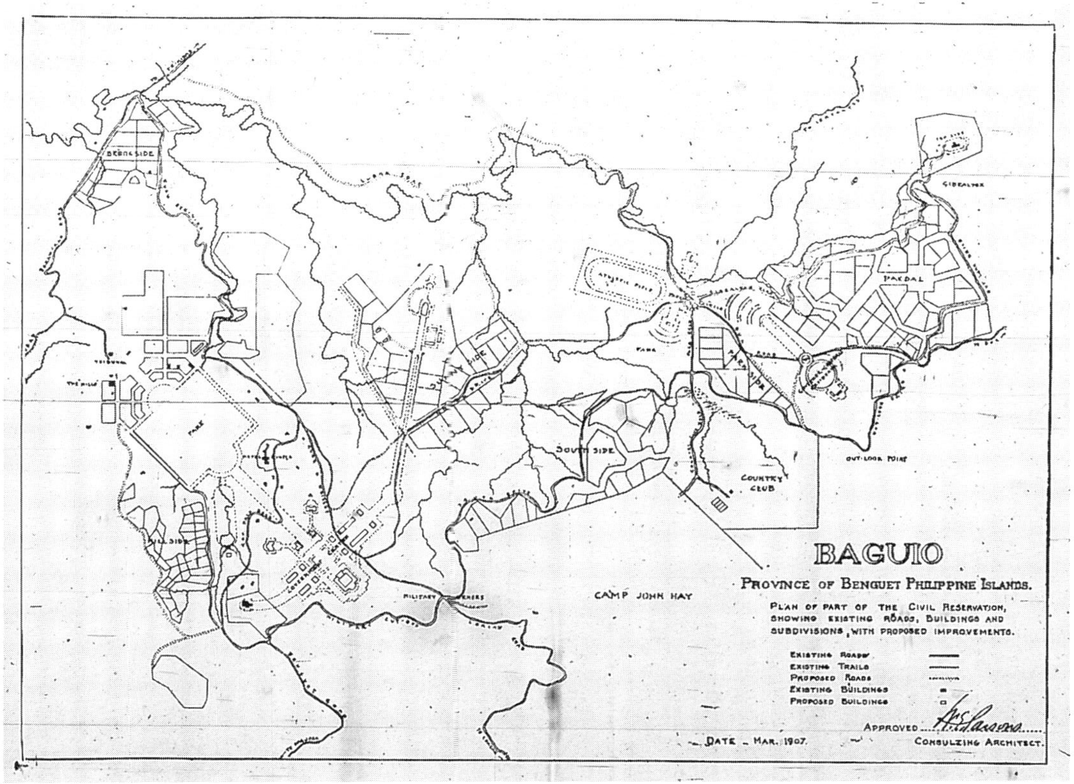


3.2 Plan of Baguio, 1907, published in the *Manila Daily Bulletin* 103 (16), November 3, 1907.

occupy the other end. Burnham created a new geometry on undulating pastureland.

Burnham used a similar strategy in his drawings for the Manila capital. There, he massed the "Government and National" buildings, which included the Capitol and departmental offices, allowing the former to occupy the center. He situated these structures to "form a hollow square opening out westward toward the sea."[126] The design turned Manila to the South China Sea, seeming to orient the city toward the function Americans anticipated for it: a trade entrepôt as it been centuries earlier. In Baguio, Burnham arranged for buildings to turn not toward the surrounding hills and mountains but to face each other across a long "open esplanade with a central *tapis vert* or greensward," a grassy mall, that Burnham planned for the space between the two government building clusters.[127] The residence of the governor general would be built of local rock and cement on the first story and timber on the second and offer a shingle roof.[128] In the original design, this and the major general's residence, positioned close to the national government buildings, were to face each other across either side of the axis. If the design for the Manila capital spoke for an imperial regime that looked beyond the archipelago, the Baguio plan seemed to encourage dwellers to turn inward and gaze upon each other and the green between them. Arguably, the "Intramuros" section of Spanish colonial Manila operated similarly; for security reasons, only Spaniards were allowed to live "within the walls" until the nineteenth century.[129] Burnham's arrangement of buildings fashioned a similar enclave effect in Baguio, something Americans had not attempted to create in Manila.[130] And this design, through which Americans congregated themselves, suggested the very purpose of the hill station: a retreat for regenerating American colonial labor.

For several years Burnham had been rehearsing the style upon which he drew for Baguio and Manila, the "City Beautiful." It appeared in his work on the White City and in commissions that followed in Cleveland and in the project that coincided with his Philippine work, a plan for San Francisco. The most recognizable element of the City Beautiful was its monumentalism, a grandeur achieved through the massing of buildings, the axial design of streets, and the inclusion of parks and processional spaces.[131] Through a particular aesthetic, generally the neoclassic, and a practice, namely, planning, the City Beautiful movement's largely middle-class proponents sought to bring new form and function to the city and also to cultivate new attitudes among its dwellers. In the face of cities' unruly growth, they proposed "architectural unity, harmony, and control."[132] Through the very design of public buildings and spaces, they

3.3 "Government Center at Baguio," in Dean C. Worcester's *The Philippines Past and Present, Vol. 1* (New York: Macmillan, 1914), 272–73.

sought to generate conditions for reconciling "individual and class interests" and even for "transcending commerce," the business of the city.[133] "When they trumpeted the meliorative power of beauty," William Wilson writes, "they were stating their belief in its capacity to shape human thought and behavior."[134] Their objectives, then, were both political and cultural. If such ambitions found followers on the mainland, they also attracted colonialists in the Philippines, where the United States had set out to build a nation and teach Philippine pupils the art of democratic citizenship and where Burnham aimed to give both "adequate expression of the destiny of the Filipino people" and "witness to the efficient services of America."[135]

In Manila and Baguio these goals and the City Beautiful approach were manifest in the massing of buildings to communicate the "preeminence" of colonial government institutions.[136] Burnham believed the technique conveyed a "dignity" to these offices.[137] As important as the clustering of buildings for creating a dramatic, even imposing, presence was the provision of open space around them. Burnham noted the effect in his plan for Manila: "open spaces" help "to dignify important buildings."[138] Where many Filipinos and Igorots lived in cramped quarters and portable nipa huts, colonial offices were the audacious occupiers of large, permanent spaces, as if the conspicuous consumption of space proved, even generated, the eminence of the colonial government.

Burnham also used Baguio's natural topography, drawing on it to convey and enhance the colonial government's literal and figurative position. In his notes on a preliminary plan for the hill station, he wrote that the group of national buildings—positioned on the hillier sections of the reservation—"will . . . frankly dominate everything in sight of it." They would be at the apex of "a crown of monumental buildings." His placement of the residences of the governor general and commander general, meanwhile, were to make the two "formally a part of the visible government powers."[139] Burnham employed the hills enclosing the Baguio valley, then, to elevate US colonial power, literally. The street system for Baguio was to perform in a similar way. Interests in efficiency, good ventilation, and sunlight informed the street layout, but the aim was also to create sightlines that privileged those in colonial offices. Burnham wrote that his "intention [was] to carry through the lines of the streets to commanding points on the hillsides and thus permit the location of monumental buildings where they command a view down neighboring streets."[140] As he would later write in the *Plan of Chicago*: "There is a true glory in mere length, in vistas longer than the eye can reach, in roads of arrow-like purpose that speed unswerving in their flight; and when and where the opportunity of level ground permits, this glory should be sought after."[141] Those who occupied government buildings looked down the length of these streets, while those whom they administered and colonialists themselves could gaze up to the public offices. Burnham recognized that this street pattern meant "lay[ing] down a geometrical scheme" "to the ungeometrical contours of the Baguio Valley." This would entail, he noted, "problems of grading and filling."[142] But these were worthwhile expenditures. Long and wide processional spaces together with the generous aprons surrounding government buildings would together produce the "monumental possibilities" of Baguio. In these ways, Burnham used man-made elevations and the natural topography of Baguio to communicate to American workers and Philippine peoples alike the dignity of colonial government, one committed to teaching natives the art of self-rule.

Burnham's invocation of Washington, DC and L'Enfant's plan and the comparison this implied hint at his sense of the colonial project's import. These also remind us of the European inspirations for his designs. By the late nineteenth century, large and impressive buildings sponsored by the state were rising in Shimla, the British hill station romanticized by Rudyard Kipling, and in Darjeeling,[143] and they suggest these hill stations' political significance.[144] But they receive no mention in Burnham's plans; instead, the Chicago planner found models for the colonial Philippines in US and European metropoles themselves and in the work of Americans'

3.4 Burnham Park, ca. 1925. Ammann Irminger Collection, Filipinas Heritage Library.

predecessor on the archipelago. Burnham admired Spanish colonial architecture in the Philippines and believed it ought to be preserved. He particularly appreciated Spanish tile roofs. He also wrote favorably of the architecture of Spanish churches and government buildings and believed that their beauty and "practical suitability to local conditions could be profitably taken as examples of future structures."[145] "The old Spanish buildings with their relatively small openings, their wide arched arcades and large wall spaces of flat white-wash possess endless charm and as types of good architectures for tropical service," he wrote, "could hardly be improved upon."[146] In this respect, American colonial architecture evinced something of the "associationalist" ethos of contemporary French architecture in Indochina. There, after 1909, at a time when the efficacy of the colonial regime had come under fire in the metropole, architecture began to communicate a colonial policy aimed—at least rhetorically—at greater participation of the Vietnamese people in government. Incorporation of "indigenous motifs" in public buildings began replacing "grandiose and vain acts of assertiveness" in design, the initial aesthetic response to the "instability and weakness of political authority" in the early French occupation.[147] While Burnham did not appropriate nipa house design, he did recognize the wisdom of Spanish adaptations to local, "tropical" conditions.

Burnham's wider urban planning also shared kinship with Spanish city designs. His work and his Spanish predecessors' showed the influence of Italian Renaissance urban plans in "the gridiron arrangement" of streets.[148] But in others respects, Burnham departed significantly from the Spanish, and through his urban plans he conveyed the distinct character of US imperialism. The focal point of a population center successfully "reduced" by the Spanish had been a church and its surrounding plaza.[149] The church, which supplanted the native *datu*'s (chief's) house as the center of the community, was an imposing structure, rising above the colonial tribunal or town hall also located around the plaza. These tribunals followed the construction of churches.[150] While ordinary Filipinos lived outside the central complex in nipa houses, the mestizo elite sought property near or on the church-dominated plaza.[151] Similarly, at the center of British hill stations loomed the Anglican church, "that essential symbol of traditional English values."[152] In the American colony, this hierarchy was reversed. Burnham grouped churches with other "subsidiary" buildings in his Baguio plan, and with this, the pastoral care of the Spanish priest, who had supplanted the Philippine *datu*, had been symbolically succeeded again—by American civil administrators executing benevolent assimilation in government offices.[153]

In these ways, while the colonial building architecture recommended by Burnham demonstrated appreciation for Iberian designs, his urban plans ultimately owed less to Spain, an imperial power in decline, and more to those European imperial metropoles with slightly firmer holds on power. Indeed, Burnham's urban plans had traveled to the colonial Philippines only by way of the United States. The influences on Burnham's designs for cities like San Francisco, Cleveland, and Chicago had made an earlier "Atlantic crossing." While American cities did offer points of comparison in his plans, Burnham's other references called up "[t]he hill towns of Italy and France, not to mention those of Japan."[154] His Plan for Manila nodded approvingly at Roman emperors and popes who valued the feature of "flowing water." Burnham identified the streets of Paris and the arrangement of buildings in Rome and Versailles as models. And he concluded his plan for Manila by likening the Pacific capital to great European cities: "possessing the bay of Naples, the winding river of Paris, and the canals of Venice, Manila has before it an opportunity unique in history of modern times, the opportunity to create a unified city equal to the greatest of the Western world with unparalleled and priceless addition of a tropical setting."[155]

City Beautiful advocates like Burnham surely owed much to European architects and planners and, as their use of neoclassic architecture and urban design testifies, to the distant Greeks and Romans.[156] Burnham was unabashed in his affection for imperial cities and their makers. He especially venerated Baron Haussmann, who had been commissioned by Napoleon III to reconstruct Paris; his work had inspired colonial public buildings in Indochina later in the nineteenth century. Burnham wrote glowingly of Haussmann in his *Plan of Chicago*: Haussmann masterminded the "cutting [of] new streets and widening [of] old ones, by sweeping away unwholesome rookeries, and by opening up great spaces in order to disengage monuments of beauty and historic interest."[157] Burnham eagerly took up Haussmann's mantle in the United States, believing that those efforts in Paris were akin to "the work which must be done for Chicago, in order to overcome the intolerable conditions which invariably arise from a rapid growth of population."[158] Burnham paid tribute to Haussmann through his plans and also in his selection of an architect to execute his designs in the Philippines, William E. Parsons. Parsons had received his training at the École des Beaux-Arts, the French school that promoted the neoclassical architectural style, and Parsons would develop Burnham's vision as the consulting architect to the colonial government until 1914.[159] If "the core values of a society" were written across a city's "street designs and public buildings, its shelters and its cityscapes"[160] by Progressives like Burnham, they employed a democratic and republican as much as an imperial and counterrevolutionary idiom, for while the neoclassic may have evoked Greek democracy and the Roman republic, it also referenced empires far and near in time. In the Philippines, teachers and pupils of self-government were to occupy halls that quietly evoked the architecture of imperial Rome or France.[161]

## Imperial Pastoral

The creation of "commanding views," the scoring of the town space for lengthy, processional avenues, and the massing of government buildings were not the only elements of the City Beautiful that Burnham brought to Baguio. He also designed what he called a *tapis vert*, or greensward, to extend between the clusters of government offices.[162] Parks and nature were as central to the landscapes he intended to offer Americans and Philippine peoples as civic structures, and in this regard, they share something with British hill stations. There, residents did not impose the grid but instead "embraced the sinuous contours of the rugged landscape and constructed their cottages along the crests of ridges and around the shores of lakes without apparent premeditation or planning." English trees and flowers trimmed roads and walking paths while English vegetables sprung in backyard gardens.[163] While Burnham wed himself to the geometry of the neoclassic at least through central sections of the hill station, his plans also embraced nature. Parks were priorities for both capitals. In Baguio he identified areas appropriate for golf and fields for play. He also advised against the construction of buildings, apart from government offices, on Baguio's hills, believing that structures would be better situated "against a solid background of green foliage" than silhouetted by the sky.[164] Just as he would in his plan for San Francisco, Burnham suggested that the city claim these hilltops as "public reservations"—that the colonial state enclose this nature so "that their cresting of green may be carefully preserved."[165] Given the place of nature in the landscape he hoped to create, he also recommended the preservation of "woods and other plantings" and warned against their exploitation by "energetic lumbermen."[166] The state ought to rein in the forces of commerce and protect nature for the public—for all to enjoy. In embracing and translating Burnham's vision, William Parsons would come to understand Baguio's development as "a problem of landscape gardening and architecture rather than one of engineering."[167]

In Manila, the creation of a waterfront and system of parks also topped Burnham's list of improvements. Here, like in his work for San Francisco and in a foreshadowing of his *Plan of Chicago*, he urged the government to reclaim private areas with bay and river access so that these might be enjoyed by all of Manila's dwellers. He recommended an extension of the boulevard along Manila Bay, calling on "palms, bamboo, and mangoes" to shade sun-worn travelers.[168] So that city dwellers might maximize their time in these green spaces, he suggested constructing extensive boulevards that would link parks across the city and enable "a continuous journey entirely around the city from park to park without losing at any point the refreshment of green foliage." His models were French cities where one could veer off "a narrow and ill favored street" for a more pleasing parkway and "enjoy a journey of some distance along a well-shaded parkway before again plunging into a less attractive quarter."[169]

Burnham's attempt to maximize Manila residents' time on parkways and in parks was consistent with assumptions of many City Beautiful proponents. To these reformers, parks served several objectives. As an "architectural accessory," they lent grandeur to urban edifices, surrounding often neoclassical buildings with wide, open space—something in scarce supply in cities.[170] They also gave urban dwellers vital "breathing spaces" that countered "disorder, vice, and disease," the supposed effects of the city's tight squeeze.[171] "Experience has shown," Burnham wrote in his plan for Manila, "that [parks] almost entirely eliminate certain classes of crimes and that their general effect is a marked improvement in the moral tone of the neighborhood."[172] Parks with swings, swimming pools, and running tracks provided "wholesome resorts" that proponents touted as good for bodily and moral health.[173] Further, parks were democratic and democratizing spaces, places where people of all classes could converge on equal ground.[174] The promise of peace and order and democratic culture inspired through landscaping and urban design compelled middle- and upper-class reform-minded urbanites in places of labor unrest and ethnic and racial diversity like Chicago and New York, and clearly it also attracted Progressives in the Philippines. Green space, paired with stately structures housing government offices, might reeducate and soothe Americans growing forgetful of home, create conditions for lessening social and political turmoil, and even edify Philippine peoples, giving them a place to perform democratic citizenship through a common consumption of nature.

In his *Plan of Chicago* several years later, Burnham would more fully articulate this task for nature and also a pastoral ideal. He contrasted "natural scenery" to the "artificiality of the city" and advised all people to "often run away from the works of men's hands and back into the wilds, where mind and body are restored to a normal condition." Again expressing the worldview of many reform-minded Progressives, he held that such escape allowed for eventual return to "the burden of life in our crowded streets and endless stretches of buildings with renewed vigor and hopefulness." "[T]he country" afforded refuge for the very fortunate; for others, there could be city parks. "Should not the public see to it that every one may enjoy this change of scene, this restorer of bodily and mental vigor, and will not citizenship be better thereby?" he asked.[175] This builder of skyscrapers, harbingers of capitalist modernity, proposed an amelioration to the very problems associated with them—from social, economic, and political upheaval to "neurasthenia, . . . hyperesthesia, . . . precarious instability" to Louis Sullivan, one of Burnham's harshest critics.[176] To these hazards of modern life, Burnham put forward good planning and also urban parks—alternative landscapes and separate spheres within the commercial city.[177] Raymond Williams has identified in the creation of oppositions like these—"natural scenery" versus the "artificiality of the city," the country versus the city—"an ideological separation between the processes of rural exploitation, which have been, in effect, dissolved into a landscape, and the register of that exploitation, in the law courts, the money markets, the political power and the conspicuous expenditure of the city."[178] The contrast between country and city was wishful thinking, a rationalization that obscured and left untouched their complex transactions.

This cultivation of separate spheres, which would appear in his Chicago designs, figured into Burnham's Manila plans and not only in his call for parks. Take for instance, his discussion of the courthouse in Manila, a structure that was to herald the rule of law in the colony. Rather than place it with other government buildings, Burnham set the Manila courthouse apart. He rejected the practice of grouping the courthouse with the post office, as he said was common in the United States. It might be practical, but it was otherwise "unfitting." The post office, he wrote, was "a business machine affecting public interests. Its character, consonant with its practical necessities, is commercial." By comparison, the Hall of Justice, "far from being solely a business machine, represents sentimentally and practically the highest function of civilized society. Upon the authority of law depend the lives and property of all citizens." "[T]he buildings which constitute the visible expression of law" should be accorded "the utmost beauty in their location, arrangement, and architectural treatment, and approaches." "A Hall of Justice," he continued, "should be treated as a thing apart, a thing majestic, venerable, and sacred. It should be above all free from the clatter of commerce and in its architectural expression should speak the greatness of its function."[179] Burnham countered "the clatter of commerce" with the "venerable" law. The contrast, an attempt to defend pursuits of justice from the field of business, may have been well intentioned; it was also ideological.

Burnham's vision for the Baguio hill station was somewhat different. It expressed the hope of resolution of seemingly opposed forces—country and city, "the highest functions of civilized society," and commerce—in a "single, appropriate landscape"[180]—one only conceivable, perhaps, given

Baguio's limited built environment and population. "[I]magine certain transverse axes, crossing the town and leading up the inclines to important buildings on the flanking hills," he wrote, "look for the green play fields here and there and picture the entire composition hemmed in by the pine ridges of the highest hills, and one will have before him an architectural group of unsurpassed effect and a business machine of the utmost efficiency."[181] Burnham presented the hill station as a landscape, a "medium of cultural representation." "[I]magine" and "look," instructed Burnham; "we will have before us," he wrote, commanding readers to the view.[182] The view was one in which nature, the colonial state, and the "business machine" were reconciled in a unified "composition." A vision of an ideal colonial town, it might be taken as "the 'dreamwork' of imperialism," one that reveals in part, "utopian fantasies of the perfected imperial prospect."[183] Burnham rendered Baguio a middle ground, where elements of the country and the city, nature and civilization, were balanced, affording those features of the classical pastoral, "peace, leisure, and economic sufficiency."[184] In this respect, his design approached that dream of forging something "unusual among cities."

No such perfect composition was possible in Chicago. In his 1909 plans for the city, Burnham reflected on the country and city as opposing forces and implied the victory of the city and elegized nature. "The quick advance of commerce and manufactures, the rapid building of railroads and factories, and the hastily constructed homes of operatives," he wrote, "crowded out nature's parterres of flowers." But Chicago's motto *Urbs in horto*—"a city set in a garden"—"lingered in the minds of men," he wrote, "and in 1839 the struggle began to secure for the fast-growing population park spaces which should at least recall the gardens that of necessity had been sacrificed."[185] The City Beautiful's parks and parkways became, then, a memorial, perhaps even a mnemonic, for a green Chicago consumed in the city's industrialization.[186] If in crafting a pastoral out of Baguio pasture, where grassy meadows and hills—fodder for the natives' cattle—became part of a picturesque landscape, Burnham was designing a mnemonic for the Ibaloi, he seems to have done so unwittingly. Yet sacrifices had been made here, too; Ibaloi pasture had been given over to American play fields and monuments to civic life and self-government.

Otto Scheerer seemed to recognize the cultural dispossessions entailed in this transformation of Baguio, losses in the creation of that imperial pastoral. Perhaps this was because in selling his land Scheerer, himself, was party to these costly transmutations. In 1933, Scheerer remembered "Baguio" before Americans' arrival. Then, he recalled, much of the town had been known as Kafagway, a name that "alludes to the center of the

Baguio basin, a piece of prairie that was more conspicuous before the creek meandering thru it was expanded into the present Burnham Park Lake. As a matter of fact," he continued, "all topographic designations used by the natives go very much into detail, covering often but a few hectares of land, and, with the frequent desertions by the Ibaloi of their settlements—as often as not a silent protest against outside interference—also the names of their former dwelling-places and adjoining grounds fall into oblivion."[187] Even earlier, in 1904, Angel Pérez had written that among Igorots of Benguet it was not unusual to hear one claim that if colonialists were to leave them to their own wishes, "we would quickly return to that happy life enjoyed by our ancestors."[188]

The Baguio that colonialists built in the years following Burnham's proposal did not perfectly adhere to his plans. They altered the position of the town's park and the business section in light of the land's topography.[189] Further, contrary to Burnham's original plan, the governor general's mansion would ultimately rise not in close proximity to the national government buildings but in the far eastern part of the town site. Still, as a whole, builders advanced the processional feel of the space by keeping the national and municipal buildings on either end of that long central axis.[190] And, as we will see, Baguio's pine-clad hills and parks would indeed come to constitute its chief draws. "By generally avoiding, at least in the earlier buildings, the grandiose Beaux Arts neoclassicism that marred Burnham's other plans," historian Thomas Hines has observed, Burnham's protégé in the Philippines William Parsons "helped ensure, in the early years at least, that the City Beautiful Movement would realize its greatest architectural success not on American but on foreign, colonial soil."[191] Just as colonial roads were thought more stunning than those on the mainland, so it was for the City Beautiful.

## Anticolonial Imperialism

In 1903, Frederic Coudert, one of Cariño's attorneys before the US Supreme Court who earlier had argued several of the Insular Cases, wrote an article published in the *Columbia Law Review*: "Our New Peoples: Citizens, Subjects, Nationals or Aliens." In his essay, Coudert made the case for "national" as a fitting term for those persons living in territories recently acquired by the United States.[192] A unique "imperial problem" required this terminological invention, namely the "domination over men of one order or kind of civilization by men of a different and higher civilization." The "imperial problem," in short, was a race problem. Coudert wrote that

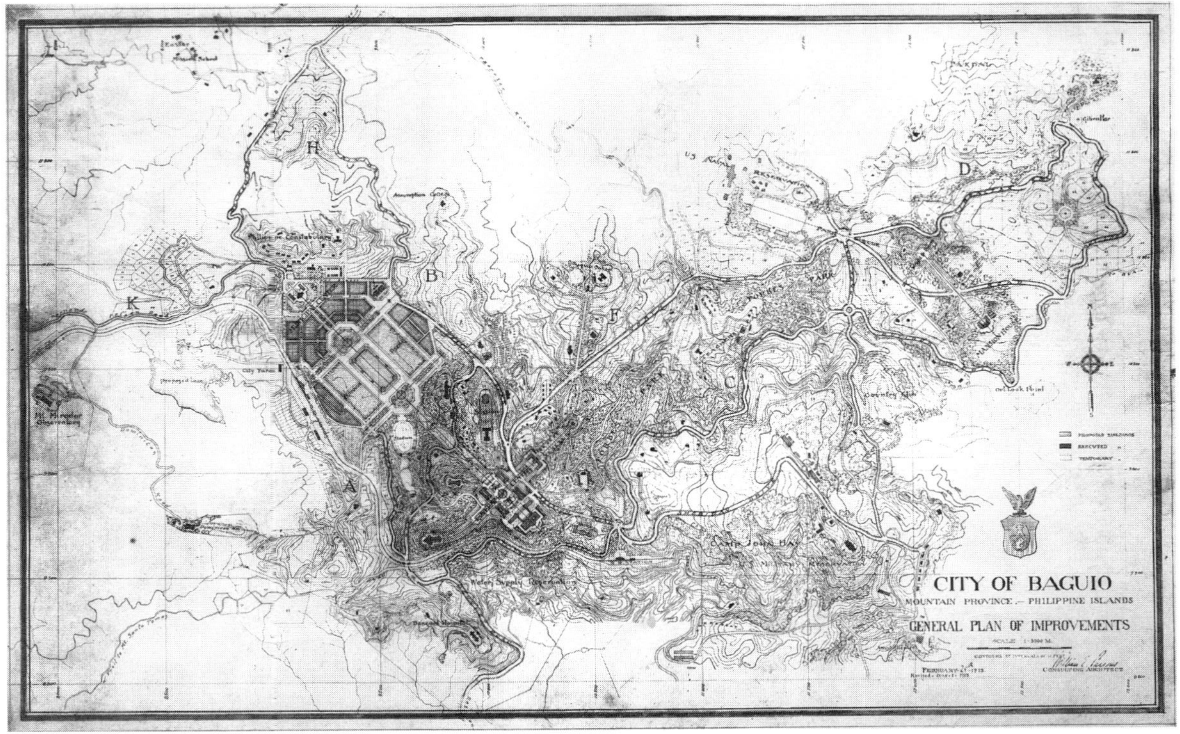


3.5 General plan of improvements, Baguio, Philippines, 7/2/1913. Daniel H. Burnham Collection, Ryerson and Burnham Archives, The Art Institute of Chicago. Digital file #194301_150506–001.

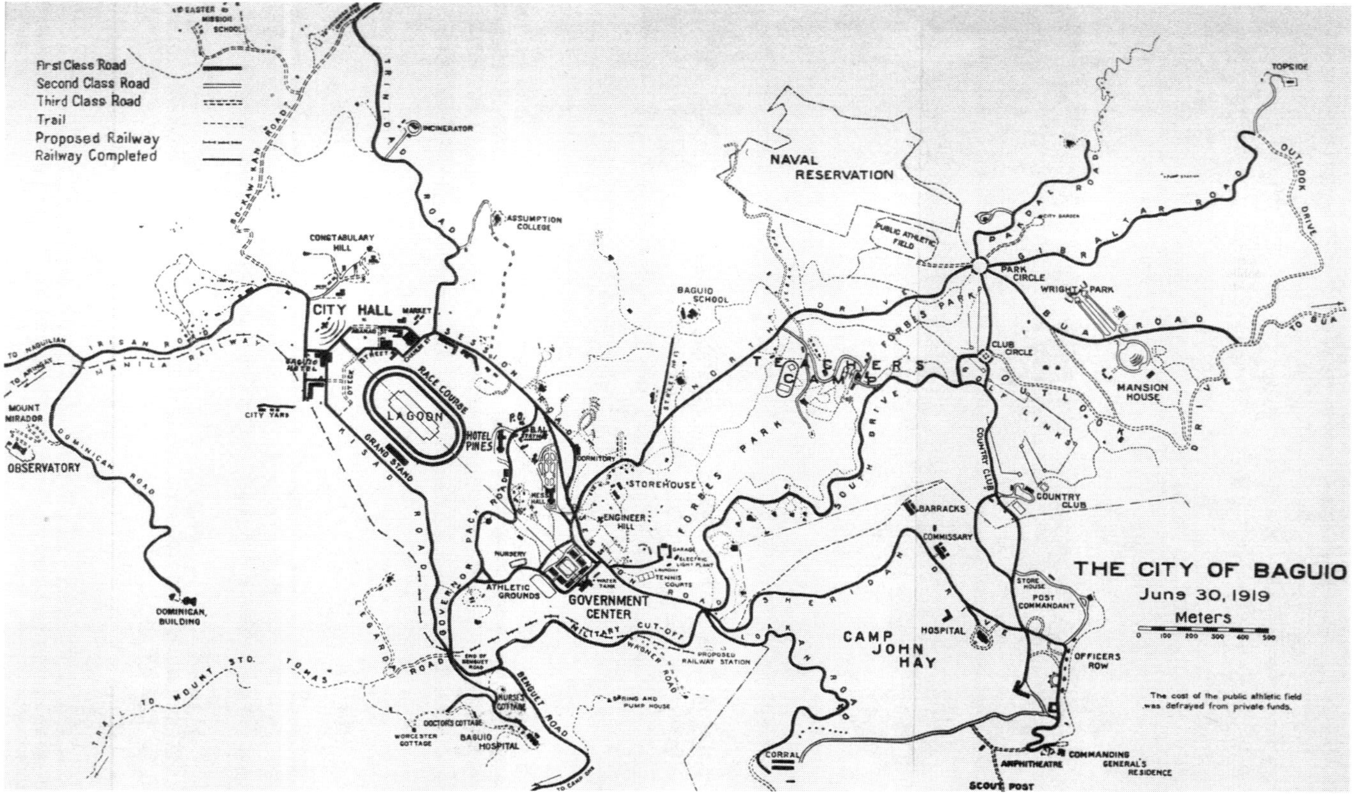


3.6 The city of Baguio, 1919, in *Baguio: The Mountain Resort of the Philippine Islands* (Manila, Philippines: Thos. Cook & Son, 1919), 22–23. Courtesy of the Department of Special Collections, Memorial Library, University of Wisconsin-Madison.

the United States' acquisitions of territory from France and Mexico offered little help in resolving the challenge. The conquest of the US West was no better a reference. The fates of the "Nomad tribes of America" and their lands, he believed, were never in question. "North America could not for mere sentimental reasons remain as a game preserve forever." "Fire water, gunpowder and well-intended but unwise policy" made the problem presented by "red-skinned hunters" "a minor question." Contrary to what he would argue on behalf of Cariño several years later, he wrote that one could not look to former acquisitions of territory to understand this imperial problem. Instead, one had to plumb another history, that of free and unfree black men in the United States. It was Dred Scott, Coudert argued, who best approximated the status of "our new peoples."[193]

In the Dred Scott case, Coudert wrote, "for the first time in our history we had a judicial declaration that there might be subjects who were not citizens." These were men who were "owing allegiance" to the United States but were still somehow "alien" to it. Coudert explained that the term "subject" suited this status well, but the word carried an unsavory ring "because of its usual reference to a monarchical form of government." Coudert proposed something more propitious, a term and concept that have endured: "national." "National would include all persons owing allegiance to the United States" while eliminating their risk of "any injurious interference in the destinies of the nation." It would be burdened with "no unpleasant inference of political inferiority or servitude to an individual." Making plain the new hierarchy generated by the "imperial problem," Coudert wrote, "All citizens must be nationals, but all nationals may not be citizens."[194]

Coudert was just one of many trying to define the relationship between the United States and its new subjects and territories in the years after 1898. Like most others, he attempted to reconcile the United States' status as a democratic republic and its status as an empire. Contemporaries made contortions as exhausting as Coudert's to resolve the tension. For example, George Malcolm, an American lawyer who wrote Baguio's city charter (and was later appointed by Woodrow Wilson to the Philippines' Supreme Court), began by explaining what the Philippines was not. "From a negative standpoint the Philippines occupy a relation to the United States different from that of other non-contiguous territory; not a foreign country; not sovereign or semi-sovereign; not a state or an organized, incorporated territory; not a part of the United States in a domestic sense; not under the Constitution, except as it operates on the President and Congress; and not a colony." He offered all this before concluding that the country was "a dependency—an unincorporated territory—belonging to the United States and under its complete sovereignty—a part of the United States in an international sense."[195] In 1914, the same year that Malcolm wrote, an officer of the Bureau of Insular Affairs (BIA), the Washington, DC agency charged with organizing the United States' new territories and dependencies, warned that "the word colony must not be used to express the relationship which exists between our government and its dependent peoples."[196] That was a sentiment that the Philippine Commission had communicated and that Secretary of War Elihu Root shared years earlier when describing the BIA. The BIA, he wrote, "performs with admirable and constantly increasing efficiency the great variety of duties which in other countries would be described as belonging to a colonial office, and would be performed by a much more pretentious establishment."[197] The British had their "Colonial Office," and the French and Italians ran their respective "Ministry of the Colonies." To Root, the BIA was not a colonial office by another name. The BIA was a mere "clearing house," as its chief Clarence Edwards once described it.[198] The United States was an empire of a different and superior kind, and the name of its colonial bureaucracy was chosen to convey that.

In creating an aesthetic for US sovereignty in the Philippines and crafting a pastoral landscape for Baguio, Daniel Burnham also shared in the work of defining US imperialism much as Justice Holmes had in his decision in the Cariño case. Burnham's plans monumentalized American power, giving governmental precincts a "dignity" that was a hallmark of City Beautiful planning. He designed commanding views and an imposing presence for those conducting colonialism, and he embedded government buildings in the green hills, situating these structures amid parks and playfields. He called for the preservation of trees and for hilltops free of development and envisioned a reconciliation of forces—city and country, culture and nature—in a picturesque landscape, an idealized middle ground that gave aesthetic form and spatial shape to the US occupation and to colonialists' aspirations. T. J. Mitchell writes that landscape is "an art that conceals its own artifice."[199] In Baguio, this pastoral landscape concealed acts of imperial dispossession.

FOUR

# "*Independencia* in a Box"

In 1904 William Cameron Forbes, the Boston businessman then serving as secretary of commerce and police, wrote to a friend, relaying some colonial gallows humor. A couple Filipinos, it seemed, were scheming to get rich off the idea of independence.[1] One claimed to be the "Black Jesus" and another, the pope; both were peddling "'*independencia*' in a box." Forbes, who had been appointed to the Philippine Commission that year and would ascend to the position of governor general in 1909, was bemused by another account—of a captured insurgent on the island of Samar: "[W]hen asked why he hadn't surrendered before he said he was waiting for '*independencia*.' He didn't know what it looked like but he did know it was a good thing and his captain had gone over to a neighboring island and was to bring it back with him."[2] In these brief anecdotes, grand, imperial promises of uplift and liberty were materialized and shrunk as transportable goods—the object of democratic politics dropped into a box. Forbes, perhaps, had a good chuckle at these stories, examples of what anthropologists call "cargo cult."[3] Across the southwest Pacific where the phenomena has been studied most, "cargo" has signified everything from refrigerators and guns to "a new social and moral order that would ensure local sovereignty and the withdrawal of colonial rulers."[4] The latter, the work of revolution, seems just what the pope, Black Jesus, and the captured insurgent hoped to realize.

The notion of "*independencia* in a box" bespoke millennialist anticipation, a kind known to emerge especially among the disenfranchised in times of rapid social and economic change.[5] General Elwell Otis reported that in late 1899, a fanaticism, possibly religious in orientation, could be found across central Luzon; "[s]elf-declared prophets were revealing and proclaiming new creeds . . . and were collecting converts." Otis offered several possible interpretations of these "disturbances": war might be keeping natives in an "excited state"; this fanaticism could be a reaction against the religious orders; or perhaps natives "were relapsing into former superstitions and beliefs."[6] In *Pasyon and Revolution* (1979), Reynaldo Ileto offers another explanation for visions and events like these, understanding them "as peasant attempts to restructure the world" in ways that aligned with their "social forms and modes of behavior." The image of independence launched from a box resonated with some revolutionaries' views of *kalayaan*, that Tagalog term meaning independence. In revolutionaries' imaginary, *kalayaan* would usher in conditions of "brotherhood, equality, contentment (*kaginhawaan*) and material abundance (*kasaganaan*)." *Kalayaan* did not just mean political sovereignty, as some Filipino elites and Americans would have it; it meant a fundamental restructuring of the social order, as the agitation of Filipino labor radicals suggested. The notion of *independencia* sprung from a box was "consistent with Katipunan images of *kalayaan* as a personified condition: Mother Filipinas . . . lying in a state of limbo or sleep, awaiting the day of final liberation that would bring about prosperity, comfort and knowledge," a version of the resurrection story taught by Americans' predecessors on the archipelago.[7] This idea reflected no relapse into old superstitions but rather an appropriation and reorientation of Catholic theology.[8] Whatever their origin, Americans, like many Filipino elites, did not and would not countenance such radical visions that amounted to an upending of the social, political, and economic order. But Americans did believe that they could deliver a form of independence as fulfilling and unifying as that conveyed by *kalayaan*. It was a reward of buying and selling in the market and not immersion in the sphere they deemed "politics."

In 1911, Republican Secretary of War Henry L. Stimson writing to then-Governor General Forbes emphasized the importance of the market, or, more narrowly, business and commerce, and argued that these must preoccupy Filipinos under American tutelage. "The temptation of the Latin-American people has nearly always been toward overattention to political thought and discussion and underattention to commerce and business," he wrote. "In such a problem as we have in the Philippines and in our other insular possessions," Stimson continued, "it is of great importance that the attention of the people should be directed to material improvement and better business methods."[9] The assumption that underlay Stimson's assessment was consistent with that governing the colonial regime, especially by this point in the US occupation. Paul Kramer has noted that through the 1910s, "US colonialists attempted to displace nationalist demands onto a depoliticized economic sphere, claiming that a 'material development' policy of infrastructure, export, and extraction was needed before any 'political' considerations could be entertained."[10] Not only did American administrators depoliticize colonial economic relations, but trade, in supposedly transcending politics, became US colonialism's *deux ex machina*. It served as an agent of nation building and an indicator of civilization.

Economic behavior had in fact constituted an important aspect of "self-government" since the concept of civilization was largely an economic notion. "Civilization" assumed "a hierarchy of evolutionary economic stages (industrial production over hunting and gathering; capitalist individualism over communal or clan ownership)." Designating a society's position on a scale of civilization could justify "interventions into 'savage' society on the part of 'civilized' nations."[11] In the political imagination of Forbes and Stimson, business and, more broadly, sound economic exchange, or trade, brought forth "free agents," who were to be the equal, independent, and civilized actors constituting a democratic nation. The freedom first felt in the humble marketplace, where one alienated and then exchanged the products of his labor for another's, was the primal experience of freedom.[12] And not only did trade develop the right sort of self-possessed individual, turning subjects into citizens, but according to some American administrators, it eradicated seemingly intractable differences among peoples. Trade, its very phenomenology, could knit the "variegated assemblage of different tribes and peoples" in the Philippines into a coherent and recognizable political unit.[13]

This chapter turns to these romantic views of economic relations, of trade, and of the institutions that housed and conditioned them, marketplaces, the one in Baguio especially. Contrary to Americans' hopes, the social, political, and even ontological yields of trade remained largely illusory through US colonization.[14] In the Philippines, Americans came to use trade and the marketplace as much to unite as to exploit difference and entrench divisions among Philippine peoples and between colonialists and colonized, too. As much as they built marketplaces to inculcate and spread a market ethic, they also engineered them to contain the market—to control how goods were sold and to limit what was for sale and to whom. They then assessed and governed Philippine peoples according to their obedience to or violation of these rules. In drawing boundaries like these, American colonialists created hierarchies instrumental to justifying the US occupation and retaining hold of the colony, a future that was very much in question at the time the secretary of war wrote Forbes of Filipinos' "inattention" to business.

But in promoting market activity as a measure of civilization and readiness for self-government, colonialists also gave Filipinos a convenient yardstick by which to assess their patrons. Further, market exchanges raised fundamental epistemological questions. For despite the mutual trust that administrators imputed to the experience of trade, for all that it could diminish and transcend, for all that it ensured and made certain, there seemed something in the experience of exchange that could also exacerbate feelings of doubt and risk. No deal and no buyer or seller could be fully trusted, and no colonial hold on knowledge or power sure. Both the instrumental use of trade and its tendency to produce ambiguities like these would be evident in the Baguio marketplace and across the hill station, host to "the greatest dog show on earth" and home of the notorious dog eaters of the Philippines.[15]

## Markets and Marketplaces

As they undertook the "Burnhamizing" of the hill station—colonialists' gerund for executing the planner's vision[16]—administrators were also doing their part to orient Benguet and its residents toward the market. In his reports while serving as Benguet's provincial governor, William Pack promoted Benguet's potential to produce goods not only for subsistence and local consumption but also for sale. He envisioned the province as a "suppl[ier of] the beef market of Manila"[17] and a producer of coffee[18] and vegetables like potatoes, even celery ("equaled in every respect [to] the best Kalamazoo variety"[19]), for eager consumers in Manila. This would mark a major change for most Ibaloi, who had long participated in but were "not of the market."[20] While administrators attempted to increase local production beyond local need, the Baguio marketplace was becoming what one visitor called a "Sabbath mecca."[21] By 1909, the hill station functioned as the center of trade with lowland communities, replacing the Ilocos town of Naguilian.[22] The hill station had also become the principal site of trade among Ibaloi, Bontoc, and Ifugao Igorots.[23] In the early 1920s, a visitor described the "rare picture" of this market as it came alive: "Hundreds of the mountaineers, both men and women, come in over the myriad trails from the adjoining country, carrying baskets of produce on their backs and heads. These are mostly packed in larger, stronger baskets to be shipped to Manila. Scores of picturesque covered bull carts that have come up from the plains, exchange their rice and coconuts, fish, chickens, and eggs, for the varied products of the hills."[24] Another traveler described "[t]housands of people of various costumes and complexions swarm[ing] about the place. Almost every conceivable sort of merchandising fills the building and overflows on the ground outside. Groups of ox carts, huge baskets on wheels topped prairie-schooner fashion and filled with rice, mangoes, bananas and various other produce are active centers of traffic. One may be fish, flesh, and grain, raw, cooked or cooking over small fires in curious shaped Filipino earthenware or in Standard Oil cans—the rivals of bamboo for universal utility in the Orient."[25]

Marketplace buildings like Baguio's garnered significant attention by the end of the first decade of the twentieth century. A primary objective of the US acquisition of the archipelago, after all, had been to cultivate trade, markets for American goods, and also opportunities for American capital. Back in 1900, Senator Lodge gave voice to the former ambitions, claiming that "[t]he mere fact that we hold the Philippine Islands increases our trade with all the East—with China and Japan alike. Trade certainly has followed the flag."[26] Lodge had even speculated that the Philippines could eventually consume more American exports than Cuba, Puerto Rico, and Hawaii altogether, something which, at least by 1913 and Forbes's departure, had not come to pass.[27] Later in 1900, the Philippine Commission anticipated colonial markets in agricultural machinery, wheat flour, preserved meats and canned fruits, jewelry, watches, and cottons, and it admonished American exporters to "familiarize themselves with the tastes and prejudices of the people rather than to ship their goods here haphazard and take chances on their proving acceptable."[28] Four years later, the commission observed that "The Filipinos are imitative, take quickly to new things, may easily be taught, as their wealth shall grow, to regard American products, which are now luxuries to them, as necessities."[29]

Another prospect that had compelled pro-imperialists was turning colonial acquisitions into outlets for "idle" American capital.[30] Before Hobson had argued that the quest for new markets drove imperial expansion and before Lenin had defined imperialism as "the highest stage of capitalism," the American finance expert Charles A. Conant had written of the "menace to the economic future of the great industrial countries" posed by "congested" capital, that is, the "excess of saved capital."[31] He proposed "imperialism" as the solution. "[T]he great industrial countries should turn to countries which have not felt the pulse of modern progress," Conant held, and invest in projects like railroad and canal building. "Existing commodities now imported from the interior of these countries at great cost will be swept on paths of steel to the seacoast, with the result of reducing their cost, increasing their consumption, and benefiting at once both producer and purchaser."[32]

Conant had supplied cause and justification for expansion, dismissing the question of whether the United States should acquire territory, arrange "captain generalships and garrisons," intervene to protect "sovereignties nominally independent," or simply exercise "free commerce" in the East through naval bases and diplomacy; all this was simply "a matter of detail."[33] But Conant did immerse himself in the colony's currency and banking issues once Americans had chosen their course in the Philippines. To help stimulate the "pulse of modern progress,"[34] Conant recommended setting the Philippines on the gold standard by minting a new silver coin pegged to the US dollar; the new peso would trade for half of the latter.[35] The "Conant peso," as it came to be known, replaced the Mexican silver dollar (also referred to as the peso) that had circulated at the beginning of the US occupation along with "Spanish-Filipino pesos, silver and paper, and fractional silver and copper coins."[36] Conant believed that training the Philippines to the gold standard would make the colony more hospitable to investors, who would seek protection from the "fluctuations in the gold value of silver."[37] He also advised permitting US banks to open branches in the Philippines. This would further smooth the path for American capital and loosen the grip of powerful foreign banks in the colony, British especially.[38]

William Cameron Forbes, a Massachusetts native like Conant, was particularly well qualified to execute these policies as the colony's commissioner of commerce and police and, later, governor general. Journeying east and building empire had been a Forbes family tradition, and William Cameron traced paths navigated by his early-nineteenth-century forebears, who amassed great wealth in the China trade. Through the nineteenth century, the Forbeses had gained entry to business in China and eventually partnered with Russell & Co., one of the most powerful firms conducting trade there through the nineteenth century.[39] John Murray, William Cameron's grandfather, turned his attention to the continental empire, developing midwestern railroads in the nineteenth century. But as a young man, he, too, had trained in the East in the 1830s. As his brothers had, John Murray worked with Howqua (Wu Ping-chien), a leading Hong merchant in Canton who had made a $26 million fortune, a degree of wealth that placed him "among the richest men in the world."[40] Ralph Bennett, John Murray's brother, described Howqua as "the life-long friend of Russell & Co.," a trusted partner whose "word was his bond."[41] John Murray managed Howqua's correspondence with foreign associates and his trade in tea and silks.[42] William Cameron's family was thus no stranger to forging close ties with the rich and powerful in the Far East toward the ends of profit making.

While William Cameron's grandfather brought home his overseas experience and expanded his wealth by laying down railroad tracks through Indian country and helping to consolidate the continental empire, William Cameron's father, William Hathaway, chose to conquer the kinds of distances traversed by his ancestors while staying in Boston. He invested family monies in communication, becoming an early backer of William Graham Bell and eventually president of the American Bell Telephone Company.[43] This family history of Forbeses who had profited in the ports and laid the tracks of empire had, perhaps, recommended William Cameron to Roosevelt, who appointed him to the commission. And if not this, it was William Cameron's own work for the Boston-area engineering firm Stone & Webster that impressed the president. Here was the next generation of Forbeses—of liberal internationalists—to help advance the United States' trade with China. William Cameron originally aspired to a seat on the Panama Canal Commission, but President Roosevelt offered him a job on the Pacific frontier instead, one which Roosevelt considered as important and "ranked with a place in his own cabinet or on the Supreme Bench," as Forbes noted in his journal. Once settled into the position of secretary of commerce and police, Forbes assumed responsibility for supervising transportation—from roads and harbors to railroads—communications, the organization of corporation law, the police, and the formation of Baguio as the summer capital.[44]

Forbes's task was developing the colony's capacity for the kind of investment, trade, and ultimately capital accumulation envisioned by the likes of Lodge and Conant. Americans' ambitions were not lost on editors of *El Renacimiento*; they warned in 1905 that in the United States "commercialism has become so pronounced a factor that the mere whisper of commercial interests is sufficient to force legislation through Congress, even when such legislation cannot be otherwise than dangerous to political precedents and constitutional limitations."[45] Indeed, Americans like Forbes did seem to have a romance with business, to play on the title of Forbes's 1921 primer *The Romance of Business*. The commissioner pursued various means to stimulate economic development in the colony apart from those recommendations of Conant. Not long after he arrived in the Philippines, Forbes promoted a postal savings bank. According to the first US colonial census in 1903, only nine major banking institutions conducted business in the Philippines; while "numerous small financial establishments" operated across the archipelago, they held little capital, "made large profits," and were frequently preoccupied in money exchanges, given the various currencies circulating in the country. The census also noted that while the number of depositors was small, deposits were large, signaling that ordinary Filipinos were not generally using banks; these large deposits belonged to commercial interests.[46] Manila and the islands of Iloilo and Cebu had branches of major banks, but elsewhere, access to and circulation of money depended on local merchants or "usurious money lenders," or upon commodities dealers, who had agencies to manage the sale of sugar, tobacco, copra, and hemp.[47] Forbes's postal savings bank, an institution operating in Hawaii by 1886, was thus intended to give poorer and rural people access to banking services by allowing anyone, even children, to buy stamps, paste them on cards, and bring them for deposit once the collection of stamps amounted to a whole peso.[48] Forbes believed the postal savings bank would encourage "the virtues of economy and thrift."[49] By 1918, the postal savings bank had branches in every province.[50]

By that point, Americans had also turned attention to another means for advancing economic activity: incubators and containers of markets—marketplaces like Baguio's. Besides modeling what the Philippines itself might become, these structures were understood to serve the needs and develop commerce especially in areas remote from population centers, just as the postal savings bank was to do. Colonial market buildings, some believed, helped to "build up centers and arteries of trade" and "encourage increased plantings of marketable products; they supply motives for more extensive intercommunication, and thereby tend to introduce new commodities and raise the standard of living."[51]

In 1912, the Philippine legislature (by this time composed of Filipinos and Americans) began accelerating construction of new marketplaces through the provision of loans.[52] Older marketplaces often lay on low land and were subject to flooding in the rainy season. And further, these sites for buying and selling tended to be small and impermanent, constituted by a series of bamboo or thatch booths or stalls. Merchants generally raised the simple structures themselves. Observers of these marketplaces noted the fire risks attached to them, especially when they were sandwiched between public and private buildings. A small spark in a thatch or bamboo stall could ignite a conflagration engulfing whole sections of a town.[53] In response to the perceived dangers of older marketplaces, the colonial Bureau of Public Works required that new marketplaces sit on at least one hectare of land, but still land situated near the center of town and preferably a major highway, railroad station, or river bank accessible by small boats.[54]

Engineers produced three "classes" of marketplace buildings, with

4.1 Baguio market, ca. 1915. Bureau of Public Works Collection, Filipinas Heritage Library.

"main market buildings" the most solid and permanent of the three. *Tiendas* with either open or closed sides were the second and third classes of buildings and involved "the lightest possible construction."[55] Main market buildings were distinguished by at least one longitudinal bay formed by concrete columns and a corrugated iron roof; these roofs were to extend one and a half meters beyond the column line, maximizing the space useable in inclement weather. Engineers prepared four designs. The simplest appeared as a single bay formed out of two lines of columns and a trussed roof; it could stretch to widths of twelve, fifteen, or eighteen meters depending upon the needs of a municipality. A more complex marketplace design created an open court by arranging four bays in a square.[56] The Baguio marketplace represented the closed *tienda*-style marketplace, which was typically anchored by concrete or timber columns topped with a light, galvanized iron roof. Each "block" or section of the *tienda* could be closed using boards that locked into place, and both sides and each end of the structure featured "a combination shutter and lifting counter" that closed to form sides of the building at the end of the day.[57] No matter the class or design, all of the new colonial marketplaces contained sloped concrete floors. A promoter of Philippine trade claimed that "[t]he ventilation, drainage and facilities for cleaning are as perfect as they can be made."[58] An engineer with the Bureau of Public Works also noted that *tienda* and market building construction had been "simplified" so as to render "skilled" labor unnecessary.[59] The uniformity of these simple structures across the colony and the regularity with which they were used may have helped to convey their authority and the promise of predictable, sound exchange. The structures eliminated some of the particularity and uniqueness that had dominated older marketplaces in addition to making the colonial state the entity now literally structuring trade.

Marketplace designs, which minimized construction costs, also maximized the income they could produce for a municipality. New market buildings were, in short, generative. As a "public monopoly," they grew municipal revenue by capturing the rent traders paid on booths, booths that sellers had formerly owned and set up independently.[60] Municipal income may have been one reason why the ordinance reforming public markets in Manila "prohibit[ed] their establishment by any other person, association, or corporation other than the city itself."[61] In Baguio, a new marketplace went up in 1908 at a cost of 6126.37 pesos. It opened for trade in March, and revenues from April, May, and June—the summer season—were 299.10 pesos.[62] A 1914 *Quarterly Bulletin* noted that municipal revenues doubled in five towns where new market buildings had been constructed. Revenue like this led the Bureau of Public Works to advise that marketplace building precede the construction of schools and *presidencias* (municipal administrative offices). While the latter two plunged a town into long-term debt and postponed further municipal development, marketplaces helped pay for themselves and began the process by which tax monies could be accumulated toward other municipal projects.[63]

This was an attractive feature of marketplaces, as municipal and provincial offices had limited sources of income. The colonial Internal Revenue Law passed in 1904 channeled 75 percent of taxes to the insular government, rendering the relationship between provincial and municipal governments and the insular government "essentially that of dependent local instrumentalities and a superior foreign occupation government."[64] More generally, in creating conditions by which municipal revenues multiplied, the colonial regime paid for itself. This was consistent with Lodge's vision of the islands as "self-supporting" and imposing "no burden of expense at all on the people of the United States" once pacified.[65] But the yields of the marketplace were also imagined to exceed the calculable. In Baguio, the marketplace, said one American, had become the "Sunday Mecca." Indeed, the blessings of marketplaces were transcendent. Or, rather, US officials promoted them by identifying what they allowed peoples to transcend: inequality and conflict.

## Market Pastorals

American administrators in the Philippines maintained that marketplaces stood as sites of liberation. "Before the new markets were erected it was often impossible for the fisherman with a few fish or the farmer with a little fruit or the weaver with a few bolts of cloth to dispose of his wares at a reasonable price, for the owners of the small native stores usually insisted on a wide margin of profit," wrote a contributor to the *Quarterly Bulletin*. "[T]he small producer needed a location where he could, for a reasonable price, obtain space in which to expose his products for sale."[66] Marketplaces were made heroic agents, rescuing the most vulnerable, namely vendors with small market share. In marketplaces, the "small producer," long imagined as the bulwark of the American republic, could become a better advocate of the value of his goods and better resist being strong armed by the middleman, the profit-seeking merchant.[67] The marketplace was, in this way, a space within which individual freedom could be cultivated.

Forbes certainly subscribed to this. He had written of the "freedom of movement" "to buy and sell that makes the poor man a free agent."[68] And he saw the United States' objective in the Philippines as the "development of the individual;" this, Forbes wrote, "the United States is endeavoring to give to other countries."[69] In the same spirit, John P. Finley, an army major stationed among Muslim peoples on the island of Mindanao, argued that in a fair trade, one practiced "freedom" and "self-control" as one exchanged the products of her time and her self. In the process of trade, he wrote, buyers and sellers and former slaves "felt the first thrills of freedom and the quickening impulse of self-control, in the possession of that which was lawfully and rightly theirs, as the product of their own ingenuity and labor."[70] According to Finley, the experience of a fair trade, like wage labor on Benguet Road, cultivated self-disciplined and autonomous political actors. By willfully alienating a product of one's labor, one became self-possessed and in that way free—no matter the colonial status of the country.[71]

Contrary to the understandings of some politically inclined Filipinos clamoring for greater legislative power, it was US sanctioned and structured trade, the market more broadly, that was the realm of freedom. This resonated with a view of market society already in circulation in the United States. There, during the late nineteenth century a "social self" began to displace the idea of a self-interested, "sovereign self," whose political identity was rooted in individual ownership of productive resources. The new social self engaged in personal development and self-fulfillment independent of his ownership of productive resources and traditionally associated political rights. An individual's "liberty" and "equality" became understood as derivative of "popular participation in the spiritual and material rewards of industrial progress"—that is, in participation in the market.[72] This was a version of what Forbes recommended to Taft in 1908: that "certain individuals should have the opportunity to give up some of their civic rights in order to get the material advantages which accrue from subjecting themselves to mild and benevolent paternalism."[73] The division of interests into the political and the economic, and the elevation of the latter, yielded something greater for Filipinos, he held.

Henry Stimson, then, was not alone in advising Filipinos to attend to commercial development and its material infrastructure over formal politics. This was how to build a nation. Taft had communicated a similar message earlier, arguing that "Civilization follows material development." "If the construction of railroads, the inauguration of steamship lines, the construction of highways, or building of port works comes under the definition of 'exploitation,'" he wrote, "then that kind of exploitation is wholly consistent with the principle of 'the Philippines for the Filipinos,'" Taft's mantra as governor general.[74] By this logic, paeans to great democrats and revolutionaries, who defied the arbitrary power of a king or priest, ought to be replaced with another iconography: salesman as public servant and business and engineering as public service. For to commentators and ideologues like Forbes, market participation not only remade people into civilized agents able to act freely; it also brought those very selves together, forging wholly new social relations. Marketplaces not only liberated the slave and piecework weaver. They also bound together buyers and sellers into new communities. They enabled foreigners to feel like neighbors, even for enemies to feel like friends. And, more grandly, as peoples congregated and haggled over wares beneath a single corrugated roof at that "Sunday Mecca," a certain sort of unity—one that vastly exceeded the time and space structured by the marketplace—was made possible. The marketplace facilitated democratic culture and, in turn, a nation.

In his 1913 article for the *North American Review*, "The Commercial Awakening of the Moro and Pagan," Major Finley wrote that despite their zealous proselytizing, Spanish missionaries had failed to assimilate the Moros and pagans to a shared faith. The Spanish, wrote Finley, overlooked "the only feasible plan of amalgamation through harmonizing the trade relations of the factional races." Commerce, the new faith to which both groups were "awakening," would bridge the divide between Moros and pagans. In his article, Finley described the creation of the "Moro Exchange," a periodic market forum for the exchange of goods between the two groups. Through the medium of commercial association, "long-time enemies have found a common ground on which they can get together, and profit more by friendly association and business ventures than by the process of slavery and extermination."[75] By this account, American-structured and American-managed trade, and the very experience of trade, obviated against conditions of war and inequality and stood to forge a new political order. Commercial encounter displaced violent confrontation. Conducting trade meant creating a language, a common currency, by which strangers could reach shared judgment on the value of an object. Trade negotiation required or at least offered the prospect of sympathetic identification and discovery of mutual interest. With this, differences between formerly warring "tribes" might be resolved and peace might prevail long enough for social bonds and a coherent nation to be manifest.

As his initial interest in work on the Panama Canal Commission may suggest, Forbes's vision of connection went beyond the nation and embraced the globe. The Boston Brahmin waxed poetic about a transnational existence formed by exchange, the production and consumption of commodities, and their movement around the world. "We find ourselves living in a world where trade is so easy that we brush our teeth with bristles brought from Siberia, eat sugar from the tropics and pepper raised in Sumatra or the south of China, wash with soap made in part of materials that come from Sumatra and scented with herbs from France, and wear shoes made of leather from the Argentine," he wrote in *The Romance of Business*. "We carry money of silver from Bolivia; our teeth are filled with gold from Alaska. We wrap ourselves in furs brought from Labrador and wear a watch made in Switzerland of steel from Germany and gold from South Africa. In fact, everything we wear and use is probably carried anywhere from ten miles to twenty thousand miles, and is the product of so many processes that we are utterly at a loss to know its origin. If we wanted to go through to the bottom of our daily comforts, we should find it perfectly impossible to trace them—so many and so scattered they are. This has transportation done for us!"[76] Awe propelled Forbes's ode to the well-traveled commodity. As he took up a toothbrush and soaped his face, places as far apart as Siberia and Sumatra became one; at once, the whole world seemed to exist in his washroom and in him—in his fillings.

Forbes's vision of unity, abundance, and peace—what we might even call a market pastoral—was not unique at the time. Englishman Norman Angell's 1909 book *The Great Illusion*, for instance, argued for the futility of war for "territorial expansion." Relations of "credit and commercial contract" had made nations of the "civilized world" so thoroughly interdependent, Angell believed, that wars for the comparative advantage conveyed by territory were, or ought to be, an anachronism.[77] In the American context, proponents of the social self maintained that "The socializing force of free exchange would naturally bring an end to the selfish struggle over scarce resources"—like political power or productive resources like land—"ushering in a cooperative commonwealth and boundless abundance."[78] "Ethical economists," who challenged the premises of neoclassical economics, expressed similar beliefs about how commercial relations conditioned social relations. Progressive Richard Ely, for example, wrote in 1903 that "the highest known relationships among men" were mainly "economic in their origin."[79] In his phenomenology of the market, economic exchange generated feelings of "sympathy" and "mutual obligation."[80] These works resonate with what anthropologists have argued—that relations based on credit and debt can forge connections both spatially and temporally. Parties to a contract "conjoin their respective futures and pasts materializing their temporal bond, as it were."[81]

## Market Hazards

While a vision of shrinking the temporal and spatial distances that separated people and things—an ideal of peace and abundance—shaped these visions of trade, US colonialists in the Philippines also seemed aware that these technologies and the realization of their full potential entailed risks. In a 1913 description of the Baguio marketplace, Emma Sarepta Yule, a professor of English at the University of the Philippines, betrayed a case of "imperialist nostalgia"—lamenting the disappearance of practices for which imperialists were themselves responsible.[82] She observed in an article for *The Overland Monthly* that "As neither the Igorot nor the Filipino understands the dialect of the other, the business operations of the market are carried on mostly by pantomime. This adds to the onlooker's interest." Yule warned, "However, as English is becoming the medium of intercourse among the younger Igorots as well as among the Filipinos, this special touch of interest will soon be a thing of the past."[83] In forging a dialogue between Igorots and Filipinos in English, this American language instructor stood to lose the easy entertainment provided by marketplace transactions. Yule neglected to note that the more friction-free trade afforded by English also implied losses for Igorots and Filipinos: they participated in an exercise that supposedly made them sovereign agents in a language not wholly of their choosing.

Forbes's description of trade also bore an undercurrent of frustration, if not a terror of the sublime. His list attempted to express the "perfectly impossible" and comprehend a totality, a seeming infinity, at once invisible and materialized in the hard, round coins jingling in his pocket. He attempted to grasp a long series of processes that make things real and set them into his hands—a cosmos of relationships that end with an otherwise mundane object atop his wash basin and in his pocket. Fathoming these travels of the commodity—the history of its production and movement—was staggering given the ease and speed of trade, so that "we are at a loss to know its origins."[84] In Forbes's rendering, the commodity embodied a universe of relationships and, for that reason, seemed to come from nowhere at all, a potentially terrifying prospect.

Indeed, American colonialists exhibited a contradictory attitude toward commerce. As much as they wanted to expand and inspire trade, believing in its social and political profitability, they also sought to control its temporal and spatial dimensions. As colonialists' close attention to drainage and ventilation suggests, their new marketplaces were designed to literally clean up trade and, in that way, control the dangerous circulations enabled by the market. Because the majority of the Philippines' agricultural yield by 1913 lay in "hemp, sugar-cane, tobacco, copra, and cocoanut oil," Filipinos turned to markets to source their subsistence foods like meat, fish, rice, vegetables, and fruits.[85] But many colonialists had come to see native marketplaces as sites where things far less sweet than mango and papaya could be acquired.[86] Isabel Anderson, an American visitor to Baguio around 1914, described squalid conditions that reportedly preceded colonial intervention in places of trade: "The local market used to be a community dwelling for all the vendors, who lived there, reveling in their filth. Their children were born there, also their dogs, pigs, cats, and chickens. It was so vile smelling that no American dared to go into it. Never being cleaned, it was the center from which disease was spread to the city."[87] In gathering together people and things from faraway origins, marketplaces stood to trade as many diseases as goods. Concerns for the health of Philippine peoples and for colonialists assigned to administer them carried the weight of past scares. The bubonic plague and cholera visited the islands early in the US occupation, with cholera alone claiming over 100,000 lives.[88] Rinderpest, meanwhile, devastated livestock.

To treat localities ravaged by disease whole town districts were set ablaze.[89] And the same tactic was applied to marketplaces deemed unsanitary. After a new one based on Bureau of Public Works designs was constructed, colonialists required that only a guard be permitted to dwell on the premises, perhaps in part to ward away any vendors seeking refuge for the night.[90] Further, when the Manila city marketplace was rebuilt, "'screen cages'" were installed to protect meat from flies. Inspectors were enlisted to guard that "stallholders wore clean clothes, kept their hands spotless and their nails trimmed, and used only clean white wrapping paper."[91] The marketplace rules recall the rituals of the Mass; like the priest dressing in his vestments, washing his hands before touching the bread and wine, and covering the sacred cup with white cloth, vendors' clean clothes, spotless hands, and white wrapping paper assisted in the transformation that took place in the market—the alienation of a good from its producer. These were the rituals that could help to assuage the fears entailed in trade.[92]

According to Anderson, Americans' sense of smell alerted them to dangers in the marketplace—the old marketplaces were "so vile smelling that no American dared" enter, she wrote. Americans who visited another nexus of the global market in the nineteenth century—the London docks—also commented on its odors. It was the smell of the warehouses that aroused American visitors to comment and attempt description. The nineteenth century "reevaluation of the senses . . . among the educated" may have had something to do with their acute awareness of the docks' perfume. Sight "came to be associated with reason and civilization," while smell was linked to "insanity and savagery."[93] These associations with smell seem similar to the ones made by Americans like Anderson who remarked on the "vile" odors of Philippine marketplaces. In shopping in the unsanitary, noisome marketplace, the border between civilization and savagery could be breached, suggesting that "anxieties about sanitation and disease" were, in the last instance, an expression of anxieties about race.[94] If this is right, then the sanitary marketplace, in managing trade, could also operate in the control of "racial boundaries."[95] Discourse around the marketplace affirmed social differences that justified imperial intervention, while the sites and their new rules insulated colonialists from the ontological risks of trade.

Health and the racial order were not the only concerns managed by colonial marketplaces. In constructing them, administrators also defined what objects ought to remain outside—what were to remain inalienable things. There were exceptions to Americans' otherwise expansive market principle, and "civilized" people recognized them. It was not just material infrastructure that marked Philippine peoples as civilized, then, but also what and how Philippine peoples bartered and sold. These practices could be used as evidence of their readiness or enduring incapacity for self-government. Differences on these grounds stood to bolster US claims to superiority and justify continued colonial occupation. In this way, markets and marketplaces were sites of racial and political formations quite apart from those visions of unity and equality advertised by their boosters. This was especially true of the marketplace in Baguio, home to the infamous dog market, a commerce that, far from signaling civilization, told Americans and Filipinos of Igorots' difference, one with political ramifications.

## Baguio's "Biggest Show Place"

Writing about Baguio in 1913, Emma Yule described the dog market as "by all odds, the biggest show place the summer capital of the Philippines has to offer."[96] Another visitor the year before termed the "stock yards" the "center of interest in this Eastern department store." He estimated some 500 dogs at market. "After the market comes the feast," he wrote. "A traveller riding towards Baguio on a Sunday afternoon meets groups of Igorrotes on foot and in ox carts usually with one to four dogs in tow. In the woods one may see an expectant group building a fire and the dog looking seriously on at the preparation."[97] This dog market would be banned in the 1920s, and in his 1920 ethnographic study of the Ibaloi, Claude Russell Moss dismissed accounts of dog eating among these Cordillerans as "exaggeration."[98] But contrary representations had long circulated.

In a 1913 contribution to *National Geographic*, Dean Worcester, a University of Michigan-trained zoologist who served as secretary of the Interior (1901–1913), wrote of the dog market as a site of the extraordinary. He had been writing for the magazine for a couple years and had sought to produce the highest quality photographs to communicate to the American public what words alone could not. But Worcester's photos often still required explanation, and he captioned a photo of the dog market this way: "Dogs are a highly appreciated article of diet, and are now brought in large numbers to Baguio from the lowlands for sale. On Sunday mornings, the Baguio dog-market presents a unique spectacle."[99] On the one hand, the dog market photo presents an unremarkable sight: a group of native men, drawing back scrawny, hungry-looking dogs (see fig. 4.2). It could say: Here are foreign peoples, our new colonial charges. They own pets, like us. They buy and sell, like us. In depicting a shared practice, the photograph offered a potential point of connection between distant peoples and places. Yet the photograph and caption of the "unique spectacle" also seemed designed to engender astonishment in American audiences: the

4.2 The Sunday dog market at Baguio. Image originally appeared in Dean C. Worcester, "The Non-Christian People of the Philippine Islands," *National Geographic Magazine* 24 (November 1913): 1200. Original caption: "Dogs are a highly appreciated article of diet, and are now brought in large numbers to Baguio from the lowlands for sale. On Sunday mornings the Baguio dog-market presents a unique spectacle."

men pictured are selling these dogs to people who will consume them, literally.[100] Philippine peoples may truck and barter, then, but not *quite* like us. They sold what we, by virtue of our position on an evolutionary ladder, would neither wish to acquire nor dare trade in such a way. Yule's *Overland Monthly* contribution also made this shift, from the familiar to the unfamiliar: "This dog market is no dog show or place where dog lovers may spend money for canines with family trees. It is a market in the sense of being a place where something to eat may be bought; the dogs brought to sell are not fancy bred, they are just dog."[101] She could see before her a similarity: both Americans and Philippine peoples traded dogs in a market. But she marked and stressed a divergence.

Yule understood American dog markets as sites where not just animals but symbolic values were traded. She implied that symbolic value was the very point of entering the American dog market: the purchase of a dog was an exercise in conspicuous consumption; it was less about possessing the creature itself, much less literally consuming it. In a textbook example of commodity fetishism, Yule suggested that the point of a dog purchase was primarily immaterial; it was a practice of dog "lovers" seeking impressive "family trees." The commodity—the dog—"abound[s] in metaphysical subtleties and theological niceties," Karl Marx might say.[102] By contrast, as Yule observed the Baguio scene, Igorots saw dogs, quite simply, as dogs: "not fancy bred, they are just dog." For the Igorot, it was about the flesh,

4.3 Dog market, ca. 1909. Schuffman Collection, Filipinas Heritage Library.

albeit for purposes Yule termed "ceremonial." She wrote, "One of the interesting customs of the Igorots, the most civilized of the uncivilized tribes in the Philippines is their practice of eating dog flesh. 'Dog-eaters' is the scornful taunt flung at them by the civilized tribes of the Philippines. They do not seem to eat dog flesh purely for food, but rather as a ceremonial meat or as a festal dish."[103] In articulating a distinction between a primitive market and a more refined "show," Yule seemed unable to avoid observing a rough symmetry of the settings. But that the "ceremonial" purpose of the canine and the fetishism of American "dog lovers" could also be more similar than different escaped her.

Yule's description of the dog market is consistent with the contradictory "rule of sympathy."[104] Sympathy in the colonial context could perform the dual work of traversing a distance, only to recreate it. Contrary to the greatest wishes of market advocates like Finley and Forbes, under these colonial conditions, sympathy could regenerate the very divergences its expression ostensibly traversed. Amit Rai explains that "[t]he difference of race, and often of gender, that divided the object and agent of sympathy was precisely that which must be bridged, effaced even, through a certain process of identification. Yet without such differences, which were inequalities of power, sympathy could not function; one might say that sympathy produced the very inequality it sought to bridge."[105] Descriptions of trade, at least in this context, seemed capable of something similar. By Yule's description, Filipinos could appear as game buyers and sellers, much as Americans. But by the same token, what Yule claimed Americans and Filipinos traded and consumed—one a symbol, the other flesh—presented an intractable difference. And this difference signaled to American readers, literate in discourses of civilization and its gradations, Igorots' and by extension Philippine peoples' ultimate inequality. Yule's representation of exchange did not dissolve hierarchy but reaffirm it.

In 1913 the notorious dog market was supplanted by another sensational trade, which also involved the archipelago's "non-Christian tribes." This case, in which what, or rather, whom, Filipinos bartered became instrumental to colonial politics and an object of consumption itself, was the much publicized story of domestic slavery and debt peonage. The case was made by Dean Worcester, "the P. T. Barnum of the 'non-Christian tribes,'" as James Blount, a US District Judge on the archipelago from 1901 to 1905, called him in a 1912 critical analysis of US colonialism. Blount claimed accurately that there existed "a very cordial mutual hatred between [Worcester] and the Filipinos" for his work in "discovering, getting acquainted with, classifying, tabulating, enumerating, and otherwise preparing for salvation, the various non-Christian tribes."[106] If the dog market confirmed the Igorot as uncivilized, in Worcester's hands, the slave trade implicated the Filipino as immoral merchant and consumer. Here again, Americans judged Philippine peoples based on exceptions Americans had established to the rule of markets, and through these limits they set to markets, they generated justifications for continuing to deny Filipinos independence.

In 1913, Worcester produced a report for the Philippine Commission called *Slavery and Peonage*. Designed to nudge consciences and stimulate action, the frontispiece of Worcester's report was a photograph of two skinny, bare-chested boys, standing on either side of the entry to a stone schoolhouse. They were identified as Ifugao, an ethnolinguistic group of the Cordillera, dwelling in a province that borders Benguet. The caption indicated that the boys were among the children who had constructed the building, and it asked: "Should it not be made a crime to sell or buy such children as chattels?" In *Slavery and Peonage*, Worcester reported the sale of men, women, and children by Moros and by Igorots and their purchase by Chinese traders *and* Filipinos. Some were to be chattel; others would exist in debt peonage, "the greater evil . . . because of the very large number of persons who suffer from its prevalence."[107] Worcester had learned of several children who had been purchased by Chinese traders and later returned to administrators of a local province. He quoted the report of an assistant director of education: "The youngest was three or four years old. Some of them already had their head sheared, so as to resemble Chinese [children]. . . . [The parents] seemed to love their children, but felt compelled to sell them in order to provide food for themselves or to meet certain debts which they had incurred." Responsible for this business of cleaving children from parents were ruthless Philippine caciques, the "rich and powerful" leaders in the provinces. The caciques induced the poor to sell their children by saddling families with overwhelming debt.[108]

Worcester all but implicated Filipinos in the practice, insisting that "apart from a false and foolish pride . . . the chief reason why [Filipino] Assemblymen object to the law which they have tabled is that it would not only prohibit and penalize slavery, but would necessarily also prohibit and penalize peonage."[109] His aim in publicizing the sale of women, men, and children was to induce the Philippine Assembly to pass an antislavery bill that had been produced by the American-led Philippine Commission. While the US Congress had outlawed slavery on the archipelago, to that point there had been no law in the United States or in the colony that actually treated the practice as a punishable crime.[110] As Worcester knew, these charges of slavery had implications beyond humanitarian concern. Worcester's work emerged as the US Congress and Filipinos debated a measure to deliver a clearer timeline for Philippine independence, the Jones Bill. Worcester's report served as an indictment of the country's putative leaders.[111]

Filipinos saw that the independence of the Philippines was at stake. In his report, Worcester quoted responses to his allegations from *La Vanguardia*: "Before the Americans came here, our civilization was quite advanced and humanitarian ideas regulated our social existence. As slavery does not exist here at present, only a madman or a visionary can fear its existence in the future. To do so would be equivalent to affirming that the Filipinos, instead of progressing, think of lapsing back into the barbarism of primitive life."[112] Filipinos were caught in a catch-22. If they legislated against slavery, they tacitly acknowledged that such an odious trade was still conducted on their shores. If they did nothing, they appeared to condone all that had been rumored.

The charge that Filipinos were involved in bartering lives was difficult to substantiate. It was true, however, that Igorot peoples had engaged in the practice until the nineteenth century and that it had seen a "brief resurgence" in the early twentieth century.[113] Historically, wealthy Ibaloi, for instance, had taken slaves. And poor Ibaloi, the *abiteg*, were described as "'like slaves' to the *kadangyan*." But the *abiteg* could also choose to leave a *kadangyan*'s settlement and live as independent swidden agriculturalists.[114] More relevant than the veracity of Worcester's claims is that the case made what, whom, and how people traded and sold a problem with political implications. Worcester's analysis presented Filipinos, those aspirants for the recognition that would put political independence within reach, as nothing more than oppressors of noble savages. Those who commodified the incommensurable and traded the inalienable violated Americans' exceptions to the rules of trade. Meanwhile, the enslaved were not free agents and self-possessed. The woebegone victim—"non-Christians," an image reproduced in newspaper and magazine articles, books, photographs, and in entries to World's Fair exhibitions—were again legitimized as the proper subjects of colonial care.[115]

As Blount's moniker for Worcester, the "P.T. Barnum of 'non-Christian tribes,'" suggests, the zoologist had become adept at this work. Since at least 1911, Worcester had been attempting to develop a means for producing and distributing films about the Philippines' indigenous peoples to enlighten American audiences and turn a financial profit, too. In June 1911, he traded letters with an American writer named William Dinwiddie about such a venture. Worcester had previously enlisted Dinwiddie in a "little publicity campaign" "to show up certain members of the Anti-Imperialist League"; Dinwiddie had referred to it as a "beautiful campaign of education" to challenge the league's "propaganda."[116] That summer, Dinwiddie was also pursuing film opportunities and communicating with Solax, a United States-based film company founded by Alice Guy, a French director whose career in the movies had begun under the pioneering producer Léon Gaumont. Worcester had proposed filming his annual inspection tour of Igorot groups across northern Luzon with the help of a government photographer.[117] A letter from a Solax manager offered Dinwiddie instructions for "taking Philippine pictures." He advised hiring a director who could capably oversee the production of a narrative, a form that would entail selecting "principal characters"; pushing the camera close to subjects to generate close-ups; and the rehearsal—the staging—of "action" scenes like "attacks upon a village," "hand to hand encounters," dances, or indigenous celebrations.[118] That fall, Worcester would correspond with a representative of Loews who had written him about rights to show Worcester's Philippine films. Game to explore the possibility, Worcester noted that he wished to retain the right to use the films in lectures across the United States. Through this work—effectively docudrama—he aimed to target "sober-minded people . . . in the importance of continuing the policy thus far followed so successfully in dealing with the wild peoples of the Philippines."[119]

In the 1913 *National Geographic* article, Worcester discussed one aspect of this policy and celebrated "government exchanges"—markets similar to that described by Finley. These, he wrote, "are nothing more nor less than government shops, where the wild man may purchase at a fair price the things which he needs and may sell his produce at its real market value."[120]

They were designed to allow the "wild man," including Igorots, to be rescued from a history of abuse at the hands of Filipinos and his "civilized" brethren. Worcester described an invidious order that prevailed during the Spanish colonial period: "In the old days the wild man was invariably shamefully cheated when he attempted to barter the products of his native mountains with his 'Christian' neighbors in the lowlands for salt, cloth, steel, and similar necessary articles."[121] While Finley's "Moro Exchange" was designed to forge peace among Philippine peoples, the government exchanges described by Worcester seemed intended to entrench difference and further division. In the exchange described by Worcester, the colonial state broke off trade relations between Cordillerans and lowlanders, making itself the fair broker and source of equality and justice.

Back in 1908, the editors of *El Renacimiento* had condemned Worcester for his instrumental use of Igorot peoples; Worcester responded with a libel suit, and the editors were sentenced to two years in jail and a steep financial penalty.[122] To be sure, *El Renacimiento* could have taken much of the colonial government to task, as Worcester's stories about Igorots and Filipinos were not unique. In 1907, the provincial governor of Benguet, William Pack, described his Igorot charges as "poor, timid, oppressed barbarians." They existed "without both the consolation and the protection of the church." "[B]eing nearly naked barbarians these Igorots had been as far back as tradition recites the natural prey of the more enlightened and Christianized tribes," he wrote. "No attempts were made to redress wrongs committed against these people, and outrages of all kinds and abuses from private and official parties were so common that safety and peace seemed to be found only in poverty and solitude." Pack celebrated the square deal that Americans were extending Igorots; only Americans, he wrote, while Cariño's land dispute continued, had "won the apparent confidence of these natives."[123]

These representations of Igorots as warranting the paternalist care of white chiefs like Pack and Worcester assumed a revision of history. As discussed in chapter 1, Igorots had managed to maintain a measure of autonomy from Spanish rule for generations. In the case of Benguet's Igorots, they had done so precisely by maintaining trade relations with Ilocanos on the coast and lowlanders from Pangasinan. Coffee, gold, copper, and tobacco had been among the goods that Igorots had traded in exchange for cotton, dogs, rice, and pigs. Their trade of gold for cattle facilitated the *baknang*'s ability to cultivate and lay claim to land and in turn enjoy "a capacity to feed, employ, and command people, kin, and village-mates," as Cariño and his ancestors had.[124] Contrary to Worcester's depiction, all Ibaloi had hardly been "prey."

It is true, however, that by the time of the US occupation, Filipinos did sharply distinguish themselves from Igorots and believe themselves superior to the mountaineers. In the introduction of his book *El Folklorico Filipino*, Isabelo de los Reyes, the Ilocano founder of the Union Obrera Democratica, "described himself as '*hermano de los selváticos, aetas, igorrotes y tinguianes*' (brother of the forest peoples, the Aeta, the Igorots and the Tinguians)."[125] But de los Reyes was an exception. Filipino nationalists treated "non-Christians—and the territory they occupied—as integral to the Philippine nation but subordinate to the political agency of Christians."[126] They adamantly rejected Worcester's representations of the non-Christians, but unlike de los Reyes they did so for largely pragmatic reasons. They called on the colonial government to make good on its claims to paternal care—care for themselves as Filipinos. Commenting on a proposed exhibition of non-Christians in the West in 1905, *El Renacimiento* wrote, "[W]hen the exhibition can harm an entire people, an entire race, it is natural that the Government, in its guardianship [capacity], come to the defense of the reputation that human curiosity and ambition put in imminent danger."[127] The following year, *El Renacimiento* condemned the form of "advertising" Americans were conducting by exhibiting Igorots in the United States. A 1906 article took to task Truman Hunt, a medical doctor and former governor of the Cordillera's Bontoc Province; Hunt had been in charge of the "Igorot Village" at the 1904 St. Louis World's Fair. The village was part of the forty-seven-acre Philippine exhibit, and it had drawn the largest crowds, men and women eager to observe the dog eaters of the archipelago.[128] Two years later, Hunt gathered another group of Igorots for exhibit in North America, leading a writer for *El Renacimiento* to ask: "who is more savage—this Hunt or the wickedly exploited Igorots[?]"[129] The effect of these displays, *El Renacimiento*'s writers believed, was to scare away capital from the Philippines—investment that Americans like Forbes had been insisting as crucial to the colony's development. The exhibits, *El Renacimiento* wrote, tended "to convince thousands upon thousands at St. Louis, Portland and San Francisco that the clothed Filipino is the exception." The newspaper reasoned that "sensible American investors" would be only wise to bank on Cuba, Mexico, and other countries in South America rather than on a land distinguished by its "naked savage inhabitants."[130]

In 1916 Maximo Kalaw, then an aide to Resident Commissioner Manuel Quezon, challenged the exploitation of Igorots by proponents of US retention of the Philippines. He mouthed and in that way mocked the logic of retentionists: "Those people are semi-civilized, if not entirely savage. Some of them eat dogs, and for proof we refer you to the St. Louis Exposition, where Igorrotes were exhibited engaged in that toothsome pastime." "It would be a pity to set this people adrift."[131] By this point, however, Worcester's charges of slavery in the Philippines had done the work he had intended. News from the Philippines in late 1913 had been grim: "Girls Sold in Islands," "Philippines Rife with Slave Trade," "Sell Girls for $100: Filipino Parents Do Not Think Such Slavery is Wrong" made headlines in the *Washington Post* and *Chicago Daily Tribune*.[132] By asserting the persistence of slavery and debt peonage and the failure of Filipinos to address the evil, Worcester helped to neutralize notions that the archipelago and its peoples were prepared for self-government. The hand of the United States was still necessary as fair broker to the Igorots and to continued political tutoring of Filipinos. In assessing political prospects around these economic practices, Americans regenerated subjects for continued colonial education. In Congress, discussions shifted from the Jones Bill to the subject of Philippine slavery. Although the Philippine Assembly did pass an antislavery bill in November 1913, when the Jones Bill became law in 1916, it included provisions to replace the Philippine Commission with an elected Philippine Senate but lacked a definite date for Philippine independence.[133] The US president retained the right to reject measures related to the tariff, public lands, and immigration. Further, the governor general held the privilege of appointing representatives from the non-Christian provinces to the legislature, and a new executive department, the "Bureau of Non-Christian Tribes," was created for "general supervision over the public affairs of the inhabitants."[134] US administrators had employed the terms of trade and representations of it to harden differences between "non-Christians" and Filipinos, retain the power to continue doing so, and keep the whole archipelago as a US colony.

## The Limits of Free Trade

Much as there was an irony in Yule's distaste of Igorot ritual involving dogs, it was the same in American colonialists' shock at the notion of slavery in the archipelago. The difference between Filipinos employing indentured servants and US colonialists' recruitment of wage workers was slim. As discussed in chapter 2, in their efforts to gather native workers, Americans had used disciplinary labor practices employed by the Spanish and in the post-Civil War US South. In criminalizing traditional subsistence practices, they robbed people of familiar means to feed, house, and clothe themselves and, in that way, practically coerced wage work. Forbes admitted in his journal that "We always suspect [road workers'] 'voluntary' labor is brought about by a somewhat free use of the municipal police, but as the result is roads, it isn't necessary to analyze too closely all the steps leading to them."[135] A thin line would seem to separate recruitment of this "'voluntary' labor" by the police force and slavery.[136] In an added irony, through wide circulation and instrumental use of texts and images—an imperial "visual economy"—Worcester and his collaborators seemed to recommodify and resell the supposed non-Christian slave.[137]

And so while Americans promoted the liberating properties of the market and aimed at its expansion, they also constructed its limits. These limits could become measures by which Americans evaluated readiness for self-government and a means for containing Filipino independence. They could also serve to contain and protect the American market and polity. To be sure, even as they sought an "open door" in Asia and free trade—something ostensibly celebrated by the military reservation on Cariño's land named Camp John Hay—Americans seemed acutely aware of the hazards this might occasion and attempted to check them. Take, for example, the politics of the colonial tariff. In the Insular Cases, the Supreme Court determined that the Philippines were "not 'foreign territory' within the meaning of such clauses in the tariff laws of the United States."[138] José S. Reyes, a *pensionado* (a Filipino who studied in the United States with government sponsorship), made the observation in 1923 that "American economic interests were united in the desire to capture the Philippine market as against other foreign traders. They were divided when it came to the admission of commodities from the islands in exchange for American goods."[139] It was not until 1913 that the United States and the Philippines actually enjoyed free trade with the termination of quotas on sugar and tobacco. Within a few years, imports into the Philippines from the United States were greater than imports from all other countries combined.[140] While the tariff policy served the United States and rich, land-owning Filipinos by nudging the colony toward production of agricultural raw materials for the US market, the policy was ultimately of questionable benefit to the Philippines as a whole.[141] Earlier, in 1909, when the Payne Aldrich bill initiating freer trade was passed (American products could enter the Philippines duty free while quotas on outgoing Philippine sugar and tobacco remained), imports from the United States had risen by 129 percent. This produced a dramatic decline in colonial customs revenues. Anticipating this loss in revenue, the colonial government levied new taxes and raised existing ones.[142] Filipinos, in short, subsidized free trade. Further, while the Insular Cases determined that the Philippines existed within US tariff walls as "unincorporated territory" (like Puerto Rico), its peoples stood outside the rights and privileges of the US Constitution. Supreme Court justices had tied themselves in knots to fix the status of the Philippines vis-à-vis the United States, finally demonstrating that American fears of the unlimited entry of Philippine commodities were indivisible from the specter of "little brown brothers" as American citizens, that is, the entry of Filipinos into the body politic. Americans wanted the benefits accrued through an open door in the Philippines, as Reyes pointed out, but neither the risks entailed in exporting the Constitution to the archipelago nor the risks in the substantial movement of Philippine persons and things to the US mainland. US tariff policy thus undermined the notion of trade as by definition a reliable agent of unity and equality. It and markets more generally were inseparable from politics—from power relations—and the politics of race, even as Americans often sought to depoliticize them.

An awareness and fear of the centrifugal forces potentially unleashed by commerce may also explain why Manila never became the "free port" like Singapore and Hong Kong that some colonialists like Forbes imagined it could be. He envisioned the capital as "one of the world's greatest emporiums of trade" and sought to use Manila much as the Spanish had.[143] From early on, Manila "functioned as a city of heterogenetic transformation, where men of different races and cultural traditions mingled to trade and exchange ideas."[144] That did not end with the termination of the galleon trade. Manila then became an open port, a status that helped stimulate an export economy in agricultural products.[145] Foreign merchants—Chinese, British, and American—and not the Spanish encouraged the production of cash crops and helped connect the archipelago's "fortunes to the booms and busts of the world capitalist economy."[146] These links to a world beyond the colony, the influx of foreign capital, and the colonial bourgeoisie's growing wealth afforded circulations with revolutionary potential. "[W]ith travel came opportunities to bring distances up close, geographic as well as social. For money enabled one to reach those at the top of the hierarchy—the governor general, the Supreme Court justices, the provincials of the religious orders." Vicente Rafael notes that *ilustrados*, those members of the class of educated Filipinos, like Pardo de Tavera, saw this potential of money and "advocated its increased circulation through commerce, [and] assumed that free trade with foreigners was the condition for freedom as such." These advocates of free trade in the Spanish colony were subsequently tarred as "*filibusteros*, or subversives, at odds with the interests of the state and church."[147] Igorots, of course, had also been identified in such a way for their ability to sustain an illicit trade in tobacco with lowlanders. A 1911 editorial in the *Manila Times* recalled the early American interest in a "free zone" "where goods might be entered without customs charge and held for reexport." But the writer noted that the Payne Aldrich Tariff Act, giving American products free entry into the Philippines, largely eliminated "much of the necessity for it."[148] In other words, Americans were already getting a good deal without further liberalizing trade. Both colonial powers—American and Spanish—wanted to encourage trade but on their terms, for they were right: the effects of trade were not narrowly economic.

## Buddha God or Solomon Levi

On the heels of the passage of the 1905 Reorganization Act, which was intended to make the colonial government more efficient and consolidate colonial bureaus and management over them in Manila, Felipe Buencamino, a wealthy *ilustrado* who had served as an adviser to General Aguinaldo and was a relatively brief member of the Federalista Party, raised some "Eminently Practical Questions."[149] He noted that from August 1898 through June of 1905, Filipinos had "paid to the occupying government, not elected by the people, the sum of 100 million pesos as taxes." Having paid this "debt" determined by the Treaty of Paris, the Filipino people, he wrote, "have a perfect right to demand from their rulers the promise to define or determine their political condition which they have been promised in the last paragraph of Article 9 of [the Treaty of Paris]." He warned that if Americans did not pay heed to this petition, Filipinos might well turn to Spain, which, as a "contracting party in the treaty—has the right to demand compliance."[150] In an article three days later, he argued that the accumulated wealth of the colony, as determined by the US-administered census, evidenced the "intelligence, honor, and work, habits of order, justice, and humanity" of the Philippine people. This fact, allied with a cohort of capable leaders and the natural resources and potential of Philippine lands, he argued, meant that the Philippine people met conditions required for self-government.[151] Unwittingly, a visitor to Baguio in 1920 had suggested that even Igorots had learned those capacities of good salesmanship and free agency that Americans deemed key to meriting self-government. "These 'little brown brothers,' as Mr. Taft used to call them, have long since discarded their ancient pastime of head hunting and have taken up the great American game of skinning one's neighbor. The almighty peso has got into the blood of the Igorrote. He gets a tropic fever at the sight of a mere *peseta* and will forever part company with his *corida* and pickaninnies for ten or a dozen of them—if he could find any one willing to make the trade."[152]

Buencamino employed both the logic of *utang na loob* and the very measures Americans had generated to manage their wards to evaluate and critique colonial leadership and launch an argument for independence well before debate on the Jones Bill. The Tagalog notion of *utang na loob*, translated literally as "debt of gratitude," expresses a state of indebtedness like that which exists between a landlord and his tenant. This condition implies the existence of a hierarchical relationship—a relation of power.[153] Vicente Rafael writes that "The hierarchy that is formed by indebtedness is based on the sensed incommensurability between the gift that is received and its return."[154] Buencamino described the relationship between the United States and the Philippines, if not quite defined by gift exchange, then as having been a contractual one. In stating that Filipinos had repaid their debt to the US occupiers, Buencamino described the Philippines as in a new state—one like *hiya*, "without any sense of indebtedness" or shame—relative to the United States.[155] With that relation of reciprocity conveyed by *utang na loob* severed by the repaid debt, the Philippines stood ready for independence.[156] The debt paid, the debtor thus redeemed.

Buencamino's essays for *El Renacimiento* suggest that as much as market relations—like sympathy—could diffuse old antagonisms and cultivate relations of reciprocity and affection, they did not preclude conflicts over intention, meaning, and the temporal reach of those bonds.[157] Indeed, as much as colonialists represented trade as binding people together in affective relations and as a solution to political problems, it still seemed that transparent valuations and consensual exchange could not eradicate the sense of risk in trade, or a feeling of incommensurability between an article and its price—a problem of power.[158] Yule's account of the Baguio dog market brought to the surface such ambivalences in the experience of a transaction. She described the actual moment of trade:

> When at last a decision is arrived at, the buyer takes hold of the rope of the dog he has chosen for his ceremonial chow, and from some invisible compartment of his girdle, produces, in coin, the price he offers. The seller, looking about as interested as a Buddha god, but really as alert as a "Solomon Levi," after due time brings his gaze to rest upon the offer. A "what's-the-use" look slowly ripples over his countenance, and he languidly, almost pityingly, shakes his head in refusal. The bargaining continues through the medium of proffered coin and languid shakes and nods, the bystanders taking a voluble part, until the deal is closed, the sale is made, and the dog is released from the canine mat and is led away by the purchaser, whose face begins to wear a peculiar smile, whether of satisfaction with his bargain or in anticipation of the ceremony or religious rite which the dog will grace, or whether only epicurean, who can say? The smile of the Philippines, whether civilized or uncivilized, is elusive, fathomless.[159]

Yule enlisted well-worn stereotypes of the Oriental, a figure who, in the words of another visitor to the Philippines, bore "an impenetrable mask who is all lies on the surface."[160] In Yule's anecdote, that mask became a poker face, assumed by the seller to extract the best return. But something more can be detected through the thick bias of this account, something that might mar an image of the marketplace as a site of liberation—political or ontological.

The marketplace was supposed to be a location where goods and vendors were "exposed" and where, for that reason, value could be reasonably assessed between honest, if somewhat cagey, brokers. Yule's account suggests that fundamental uncertainties afflict this scene of disclosure and inspection: Who got the better end of the deal? Was the seller a "Buddha god," a serene, indifferent party to the deal, or a "Solomon Levi," a calculating trader figured as Jewish? It is unlikely that indeterminacies like these were peculiar to the Baguio dog market, or to trade in and with the East, even as Yule wished to cast them as primarily cultural exotica. These ambivalences seem written into the historical experience of expanding markets.[161]

One may wonder if, in the context of an unrelenting native resistance through the first decade of the twentieth century, that "elusive, fathomless" "smile of the Philippines, whether civilized or uncivilized," made US administrators wonder whether they had bought a bunch of bad brokers—a people unknowable and therefore uncontrollable—when they took the archipelago. In a 1902 *Atlantic Monthly* article on "race prejudice" in the Philippines, James LeRoy quoted an editorial from the American-owned *Manila Freedom* expressing just this concern: "Every Filipino is an insurgent at heart, and every Filipino hates the Americans if the truth was known. They take our money, and they smile to our faces, but in their hearts they have no use for us or our government. Incapable of gratitude, they view our generosity in the light of a weakness, and at the first favorable moment betray the trust reposed in them."[162] It seemed an irresolvable uncertainty, if not anxiety, remained at the point of sale. Both Yule's account and that cited by LeRoy suggested that melding peoples to form local society, societies together to make a nation, and nations to make a universe of faithful friends would be an impossible order for free trade alone.

Those Filipinos peddling "*independencia* in a box" may have recognized this. In a 1904 report, David Doherty wrote that at least some of those revolutionaries with messianic visions were holdout insurgents, or *ladrones*. "In my judgment," he concluded, "the cause of 'ladronism' is the desire of the people for independence."[163] These peoples, it seems, remained committed to the revolutionary vision of *kalayaan* and an understanding of *independencia* that exceeded "political autonomy" to signify something far greater: "the attainment of certain possibilities of existence." "Politics," Reynaldo Ileto writes, "could not be divorced from all other aspects of life, like morality and economics."[164] Far from commercializing politics, at least some of those "popes" and prophets promoting independence were politicizing the economy and calling for a moral economy, one based not on a sympathy that reproduced hierarchy but on the emergence of a "radical brotherhood."[165] Vicente Rafael has described that "radical brotherhood" as the condition of *layaw*—"the pre-oedipal moment of perfect reciprocity between mother and child. Idealized as a state of bonding without bondage, the condition of *layaw* . . . comes from acts of giving that do not expect a return and so escape the economy of the gift; of taking without incurring a debt and so dispense with the formation of hierarchy."[166] This was a future that American proponents of market expansion and even Filipino nationalists like Buencamino who accepted Americans' terms of government may also have idealized but did not bring about. In advancing marketplace construction, administrators sought to make a market logic govern greater areas of life while insulating it from questions of power—the imbalances of power that define colonial conditions. Had he read Doherty's report, then, the laugh that William Cameron Forbes and his correspondent may have shared at the expense of the Black Jesus and the hold-out insurgent on Samar should have been an equivocal one. It was they who put exalted "*independencia* in a box."

FIVE

# Savage Hospitality

By 1911, Baguio could truly be called the colony's summer capital. Close to six hundred government employees, from the Bureaus of Agriculture and Audits to Public Works and Supply, joined Philippine Commissioners that summer season, ascending the Cordillera to seek refuge from the Manila heat.[1] The exodus included some 250 Americans and 350 Filipinos whose presence allowed the government to carry on its work in the mountains.[2] Ever the assiduous promoter, Forbes had, perhaps, encouraged editors of the *Manila Times* to publish its favorable report of government-built housing. Filipino and American employees, provided with their own accommodations, were granted separate but equal treatment. Government dorms included "modern improvements, baths, recreation rooms, pool and billiard tables, and many other opportunities for pleasure." Workers of the "higher grades," meanwhile, enjoyed detached cottages, topped with tin or galvanized iron roofs, and each with a fireplace.[3] In 1911 alone, a dormitory, over twenty "large" "cottages," and forty-five two-room cottages had been built.[4] A bulletin published for teachers during their annual Baguio vacation observed that the cottages could not be distinguished for beauty, but they did "add to, rather than detract from, the landscape," compliment enough as the landscape was one of Baguio's chief draws.[5] One made the most of a visit to Baguio out of doors on the hill station's tennis or croquet courts, playing baseball or polo, or trotting on horseback along Benguet trails.[6]

Baguio offered this much and more to nongovernment workers. The *Manila Times* noted two "first class" hotels accessible by roads that were nothing short of "perfection."[7]

5.1 "Gov. General's Mansion, Baguio, P.I.," ca. 1915–1920. George Grantham Bain Collection, Library of Congress, Prints and Photographs Division, LC-DIG-ggbain-22586.

Other city services would come to distinguish the hill station. Between 1910 and 1920, Baguio gained a town waterworks, waste collection system, sewage system, and electric light plant. The city began to manufacture ice and built a second marketplace.[8] A two-story, green-and-white city hall would be completed in 1911.[9] A mayor occupied the office and by the terms of the 1909 city charter he was to be a "nonpolitical engineer" appointed by the governor general; a native of the Philippines would not hold the office until 1937.[10] The mayor presided with the city council in a "modified city manager form of government"—a structure favored by Progressives seeking to put government above corrupt politics and in the hands of those claiming expert knowledge.[11] About three decades later, J. C. Orendain, a Filipino attorney and author, would write that "Nowhere in the country has more attention been lavished on public improvements than in the Pines City." And he would acclaim Baguio as "one of the most progressive and cleanest cities in the world. Such is the metamorphosis of Baguio under the American flag."[12]

Well before Orendain's writing and indeed by the 1910s, Baguio had achieved renown in the colony. Advertisements appeared in the *Manila Times* for the Hotel Pines, hailing Baguio as "the Switzerland of the Orient."[13] Another hawked "Stetson Hats for Baguio-Goers;" and yet another promoted "Benguet Buttermilk" and "Benguet Cheese"—"Only 'Pure' Brand 'Natural' Milk Used in Manufacture of these Products."[14] The last was a venture of Metcalfe A. Clarke, the American who had maneuvered to buy the disputed land of Mateo Cariño (the future military reservation) and had helped found the gold-mining operation Benguet Consolidated. As these ads suggest, Baguio became a signifier of things Americans longed for most while in the Philippines. These included modern amenities possible for the kind of engineering intelligence reflected and celebrated in the Benguet Road. And central to Baguio's charm was also nature—as climate and setting for outdoor recreation, fresh produce, and landscape views that seemed to transport a visitor to the northern latitudes. The globe-traveling journalist Frank Carpenter described this Baguio in 1925: "Right under me," he wrote from the Baguio Country Club, "dropping down into a ravine, is a sunken garden, with terraced flower beds that extend halfway round the clubhouse." "There are big blue hydrangeas, blood-red bougainvilleas, and with them a variety of strange blossoms that would delight any botanist. A wide veranda surrounds the club, and sitting in the dining room, which is perhaps fifty feet square," he continued, "one can look out through immense plate-glass windows on scenery as fine as that of the Riviera. The whole district of Baguio is a garden that would be the pride of any multimillionaire of the United States if he could drop it down near New York or Chicago."[15] Baguio's climate and flora, and the society and recreation found there, gave Americans in the Philippines more than a comforting glimmer of home. "I find it hard to describe this mountain capital," wrote Carpenter. "The whole scene has the effect of a great landscape garden designed by Jehovah and developed by man."[16] It was as if Americans who ascended the Cordillera had partnered with the divine to bring into being the transcendent.

Together with its modern conveniences and the felicities of the mountain setting, the hill station appealed to Americans for its peculiar and ultimately reassuring marriage of the refined and the primitive, the familiar and the strange, the domestic and the foreign. Carpenter's vantage from the verandah of the Country Club gave him a view of "an eighteen-hole golf course, where I see a foursome of Americans, with Igorot caddies at their heels."[17] Another visitor, a reporter for the *New York Times*, wrote in 1923, that "one easily could fancy Baguio, with its rustic cottages and nearby peaks covered with pines, to be in the Catskills or the Adirondacks"—"[i]f it were not for the half-naked Igorrotes seen on the roads and working the gardens."[18] One could visit the marketplace, be repulsed and transfixed by those Igorot dog eaters, and then retire to a country club and be served by one of them. One might quake at the sublime beauty of the Cordillera during a climb up the treacherous Benguet Road, and yet rest easy upon reaching the Baguio meadow. Baguio enchanted guests for conjuring a "great landscape garden" in the Adirondacks, a Switzerland south of the equator, a New England hinterland peopled by men in g-strings.

As these Baguio sketches suggest, Igorots played an important role in the hill station. Americans cast them as curiosities and landscape features on par with sensational views of nonhuman nature. In travel narratives and hill station boosterism, Igorots served as signs of the untamed mountain provinces that Americans used to justify continued colonial rule. On other occasions, Americans assigned Igorots to the "landscape garden"—the image of the Ibaloi in his g-string tranquilly tending the colonial golf green and earth. They included Igorots in a vision and representation of Baguio manifest in space: a garden tucked into the sublime Cordillera and, more metaphorically, a middle ground between the mountain wild and the civilization signaled by Americans' presence. Here, the Igorot appeared as unthreatening and as local color. These material and discursive pastoral renderings of Baguio obscured the dispossessions of Ibaloi peoples' pasturelands and the appropriations of Igorots' labor. Baguio delivered the "solid satisfactions" of the classical pastoral named by Leo Marx—"peace, leisure, and economic sufficiency"—by incorporating Igorot herdsmen into the imperial idyll with no trace of irony.[19]

If Daniel Burnham offered a sketch of Baguio's pastoral potential in his 1905 plans, it was William Cameron Forbes who brought the vision to fruition as the hill station's greatest advocate. And at the pinnacle of this home-away-from-home for US colonialists were the country club where Carpenter sat admiring the view and Forbes's personal residence Topside. These sites, built while Forbes served as the colony's secretary of commerce and police and, later, as governor general, beckoned as the most desired destinations off Benguet Road. They occupied exclusive recesses beyond the town center and stood as the greatest realizations of the Baguio pastoral. They were also the private preserves of US colonial power. As with Baguio as a whole, they remade America in the Orient for homesick and weary colonial workers and, in that way, helped to stave off "Philippinitis" and reproduce colonial labor. Like in the emergent American suburbs, the colonial class created itself through its production and use of these spaces.[20] These sites performed other work. Forbes used the country club and Topside to entertain and cultivate a class of political collaborators. In these special provinces of the "special province" of Benguet, on the golf green and in the gardens, he gathered elite Filipinos, appropriating the role of headman of the United States' colonial charges. Here, the foreign occupier assumed the role of host, and Filipinos became guests, there only for the good grace of Baguio's new chief or until they themselves purchased into the pastoral dream and acquired property. Under Forbes's leadership, Baguio became a "show place," a model of what US tutelage offered and also a stage for setting in motion new relations of social credit and debt.[21]

In these ways, while gardens, a polo field, and golf links may seem like somewhat superfluous sites of study, sitting far afield from the usual domains of imperial politics and its history, when used by the likes of Forbes, Baguio nature and lands and even the Ibaloi who had inhabited them came to play a part in the practice of politics and the work of state building.[22] The hill station gave Americans a unique position from which to build legitimacy and grow colonial power and capital. For, ultimately, Americans occupied the Philippines to build a social infrastructure for market expansion and the "political conditions for capital accumulation."[23] As Forbes put it, "the surface of the field is to be kept smooth by the Government: it is the merchants who play ball."[24]

Baguio thus appears as a "location of power"[25] as much as Manila, arguably all the more so for Americans who fashioned it into a model town and place defined by its contrast with the rest of the archipelago. In dispossessing Igorots, Americans accrued a surplus by which they could fashion their pastoral, reproduce American labor, and incorporate Filipinos into new relations of obligation. And through the hill station's gardens, parks, and even the design of the country club and Topside, Forbes associated US power with nature itself, naturalizing imperial dispossessions and the US occupation. In these respects, the beauteous nature that Carpenter attributed to the hand of Jehovah proved to be a pastoral fiction and the Cordillera mountains anything but a retreat from the work of colonizing the Philippines.

## The Country Club(s)

Daniel Burnham drafted plans to remake Ibaloi pasture into an imperial pastoral, but it was William Cameron Forbes who saw much of the work through and zealously cultivated the Baguio where visitors believed their Stetson hats grazed the heavens. When appointed by President Roosevelt as secretary of commerce and police in 1904, he was put in charge of Baguio's development. At the time, little had been done to advance the US hill station; the task was largely his, and it tried him. The Baguio resort remained controversial among Filipinos and even some American colonialists, as it would through Forbes's nearly decade-long tenure on the archipelago. In certain respects, the building of Benguet Road had hampered the town's development. As money invested on the highway seemed washed away with every rainstorm, the hill station struck many as another expensive waste, when already *El Renacimiento* published articles criticizing the "great luxuries and marvelous prodigality in salaries" of US colonial government employees.[26] Forbes worried in his diary that some members of the Philippine Commission seemed "ready to abandon the whole matter rather than subject themselves to the criticism which would certainly follow additional appropriations either for the road or for the town." "Americans were not enthusiastic," Forbes wrote. Government itself was "lukewarm; people came, liked it but did not return. It was evidence that they had to be amused."[27] To Forbes, the solution was the country club and the society that such an establishment fostered. These might provide amusements to entice visitors and spur the purchase of properties and the building of houses. Use capital to generate capital: Build up Baguio, move government offices north in the summer, then surely the town would be worth its road. Daniel Burnham employed this same logic of development when promoting his plans for Chicago in 1909. He also believed that the right kind of urban development would encourage "our people" to remain "and others will come to dwell among us—the people who now spend time and large amounts of money in Paris, in Vienna, and on the Riviera."[28]

Personal experience had led Forbes to believe in the value of a country club to the speedy development of the hill station, and in building the institution, he tapped his rich inheritance once more. In his hometown outside Boston, a club had induced some twenty members, William Cameron included, to buy "estates" in close proximity and construct homes amidst "country" "occupied only by old fashioned farms, pastures, and wood lots."[29] The Brookline-based club had inspired a suburban development. The club's founder was James Murray Forbes, a cousin of his father, who seems to have established the first American country club in 1882. A 1905 article on "Prominent Country Clubs" in *New England Magazine* recalled that James Murray's organization went simply by "The Country Club."[30] It appealed to members' passion for horses, offering a racetrack and steeplechase, and it reflected an "Anglomania," as members sought to mimic the ways of the English and Scottish upper class. Notably, it also included women and children in club life.[31] In these features, the Country Club drew on traditions of the most privileged Americans' urban and rural sanctuaries like the city club, a domain of elite white males; aspects of the country resort, spas in places such as Saratoga, New York; wilderness resorts like the Ausable Club in the Adirondacks patronized by hunters and fishers; and suburban settlements like Chestnut Hill in Philadelphia.[32]

Though the name might imply it, James Murray's was not the ur-"The Country Club." The first incarnation owed to another member of the Forbes clan and emerged an ocean away from Boston's shady suburbs: Shanghai. James Murray had visited the port city and his cousin, Francis Blackwell Forbes, a partner with Russell & Co. and a botanist who published work on Chinese flora. Francis had also been a promoter of the Shanghai Club, "one of the largest institutions of its kind in the Far East" by 1908.[33] But it was another club called "The Country Club" on Bubbling Well Road in the port city that seemed to inspire the Brookline, Massachusetts, establishment. A history of the Brookline Country Club notes "that in his later writing about our Club [James Murray Forbes] is always careful to say that he founded the first country club in the United States and not 'in the world.'"[34]

The Shanghai club treated members to billiards, tennis, and squash, and it did so in a pastoral setting that "resembl[ed] a large private country residence, surrounded by beautiful shady trees, green lawns, and well-kept gardens, the whole occupying an area of nearly 60 mow."[35] James Murray's Brookline retreat employed a similar aesthetic. The clubhouse evoked the "English country house, complete with leathery appointments and a retinue of faithful servants. . . . Like the great country homes of England, the clubhouse had both common spaces—a reception hall and formal dining room—and distinctly gendered spaces, including separate entrances for men and women. . . . The whole place bespoke tradition and venerability, belying the club's recent vintage."[36] The Shanghai Club, incidentally, planned in 1908 to construct a new clubhouse that gestured even further back with an "English Renaissance" style.[37] Like its counterpart in Shanghai, the Brookline club enticed members with the possibility of withdrawal into "an imagined Anglo-Saxon civilization." Early member lists sounded like "a roll of Mayflower descendants, which to some extent it was," writes historian Randy Roberts.[38] In formal ways and even in the society it assembled, the club expressed nostalgia for simpler times before the social, political, and economic challenges wrought by members' own occupations.[39] It offered a substitute for the "vanishing village" through a club—a "small, stable, and easily understood corporate" establishment "that, although democratic in practice, exercised nearly absolute control over access."[40]

A yearning for the "village" and a preserve of Anglo-Saxonism drove not only those Forbeses in bustling Boston but arguably also those family members transacting business in Chinese ports. The desire for the seemingly simple life in a pastoral refuge may have animated Forbeses' Chinese associates, too. Business bred (or required) genial relations between Hong merchants and their foreign partners, and the two were known to impress each other with carefully designed gardens and good food, interests which they shared.[41] In their appreciation for nature cultivated into landscapes, these Chinese merchants followed the ways of Chinese rulers since their state's inception. Emperors sought domains at a distance from the royal palaces, where they could pass the time in wonder at nonhuman nature and gaze at menageries with birds, fish, and other animals. Some eventually became "country houses called the 'suburban pleasure palaces.'"[42] The most famous of these retreats, Yuanming Yuan, known as the Old Summer Palace, outside Beijing, was destroyed by British and French troops during the Second Opium War. During the Jin dynasty, the Chinese aristocracy appropriated emperors' practice and took to building private gardens, too. These were to show off the elite's own fondness for the simple life of the country, a sensibility that emerged in this time of social turbulence.[43] In this way, The Country Club—in its Shanghai and Brookline incarnations—presented no original response to the disorientations of rapid change. The pastoral's history extends long and wide. The particular form of the country club, however, emerged in the nineteenth century, and it had found a purpose in the East among merchant exiles as well as in the West. To the East it would return with William Cameron Forbes.

The experiences of Forbeses in Shanghai and Boston did not alone prepare Forbes to drive the establishment of the country club and Baguio as a pastoral retreat. He would seem uniquely qualified in a second respect, for he was not only of the Forbes clan; he traced roots through the Emerson family tree. His maternal grandfather was transcendentalist Ralph Waldo Emerson. Emerson, himself writing at a tumultuous time in American history, believed in the therapeutic power of nature. "The land is the appointed remedy for whatever is false and fantastic in our culture," Emerson wrote. "The continent we inhabit is to be physic and food for our mind, as well as our body. The land, with its tranquilizing, sanative influences, is to repair the errors of a scholastic and traditional education, and bring us into just relations with men and things."[44] Perhaps such a salve is what owners of the estates built around The Country Club sought in living amid "old fashioned farms, pastures, and wood lots" and what merchants pursued through "the shady trees, green lawns, and well-kept gardens" of the Shanghai club. The "tranquilizing" effect of nonhuman nature might reorient those men wearied and made unwell by the scramble for profits and control, a pursuit that the Forbeses had charted internationally. What Emerson himself would have made of the imperial quest for overseas nature and adventure to atone for the "false" and "fantastic" in civilization, one cannot say. Some form of the nature remedy he had prescribed, however, was arguably an objective of those who retired to John Murray's Brookline club, as it was for visitors to Baguio.

Forbes's plan to make the country club the engine of Baguio's development—to trade on the pastoral—worked. With allies, he recruited other Americans who established the rudiments of a country club in 1906, acquiring some 100 acres in the first public auction of Baguio lands.[45] By 1911, the club counted over 150 members—all men—including military personnel like John J. Pershing, who would hold the governorship of the Moro Province among other colonial appointments, and Henry Allen, who organized the Philippine Constabulary; Forestry Bureau Chief George Ahern; judges including Charles Elliott; and businessmen like Manila Railroad president Horace Higgins, Richard Laffin, the first vice president and general manager of the electric company Meralco, and Charles Swift, a streetcar entrepreneur from Detroit and electric railway owner in Manila.[46] About twenty Filipinos were members, too; these included Benito Legarda, a member of Aguinaldo's cabinet before becoming a member of the Philippine Commission in 1901; Jose de Luzuriaga, another commissioner; Manuel Quezon, the future president of the Philippine Commonwealth; and the wealthy Felix Roxas, who served as the appointed mayor of Manila between 1905 and 1917.[47] They helped underwrite the club's development from a few "slabs" and screens.[48] Forbes himself, unwilling to accept failure if his sizeable personal resources could avert it, loaned money to complete several guest cottages.[49]

Besides the country club, those who made the journey from Manila by late 1907 would learn of other developments underway: a new hotel, the Hotel Pines, and a residence for the governor general bound by ten hectares of parkland. Commissioners followed Burnham's plans and also identified and reserved land for a town park.[50] Hill station visitors could delight in cool temperatures in these spots, fill up on fresh vegetables and fruits, and enjoy the kind of recreation that made possible the material and cultural reproduction of American life in the colony. Cool temperatures meant that the great hearth featured in the 1910 pamphlet "A Brief History of the Baguio Country Club" was not just decorative.[51] These temperatures were believed a necessary condition for white persons' health in the torrid zone. An early Philippine Commission report reflected on the salubrious Benguet climate, relating its pleasing coolness by describing the ruddiness of mestizo children's cheeks.[52] Baguio's climate afforded another aspect of good health and comfort and an aspect of the pastoral by allowing for rich gardens and the cultivation of crops familiar to Americans. Strawberries, cabbages, bananas, potatoes, and coffee sprang up in the valleys around Baguio. "Fresh vegetables of all sorts tickle palates which have grown weary of the canned goods of the lowlands," wrote Dean Worcester.[53] Baguio's climate gave Americans a chance to make it into home and into an American garden for these colonialists deprived of familiar comforts. These benefits redounded to Manila's American denizens, observed Baguio enthusiast Isabel Anderson. Americans could feast on peas, squash, and wax beans, and Anderson shared a friend's wish: "'If the time ever comes when we can have real cow's milk and cream, then our food will be as good as anywhere in the States.'"[54] Metcalfe Clarke would gratify such yens by vending Benguet buttermilk and Benguet cheese.

Refreshed and well fed, Americans could attend to their health in another way. They could exercise themselves in sports like golf and polo, activities that some feared sped the onset of illness in the hot lowlands.[55] In engineering a golf green, Americans followed not only US country club models but experienced colonialists', too. The British constructed golf courses across colonial India.[56] In his prospectus for the country club, Forbes promised that the tropical location would pose no obstacle; the course would be "equal to the finest known in Scotland."[57] In the United States, the sport had been helping to drive the establishment of country clubs by the 1890s, supplanting "horsey sports" as clubs' key attraction. Some observers even came to identify the sport with "classic American values" like "democracy and equality." A 1908 treatise on golf also pointed to nature as one of its key draws for players.[58] Polo, another Baguio offering, put players in sight of nonhuman nature, too. It was played on "a great level greensward surrounded by pine-clad hills" that "look[ed] off, at the southern and eastern extremities, to a mountain view unsurpassed in the annals of sport." This spot tripled as a space for baseball and cricket.[59]

## Reengineering Nature

To enjoy these settings for recreation and a table heaping with fresh fruits and vegetables, Americans conducted considerable reengineering of the environment, building on nature's assets to heighten their appeal. The prospectus for the Baguio Country Club described "A beautiful, purling stream" that could "be so arranged, by pumping or otherwise, as to form a lovely cascade which will fall into the swimming tank."[60] The parks that Daniel Burnham and William Parsons had planned for the hill station were "not going to be hard, formal parks with stiff hedges and flower beds, but are to have natural masses of exotic plants and tropical and pine trees, Baguio's typical and crowning feature." The municipal and government centers would enclose a lagoon offering paddle boat rides.[61] The naturalism of the parks, as much as the golf and polo grounds, were the handiwork of landscape architects, who transformed Ibaloi pasture into American playground. The government also planned to terrace the hills below the municipal offices, bridging the elevations on either side.[62] In this effort, Americans again demonstrated their engineering skills even as they, wittingly or not, appropriated a technique used by Igorots to make the mountains accommodate the production of rice. A method employed in production was reoriented toward consumption—of nature as landscape.

The journey to Baguio itself offered occasion for consuming nature as a series of thrilling sensations and sublime landscape views framed by early visitors and Baguio boosters. The Benguet Road, that "marvel of engineering audacity," offered a portal between "the exuberant Tropics, hot sizzling" and a place "where fruits, flowers, and trees known to the temperate regions grow in marvelous abundance."[63] Like the experience of Baguio, the road journey produced a feeling of sharp contrast with ascent leading visitors to a place more and more like the temperate climes of the United States. One newspaper article projected views to be had along the way, including "Bridal Veil falls": "a wall of pure white water" before which "language fails and silent admiration prevails." "It is like a great set piece worked in the mountain, its frame of ferns helping to display its grace and charm."[64] Even man-made structures like the Jesuits' retreat atop one of Baguio's hills enhanced visitors' appreciation of nature: "silhouetted against the glorious sunset sky, over and over again do those beautiful words of Coleridge ring in the traveler's ear," wrote one visitor.[65] Dean Worcester described the commanding views from his private residence, writing that "the scenery from windows and porch is simply magnificent. We have glorious sunrises and sunsets, unsurpassable cloud views."[66] The builders of Baguio fashioned nature to produce pleasing effects in material ways—by moving earth—and visitors did so discursively, naming views even as they sometimes described them as beyond words. Both manufactured a *landscape*, "set piece[s]" observed and sketched through language, that turned nature into "objects of conspicuous aesthetic consumption." To consume these Cordillera views was to claim membership in rarefied colonial society.[67]

It was not nonhuman nature alone that enticed visitors to the country club and hill station and that became objects of consumption. As Frank Carpenter's description and other accounts of Baguio visits suggest, Igorots were equally sought-after spectacles. A brochure for the Baguio Country Club advertised "wonderful tropical vegetation, magnificent mountain scenery, strange wild peoples, and the most remarkable terraced mountain sides in the world."[68] Here, the peoples indigenous to the Cordillera were distanced from the reader as "strange" and "wild" and set aside other objects in the landscape. A brochure and timetable for the Manila Railroad line to Baguio from the later 1920s seemed to offer both the Igorot and the Filipino as spectacles for visitors' diversion (see fig. 5.2). The standing figure wears Filipinas' traditional Maria Clara dress; the figure poised on the rock with the headband depicts an Igorot. The two absorb the pros-

5.2 "Baguio: Philippines Summer Resort," published by the Manila Railroad Company, November 20, 1928. American Historical Collection, Rizal Library, Ateneo de Manila University.

pects available at the Baguio hill station, while the viewer of the brochure commands the images of nonhuman nature and them.[69] Writing on "The Simla of the Philippines" for the *New York Times* in 1920, Fitzhugh Lee Minnigerode described encounter with the Cordillera's indigenous people as a form of entertainment that stood to enliven a trip to the hill station. Besides the recreational possibilities and opportunities for dancing and drinks, he wrote, there lay "always in reserve the pleasing possibility of running head on into an Igorrote head hunter, with his long hair flowing over his shoulders and his brown body as naked as the day he was born, excepting only his 'G' string, which is not worn for modesty, but only to fasten his vicious-looking 'bolo' to his side."[70] Frank Carpenter described a similar scene as a "snapshot" along the road to Baguio. "As we go we shall meet half naked Igorot men and women dragging stones from the cliffs and with rude hammers of steel breaking them inside rings of hoop iron into bits the size of a walnut. Most of the men wear only a shirt and a gee-string. I made a snapshot of one—a man as brown as a berry, who had left his shirt home."[71] A picture of an Igorot, or better yet with an Igorot, perhaps carried a particular currency after the 1904 St. Louis World's Fair that had made the indigenous peoples infamous and outraged Filipinos.[72]

This treatment of Igorots corresponds with one of two trajectories of the pastoral that Raymond Williams identifies in *The Country and the City* (1973). One path leads toward the close study of nature and hence toward "nature poetry" like Emerson's; the other pursues "theatrical and romantic" expressions wherein "shepherds and nymphs . . . are lay figures in an aristocratic entertainment."[73] Arguably the latter was the function of the Igorot in the Baguio pastoral. At the hill station, Americans sought to make real a literary landscape, coupling civilization signified by Americans' presence, their city hall and government center, their modern roads and marketplaces, and the "engineering audacity" these represented, with the mountain "wild" embodied by the Igorot in his g-string. Igorots reminded readers and viewers of the proximity of the wild even as their loss of land and relative power, at least in Baguio, indicated by their menial jobs, disarmed and positioned them safely within a pastoral middle ground. In these ways, Cordillerans, like the mountains where they dwelled, became "objects of conspicuous aesthetic consumption."

To transform the pastoral from an aesthetic form to a colonial reality, Americans employed Igorots twice over. As Carpenter's description reveals, Americans conjured and then consumed Igorots as representations of the wild and the primitive even as they employed them as workers for turning Baguio into a modern and most avant-garde Philippine city. For instance, Cordillerans were crucial to building Benguet's celebrated roads. 250 to 500 Igorots labored along Benguet Road and were paid about forty cents a day.[74] While more and more Igorots would be compensated for labor on the public works projects that distinguished Baguio, they also remained responsible for committing ten unpaid days of work in the province, often on the upkeep and repair of roads. William Pack called this "free labor" in a 1907 provincial report, the very kind of labor that had been used to build some 200 miles of passable road back in 1903.[75]

The country club and Forbes himself also employed Igorots. To keep Topside shipshape, Forbes was retaining in 1906 "four bulls, thirty Igorots, with their houses of grass, and stables, all built. Also eight Japanese masons, an American negro, and a few odd Filipinos." His household staff, meanwhile, consisted of two "Chinamen for the kitchen and the table," a "couple of Filipino boys who can talk English to be on hand to receive and interpret," and ten Igorots to attend to the property and be uniformed in "blue coats and a leather belt and a fine bolo each, also a raincoat, and . . . breechless."[76] This division of labor based on nativity recalls the organization of work on Benguet Road just as it evokes Thorstein Veblen's description of domestic service as "a spiritual rather than a mechanical function"; it was part of the stage or "set piece" that was Forbes's home.[77] And it was similar to the staffing of the Baguio Country Club, where fifty Igorots had been enlisted to build the

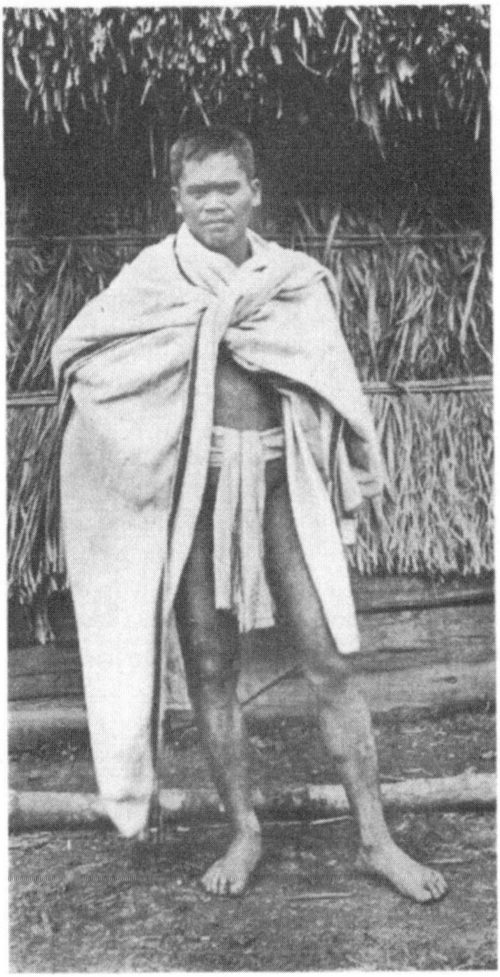


5.3 "A Caddy," in *The Baguio Country Club* (Baguio, Philippines: The Club, 1910), 21.

golf green.[78] Igorots also served as caddies there, a reminder of work they had long performed for colonialists as carriers of gear and building materials along Benguet Road; in 1920, C. R. Moss rated the Ibaloi "the best *cargadores* [carriers] of all the Igorots."[79] A pamphlet on the Baguio Country Club featured a photograph of one such Igorot caddy, a sight and experience that the *New York Times* promised prospective visitors (see fig. 5.3). "The nineteenth hole is still the popular place to get 'teed up,'" a reporter wrote for the paper, "and in place of a naked Igorrote kid doing the caddying there is a flock of Igorrote muchachos—all dressed in the popular 'G' string-style—doing the carrying."[80] In descriptions like this, Americans seemed to collect encounters with Igorots just as they collected landscape views.[81]

## Naturalizing Dispossession

The power to order the land and its indigenous peoples in this way indicated American power; to claim bolo-carrying Igorots as an entertaining view meant they surely had been disarmed. And in many respects they had been. By the fall of 1905, while Cariño continued to pursue his case against the insular government, the Baguio town site had been surveyed and a street system laid out. Soon after, under the direction of the colony's consulting architect, a section of the town site deemed residential had been carved into alienable lots.[82] Americans continued building roads and trails, recruiting Igorot labor to the task of rendering former pastureland—newly measured, subdivided, and identified with numbers—available to buyers. On May 28, 1906, 144 Baguio lots were sold at a public auction.[83] The governor of Benguet Province reported that "the prices realized average[ed] about three times the values assessed."[84] A couple months later, the colonial government further subdivided the government reservation and identified zones for residential purposes, "ecclesiastical residences, charitable, religious, or educational purposes." And they called for another public auction of lands.[85] Forbes noted in a report to the governor general that Baguio "lots immediately around the country club seemed to be very favorite."[86] Forty-four more lots were sold in Manila and in Baguio in January and April of 1908, bringing total sales to 33,369.13 pesos. That number would grow, as a "Filipino residence section" with thirty lots was then underway.[87] While Baguio's *baknang* had clearly participated in rendering Benguet lands private property—commodities—it was Americans who clearly advanced the process. They made the land legible to capital and to the state, surveying and ordering Baguio pasture into a series of zones and then into discrete parcels that could be affixed with a price per square meter. The sale of these parcels supported new colonial projects, while the rest remained as public domain.

In turning Igorot lands into commodities and into public domain available to homesteaders, Americans overlooked Igorots' customary use of the land and its unique features. Benguet Governor William Pack had noted in 1907 the incongruence between Ibaloi use of Benguet lands and the Public

Land Law. "Owing to the rough and broken surface of the country the parcels of ground cultivated by the natives are small and detached from each other. That is, one may own several cultivated areas, mostly rice paddies, ranging in extent from one to four or five acres, all small," he wrote.[88] The land law, however, allowed property holders to claim only one parcel. To remedy this, Pack urged the Philippine Commission to limit claims by acreage rather than by the number of parcels.[89] The land law and the colonial property regime posed another challenge for some of Benguet's pastoralists. The *baknang*'s responsibility and his influence spanned the reach of his cattle's grazing—the movement of his livestock rather than lines plotted on a map. Americans' system of private property ownership conflicted with this Igorot practice of "stewardship."[90] And ultimately, the sale or appropriation of Baguio land pushed cattle raising, the foundation of Ibaloi's social organization and political economy, to the outskirts of the town.[91]

By 1910, 295 town site lots had been sold in the government reservation, but only ten residences had been built, including Forbes's and his secretary's, indicating to Forbes that many had purchased land for "speculative purposes"[92]—just what the government had suspected in the case of Mateo Cariño's land dispute. Forbes acknowledged that this kind of speculation had raised property values to sometimes ten times the original purchase price. Perhaps because of these transformations of the land, ordinary Igorots worked for colonialists, standing for pictures, smoothing the golf green, and serving as caddies. Land that had been theirs, if only in an attenuated sense as a *baknang*'s real or fictive kin or as the *abiteg* laboring on the *baknang*'s land, a means of reproducing life in material and cultural ways, had been incorporated as the US government reservation.

The colonial state had created a surplus and a pastoral out of Igorot dispossession. And these dispossessions may help to explain the expedience of the particular form Forbes's clubhouse, completed in April 1908, and his private residence would take: the bungalow. Their style had been forged in another colony; the bungalow had "Anglo-Indian" origins and had found a home in British hill stations.[93] Anthony King traces the circulations of the bungalow from colonial India to England and then back to former colonies like the United States in the early twentieth century. There the bungalow served largely as a vacation or weekend home, as it would in Baguio. Imperial relations generated not just the aesthetic spark for the bungalow, King explains, but also the surplus time and capital that afforded the weekend retreat. "Just as the profits of the West India sugar plantations had provided their owners with country seats such as Fonthill or Harewood," he writes, "so the wealth of the East India trade had

5.4 "The Baguio Country Club," in Dean C. Worcester's *The Philippines Past and Present, Vol. 1* (New York: Macmillan, 1914), 288–89.

a similar impact on the landscape at home."[94] The Forbeses, architects of the United States' China trade and The Country Club, would have been no less acquainted with this relationship than the proprietors of Fonthill or Harewood.

The bungalow-styled clubhouse, while hardly suppressing these imperial relations of power, muted them; hence, its expedience. The Baguio clubhouse contained "a spacious assembly hall with a large fireplace and cozy chimney-corner seats; an office for the club steward; a buffet; a locker and a bathroom for men and one for women; a kitchen; a storeroom and a servant's room." Although the bungalow typically referred to a single-story structure, the clubhouse, like Topside, featured a second floor dotted with dormers and provided sufficient space for twenty overnight guests. Forbes wrote that "[t]he second story was so constructed as to overhang the first at the east end, making a large open-air dining room with roof and ceiling, but without sides. This is a favorite place for serving luncheons and dinners, as it commands a magnificent view of Gold Creek Canyon and the mountains beyond."[95] The bungalow style, noted for its "low-pitched, sweeping roof" and a substantial stone chimney, together with a verandah, was to embed the structure and its dwellers in nature, rendering it "closest to Nature both in form and in site" (though of course not as close as Ibaloi huts).[96] A 1911 book on the bungalow elaborated the vision: "to secure the intimate relationship between a home and its surroundings that conveys an impression of peace and stability rather than strife and unfitness."[97]

This formula describes fairly well the Baguio clubhouse. The country club bungalow could seem no jarring imposition on Baguio; it stood to convey "intimate relationship" with its setting and "peace and stability" even as it "command[ed]" landscape views and hosted and housed the colony's most powerful Americans and Filipinos. In a journal entry from 1908, Forbes wrote of baseball games among officers, civilian leaders, and teachers. "The 10th Cavalry is here, and part of the Constabulary band, and music is to be heard harmonizing with the mellow sounds of the wind in the pines, and the ringing click of the meeting of the golf stick with the golf ball."[98] In this picture of fellowship and harmony with nature—the police band playing in unison with the wind through the pines—Forbes naturalized the US occupation just as the bungalow form did. In his work on the pastoral, Raymond Williams writes that "The clearing of parks as 'Arcadian' prospects depended on the completed system of exploitation of the agricultural and genuinely pastoral lands beyond the park boundaries." These landscapes intended for consumption supplanted the landscapes of production of shepherds and farmers, and they were to be "seen from above, from the new elevated sites; the large windows, the terraces, the lawns; the cleared lines of vision; the expression of control and command." Landscape views and English country houses were underwritten by dispossession and enclosure of land in a pattern not unlike Baguio's own development.[99] Americans had thus made Baguio an imperial pastoral: they dispossessed and objectified many of its long-time dwellers, building in part on land granted and taken from Mateo Cariño; they reengineered nature into a picturesque setting, created vantages for observing it, and regenerated the space discursively in ways that naturalized their power.

By 1911, all the colonial government bureaus were sending portions of their staff to the hill station. American teachers were also retiring to a "Philippine Chautauqua,"[100] where they could "compare notes, take courses in teaching and enjoy a little American society, . . . and get a change from the heat and monotony and isolation of their lives in the scattered Philippine villages to which they were assigned."[101] The 1910 annual report of the Bureau of Civil Service suggested the economy of the move: "[T]he increase in the strength and vigor of the employees due to the three months spent in a cool climate will render their service of more value to the Government during the remainder of the year."[102] The *Manila Times* seemed to agree on Baguio's power to revivify the colonial workforce, admonishing colonial officials in 1908 to "Go, and help your clerks go. It will make better men of you and them. In the shadow of those giant mountains and 'Neath the wide canopy of a heaven which seems measurably nearer one drops his mean and petty thoughts and his heart throbs close to Nature's own. It is a regeneration."[103] A later advertisement for the Manila Railroad's "Friday night special" train to the hill station would emphasize the hill station's therapeutic powers, too: trains "make it possible for a business or professional man to attend to all his affairs all week and yet have a nice rest and week-end visit with the family. It makes an agreeable change for him and the return SUNDAY NIGHT TRAIN puts him back in his office Monday morning ready to resume business with renewed vigor."[104] Just as homes in the United States' new suburbs were produced by and helped to reproduce a domestic professional-managerial class, so Baguio reproduced its colonial analogue culturally and psychically, offsetting Philippinitis.[105]

The string of foreign visitors to Baguio suggests that some of Forbes's and even his Spanish predecessors' greatest ambitions for the "Philippine Alps" came true. "In season," wrote one visitor, "Baguio is a gay place. Titled foreigners from China, Japan, the East Indies, and Europe, may be found mingling at social functions with the cosmopolitan populace of the summer capital."[106] At the country club Americans could rub shoulders with high-ranking counterparts visiting from nearby colonies. The country club and Baguio as a whole provided a rendezvous where colonialists from stations in the southern hemisphere, perhaps all seeking reprieve from their colonial charges and colonial heat, could summit and trade notes, much as British and American merchants had done in the clubs of Canton and Shanghai.[107] "Now and then the Governor-General," wrote Monroe Woolley, "drunk with Yankee democracy, dons a baseball uniform to play the game with his Cabinet." On other occasions, "the Count of Sen-Sen, from Kobe, or a mandarin from Tien-Tsin, or a Sultan from Sulu, with a British Lord from Hongkong, form a part of the opposing team."[108] Perhaps in that comingling of colonialists and native elites Americans affirmed to themselves their status and membership in an imperial caste, if one they sought to rise above. By 1910, a writer for the *New York Times* complimented Forbes's efforts by calling his country club "the undoubted 'piece de resistance' on the club menu" of the Philippines.[109]

## Forbes's Show Place

Hosting counts, sultans, and lords was not the only entertaining that Forbes conducted in Baguio. Perhaps of even greater consequence for the survival and success of Baguio—and his own as an administrator in the Philippines—was the hospitality he extended Filipino elites at his private residence, Topside. In turning his home to such purposes, Forbes seemed to undermine the idealized pastoral, turning a place intended for leisure and recreation into the domain of express political work. Here the purpose was not only the reproduction of American labor or the naturalization of the US presence but also the creation and maintenance of a class of Filipino collaborators.

When residential and business properties went up for sale in 1906, Forbes claimed a top prize: twelve acres overlooking Baguio that he had sited for his own home Topside. "It . . . is on a hill that stands right out over a valley, a thousand or two feet drop, overlooks all Baguio, and is not more than two miles from the centre."[110] Breaking with the tradition of the Spanish *principalía* whose claim on elite status was indicated by their homes' position on the town plaza, Forbes chose to locate his house apart from the center of Baguio and even apart from the enclaves that Burnham had sketched for the hill station's wealthy future patrons. He positioned Topside above them all. In a footnote to a May 1906 journal entry, Forbes explained the house's name as "Chinese pidgin-English" "for 'upstairs.'" "[I]t seemed a very appropriate name, as it looked off over the whole surrounding country."[111] Forbes seized the most commanding view and nature supposedly approved; a poem that appeared in Topside's guestbook opined: "when to Nature's lovely work / Was added Topside's friendly glow, / She smiled again—she saw for this / Perfected Baguio."[112]

Though Forbes and his guests tended to naturalize the appearance of Topside, its construction proved a tremendous challenge. Like the road to Baguio, the project collided with the natural environment, and in a letter to his brother Ralph, William Cameron gave vent to his frustrations. Building in the rainy season drained the commissioner. The skies seemed to open up each afternoon, and rains dug "quagmires" into the roads, rendering the transport of supplies by wagon and mule next to impossible.[113] Forbes kept to his plan, however. Designed as a bungalow by Burnham and William Parsons,[114] Topside presented something of a contrast or, perhaps, a complement to the Beaux Arts style that Burnham had brought to the Philippines. Topside offered the intimacy and warmth of home. An article in the *Manila Daily Bulletin* described "the great fireplace in the living room [as] a popular place for Mr. Forbes's guests, of whom some experience for the first time in many years the pleasure and fascination of dreaming with the glow of that back log."[115]

Still, it was only nominally rustic living that Forbes intended there. Even as its bungalow form gave it a low profile, Topside was to be a "show place,"

5.5 A photograph of Topside, captioned "Baguio, Mountain Province. Private residence of the governor-general of the Philippine Islands, undated." bMS Am2212 (77), William Cameron Forbes Collection, Houghton Library, Harvard University.

an exercise in ambitious construction and "conspicuous consumption," his contemporary Thorstein Veblen might say, designed to give pause to Baguio's critics.[116] No "badly built, leaky, inconvenient, and unplumbed shack," it was made of stone as the houses of the Spanish *principalía* had been.[117] It offered modern amenities and was comprised of materials that came distances traversable only because of Forbes's power and financial resources.[118] The Boston squire made his intentions clear: "I think it very desirable there should be somewhere an object lesson to people below how beautiful this place is and how comfortable one can be. For this reason, among others, I am planning to make this place a little different from ordinary Philippine custom, and make it look more as if we had got out of the Orient and back to America, that is, so far as the interior of the house is concerned."[119] Much as the bungalow in British India came to carry "cultural and political meaning" and contrast with native Indian housing,[120] Forbes's place was exceptional, and it exhibited what Filipinos stood to gain—all the services and comforts that accompanied US progressive rule. Such a message, some held, was an important one. As James LeRoy, an early secretary to the Philippine Commission, contended, "the success of our political venture in the Philippines depends in large measure on the extent to which we can arouse in the people a desire for better homes, better towns, and better surroundings."[121]

Topside's interior displayed Philippine woods and served as an advertisement for Philippine forestry, a field that Forbes hoped to develop as a source of state revenue. These fine woods would eventually appear in Forbes's Norwood, Massachusetts, home, which was built in 1914 following his time in the Philippines. Just as William Howard Taft returned to the United States with tokens of his Philippine sojourn like carabao skins and hard woods that became Oval Office floor boards, so did Forbes.[122] His Georgian Revival mansion featured twelve species of wood from the Philippines.[123] Across its thirty rooms, Forbes may have displayed some of the "trophies, souvenirs, manufactured articles, samples of handicraft" that he had collected. His booty included stuffed birds, battle axes, bolos, baskets, woven cloths, wooden spoons, canes, and bowls from across the Philippines.[124] The Philippines became a series of objects that could sit on the shelves of his mansion and in Topside, too. In Baguio, these collectables included interior furnishings built by Bilibid prisoners, who crafted "handsome" tables and chairs "for much less than the catalog prices in the States."[125] The products of these laborers were not only good bargains but also signs of the efficacy of US colonialism: it transformed supposed social miscreants—some, perhaps, imprisoned for vagrancy or ladronism—into productive craftsmen.

In making an exposition space of his bungalow, Forbes's home seemed to undercut the pastoral ideal. It hardly stood apart from the market or colonial politics; rather it put these very things on display as if for sale, demonstrating a kind of power similar to that of visitors who "commanded," "captured," and snapped landscape views. Forbes complained about this work in his diary. "The possession of the show place in a place like this and the obligation it entails, is not all heliotrope," he wrote. That May of 1908 he wrote of a string of visitors including the Belgian consul and army and businessmen who "all came and stayed and went, or came, and stayed, and stayed. . . . Sunday I had no less than fifteen in the house."[126] Of course Forbes could only blame himself for such success and the exhaustion he felt at playing host. Not four months after he had purchased the Topside land, he was aiming to entertain the "big bugs that come up—like the archbishops, admirals, and 'highfalutin' visitors—so as to show them how easy the trip is."[127] The notion was that good hospitality would lead them to invest themselves socially, economically, and thenceforth politically in Baguio and, in that way, in the US occupation. The Baguio pastoral, represented by a bungalow atop a hill in the Cordillera, was to speak for the whole of the US regime. Forbes wanted the white light of publicity to shine right there. Forbes even invited Harvard philosopher William James, attempting to lure James finally to the side of imperialists. "If you will do this," Forbes wrote James in 1907, "I am sure that I will be able to put you in

5.6 "Igorrote Fiesta, Topside, May, 1907," W. C. Forbes Collection (picture no. III:A17:004), American Historical Collection, Rizal Library, Ateneo de Manila University.

a way of seeing what is being done out here in a way that will be delightful to you and at the same time enable you to form a correct opinion, which it is very difficult for anyone living so far away to do and particularly people living in Boston."[128] It seems that James stayed put in Massachusetts.

American and European visitors were not the only ones whom Forbes tried to romance with Topside. Visitors included his neighbors—Igorots, as a photograph from 1907 shows (see fig. 5.6). In fact, Forbes had arranged for the country club to host a feast for Igorots, a *cañao*, in 1906. He explained that "[a]n ox had been roasted whole and the Igorots were there in force, dancing and beating tom toms and their peculiar kinds of drums."[129] Isabel Anderson spoke of a *cañao* she had witnessed, too. It took place ten years later and was hosted by the then-vice governor of the colony. "A line of partly dressed dog-eaters arrived, bowing as they passed," she wrote. "They proved to be the chiefs or head men, who had put on what clothes they possessed for this occasion."[130] In hosting a *cañao*, Americans were appropriating a practice of the wealthiest Igorots like Mateo Cariño, who maintained and heightened their status through such feasts. Forbes assumed the role of chief for himself much as Kennon had done along Benguet Road. American colonialists had assumed the pasture of Cariño and his posture, too, conducting "politics by other means."[131] Symbolically, the imperial trespasser had become magnanimous host, and the native person the indebted guest.

Forbes played host to Filipino guests as well. To ensure Baguio as an American oasis in the Philippines, Forbes needed Filipinos, including political opponents. So Forbes opened Topside—a supposed retreat from Manila—to Filipino elites. Forbes did not conceal his intentions. "My object in this whole thing was to get the wealthy Filipinos interested," he wrote, "because I knew if they could come up and be enthusiastic that the success and permanence of Baguio was secured, as they could wield enough influence with the Legislature to see the thing wasn't blocked once they had property interests there." Self-interest, he assumed, would trump nationalist convictions that could complicate the American enclave's future. Once he had roads, houses, two hotels, tennis courts, and a golf course, among other amenities, he sent out his invitations, "knowing that [his Filipino guests] would feel not as pioneers—which takes an initiative not always found in the tropics, and not always found in the well-to-do—but as participants in something the success of which was assured."[132]

"The result justified my calculations," Forbes announced to his diary,[133] and he proudly recorded that even Secretary of War Taft celebrated the role of Topside in building Baguio.[134] Prominent Filipinos, awakened from their supposed tropical languor, bought Baguio lands.[135] Through Forbes's tenure, elite Filipinos and even former revolutionaries visited Baguio, purchased land and built houses there, and joined the country club. Besides Manuel Quezon and Benito Legarda, General Manuel Tinio, the youngest commander of Philippine revolutionary forces, and the Zobels, a wealthy and powerful family from Manila, made their way north into the mountains.[136] In 1906, property buyers in the second auction of Baguio lots included Manuel Araullo, a Manila judge; businessman Mariano Limjap; Rafael del Pan, an *ilustrado* lawyer; and Salvador del Rosario, a one-time supporter of the Philippine Republic.[137] Emilio Aguinaldo visited Baguio in 1909 and stayed with then-Vice-Governor Forbes. The *Christian-Science Monitor* noted that "This is Aguinaldo's first visit to the northern mountains since his retreat in that direction, with several columns of American troops in pursuit."[138] Country club members by 1911 included Gregorio Araneta, a judge and Philippine Commission member; attorney Francisco Ortigas; lawyer, writer, and Nacionalista senator Rafael Palma; the journalist and attorney Juan Sumulong; and wealthy businessman Enrique Zobel de Ayala.[139] In 1912, members of the mestizo Roxas and Elizalde families, the former owners of a major manufacturing and trading firm, the latter proprietors of the largest sugar mill in the Philippines, together with the Zobel, Brias, de la Rama, and Garchitorena clans, were reportedly building houses in Baguio.[140]

As this list shows, many of those embracing the hill station's offerings were wealthy Filipinos, some former revolutionaries, who had chosen collaboration with Americans as the course best suited to advance their own interests and the Philippines' welfare. From early on, American authorities attempted to attract these Filipinos by greasing the wheels of the patron-client system, "[h]ighly personalized, multifunctional, and affect-laden" relations that connected people with uneven shares of power.[141] The system was rooted in a relation of "debt and idealized reciprocity," or, in Tagalog, *utang na loob*.[142] *Utang na loob*, as discussed in chapter 4, structured the relationship between a landlord and his tenant, who returned a portion of his crop as tribute to the landlord. Their commerce was unequal, but their relationship endured thanks to the landlord's distribution of "material and affective surplus" that the tenant returned with "public loyalty and deference."[143] The Ibaloi chiefs distributed their surplus through feasts and parties for lesser kin and for those who tended their cattle or lived on their lands; local Filipino leaders did the same and were known to underwrite public improvements with their own monies, just as Forbes had invested his in Baguio.

Through the patron-client system, Americans tried to steer Filipinos into cooperation with the US regime—into accepting patronage, or "tutelage"—in exchange for a share of power.[144] Some received political appointments in exchange for their loyalty. Benito Legarda, Pardo de Tavera, and Jose Luzuriaga were awardees of this patronage, winning seats on the Philippine Commission in 1901. Others were granted posts in government bureaus or in the courts.[145] By the end of 1901, the colonial "politics of patronage" incorporated all but the "special" provinces of the Philippines, and just as it had been the Spanish policy, government jobs over the municipal level were appointed positions.[146] Apart from these posts, under US rule, elites were known to enjoy tax breaks.[147] Americans imitated the Spanish here; they, too, had cultivated native elites by awarding them exemptions from tribute payment.[148]

Grants of social capital served as another method for recruiting collaborators; this goal spurred the hospitality that Forbes extended to Filipino elites.[149] "It is our practice to recognize the Filipino socially, to try to win him by little acts of friendship and courtesy and of personal consideration," Forbes wrote to his sometime correspondent Rudyard Kipling, an avid observer of colonial social relations. "[W]e are accustoming [the Filipino] to certain phases of democratic simplicity of a sort that hitherto have been quite unknown among Oriental peoples I believe. For instance our highest officials, including the Governor-General, wear the same clothes as other civilians, walk the streets and enter into games and sports with natives and others on a basis of equality."[150] Narrowing the social distance between the white colonial officer and "Oriental peoples," then, and projecting the appearance of equality: that, Forbes implied, would win the affection and cooperation of Filipino elites. Forbes learned a lesson of William H. Taft in this respect. Over his tenure as governor general, Taft had come to believe "how deeply *ilustrados* yearned to be treated as the legal and social equals of their colonial overseers," and he calibrated his methods of rule in light of that understanding.[151] He had participated in Manila's "fiesta politics," Paul Kramer's term for Filipino and American elites' socializing at galas and balls.[152] Occasions like these spoke to a connection between Americans' social recognition and their political recognition of Filipinos. When Filipinos played host, these affairs gave native elites an opportunity to demonstrate their full possession of markers of "civilization" and to draw Americans into new relations of debt and reciprocity.[153] Kramer terms this kind of politics a "non-coercive mode of imperial power" that offered "not simply the buy-off of local elites, but the creation of buy-in," in this case through "the radiance and prestige that attach to asymmetric power and wealth."[154]

Prestige certainly attached to Baguio. An American writer linked hill station sojourns to one's social status this way: "The person that doesn't go to Baguio does not amount to much, in a social way, you know. That's the idea."[155] If visits to Baguio made clear one's standing in the colony, then by inviting Filipino elites into Topside, his domestic "show place," and into participation in Baguio's "material advantages," Forbes accorded Filipinos a measure of social recognition and projected the image of social equality that he had boasted to Kipling. Americans had transformed the capital appropriated from the Ibaloi into a political asset—a source of social capital that they then put into circulation with Filipinos. An episode from the summer season of 1907 shows how Baguio could serve as such an instrument for forging alliances with Philippine elites; however, it also suggests the limits of these patron-client relations for colonialists' hold on power.

## The Limits of Patronage

In the spring of 1907, the colony prepared for elections of the first Philippine Assembly, the popularly elected representative body that would serve as a counterpart to the appointed Philippine Commission dominated by

Americans. During the Baguio season, Forbes hosted Henry Bandholtz, general of the constabulary forces, and Topside architect William Parsons, together with several prominent Filipinos including Legarda, then a commission member, and Araneta, soon to become one. To this group of visitors, Forbes added a rising star from the province of Cebu: Sergio Osmeña. Osmeña governed "the biggest province in the archipelago, in point of numbers some 700,000 people, and controls them so that he can do almost anything that he wants," wrote Forbes to his brother.[156] Osmeña, Forbes believed, would be the first speaker of the Filipino Assembly. The Cebuano was part of a generation of Filipino politicians who would drive a politics that tacked between resistance and "cooperation" with Americans. He was a "Nacionalista," a member of an emergent political party that had grown frustrated with Federalistas' lock on influence and patronage and their stance on independence. While Federalistas generally accepted and sought collaboration with Americans, Nacionalistas, even though accepting the reality of the US occupation, vocalized commitment to future Philippine independence.[157] Or, as Osmeña would write Forbes in 1944, "the Nacionalistas proudly waved their banner for immediate independence with the backing of the people." Yet they also sought to "reconcile" this goal with "American sovereignty." "Some radical elements in and out of the Assembly—supported by prominent political figures with European training and experience—favored a policy of noncooperation," he wrote. "The majority of the Assembly, however, openly advocated cooperation with the American administration, as represented by the Governor-General and the Commission." Osmeña led this latter faction.[158] His nationalism was of a different character than that demonstrated by some of those *ladrones* and the Partido Nacionalista, the pro-independence party founded by Dominador Gomez, Isabelo de los Reyes, and Pascual Poblete.[159]

Osmeña's political fortunes were on the rise, and Forbes wished them to be rightly invested.[160] Were they to win Osmeña's endorsement, the hill station and the colonial regime symbolized by Baguio could enjoy a greater measure of legitimacy and security.[161] "I knew very well that the Legislature would never designate Baguio without having seen it," Forbes admitted to his diary, "and it would be difficult to get the delegates to go up, particularly as the Filipinos had heard great stories of the road and some of them were afraid of it and of the trip." And so when Osmeña asked Forbes for a meeting during the "Baguio season" of 1907, Forbes made it conditional on location: the session had to take place in Baguio. Bandholtz, together with Osmeña, Araneta, Legarda, and perhaps other Filipinos, convened, and "enough of them were converted to the delights and advantages of the place so that the opposition to Baguio was greatly decreased from that time on, and has finally disappeared entirely."[162] Ironically, Osmeña had already been acquainted with Benguet; as a former collaborator with Aguinaldo he had ascended into the Cordillera to meet the general during the war. Years later he would purchase property and build a home in Baguio and reflect that Topside "had a vital role in the building of" the hill station.[163]

During that 1907 visit, Forbes took Baguio not only as an end but also a means of politics. Forbes sought another concession from Osmeña, for in that year, more than the prospects for Baguio hung in the balance. So too was the direction of politics in the colony with the Philippine Assembly due to form. Forbes, it seems, also invited Osmeña to coax an assurance from him. Forbes wished to know that when Osmeña and the Nacionalistas came to power, they "would not endeavor to upset the present form of government—that they would be satisfied when they had the office." Surely to Forbes's relief, Osmeña so promised.[164] That October, Forbes recorded a description of Osmeña's appearance at the Assembly: "Before the desperately slender, sensitive, aristocratic-looking, and composed young man of twenty-nine arose, the whole house was applauding, and as he sat down the demonstration continued and almost grew till he had to bow. Everyone knew that here was the real leader of the Philippine Islands, a man of fine reasoning power, force over men, firm, incisive, an orator, a lawyer, frail, almost diseased in body, but sympathetic and responsive as a girl." Forbes concluded ominously, "The only danger is lest the unthinking at home shall confound the evidence of capacity which the Assembly will show as evidence of a democratic capacity instead of the evidence of a power of being dominated."[165] Forbes wished for the performance of a man, rendered "sympathetic and responsive as a girl," to be recognized not as "democratic capacity," the supposed consequence of effective political tutoring, but of political domination, an authority's exercise of power.

While Taft would write in 1907 that the convening of the Philippine Assembly and governing responsibilities "in the hands of Filipinos" tended to produce "both conservatism and an interest in the existing government" among Filipino leaders, Forbes's concerns about the vulnerability of colonial authority in light of Osmeña's leadership proved warranted.[166] Subsequent to the formation of the Assembly, Filipino politicians turned from the "institutionalist concerns" of the Filipino revolutionaries to efforts at enlarging legislators' power at the expense of the executive.[167] An article published by *El Renacimiento* in September 1906 had foreshadowed this turn; it suggested that the Assembly's first task ought to be to curb the Philippine Commission's powers. One way of leveling the power between the Assembly and the executive, the writer suggested, was for the Assembly to claim "the right to accuse the governor general or the most senior officials of crimes of infidelity in fulfilling their duties"—in effect, of failure as patrons.[168] By early 1908, just several months into its existence, the Assembly was debating the right to co-legislate in the non-Christian provinces, citing racial unity, territorial integrity, and the fact that some of the funds of the Philippine treasury were used in the administration of the Moro Province and non-Christian tribes, among other arguments for seizing authority.[169] A month later, shortly before the inception of the 1908 Baguio summer season, *El Renacimiento* questioned the government's break for the north. The article noted that the expectation that legislators accompany commissioners to the summer retreat put an undo financial burden on assemblymen. They would be forced to maintain two homes, one—in Baguio—where the cost of living was through the roof, or, in a literal translation, "to the clouds." If assemblymen did not accompany the commission, however, what would become of the work of legislating for the colony?[170]

Small episodes though they may be, these early debates of the Assembly are reminders that provision of political office and social equality did not alone satisfy Forbes's guests, contrary to Forbes's anthropology of Filipinos. Assemblymen showed they would readily domesticate the political education colonialists offered and make it a source of empowerment for themselves, if they did not dismiss it altogether.[171] Filipinos could also domesticate the longing that had informed Americans' quest for a mountain retreat. Perhaps some of these ambitious assemblymen had read Forbes's grandfather Emerson and had been as inspired as a writer for *El Renacimiento* had been. "After reading [Emerson], one feels happier, more at ease, ready to undertake great challenges, finding that it has stirred in us many dormant energies, overlooked strengths, hidden enthusiasms," he told the newspaper's readers.[172] Such empowerment might well counterpoise any signs that Filipinos were "being dominated."

## Philippinitis

Through Forbes's tenure, Baguio never totally shook its controversial status. In 1911, on the eve of Woodrow Wilson's election, Forbes's ouster, and the Filipinization of the colonial administration, and just as Baguio was truly functioning as the colony's summer capital, an American newspaper called the hill station a "white elephant": "so far as the great mass of the Filipino people are concerned, Baguio might as well be a resort on the moon. It benefits them not one whit."[173] When Forbes's successor, New York Democrat Francis Burton Harrison, arrived in 1913, the government's ritual retreat to Baguio came to an end. Harrison reportedly called the hill station "a piece of 'Republican folly and extravagance.'" Ultimately, however, he fell for the charms of Baguio, and by the time he left the archipelago, the hill station was larger and even more well known.[174] Still, if the existence of Baguio as an American retreat was never wholly assured, neither was the fate of the colonial regime.

The country club and Topside, even the hill station, spoke of American vulnerability in their intended function as retreats. Their associations with the simple life reflected the needs and weaknesses of colonialists and spoke to what Warwick Anderson, in his work on colonial medicine in the Philippines, calls "the distressed and assertive colonial culture of bourgeois white males."[175] Dean Worcester would retreat to Baguio from Manila in 1908 to "get over the effects of a carbuncle of record dimensions" that had made him quite ill.[176] But perhaps he also sought refuge in Baguio from those Filipinos who had come to loathe him and his exploitation of indigenous peoples. He wrote that at the hill station, he could operate "free from interruption" and apply himself to his responsibilities, something that seems to have been more challenging in the political capital.[177] Elsewhere, Worcester wrote of the psychological depths men could reach even in the non-Christian provinces of the Philippines: "The most serious of our difficulties have been those which necessarily result from extreme isolation and loneliness." The consequence, Worcester said, was bickering or bitterness among fellow colonial workers. Others "have completely lost their sense of perspective. Little things look very big to them."[178] Frank Carpenter also wrote of the need for respite in his 1925 book, explaining that "experience has shown that prolonged residence in the Philippines has a bad effect on the mental and physical condition of people not born and brought up in the tropics. While they may not become actually ill they lose their energy, grow irritable and forgetful, and find the least exertion burdensome. A trip to Baguio has often proved the best medicine for such cases."[179] What they forgot, what wearied them, and what irritated them, Carpenter did not say. They required "medicine" yet were not "ill." They suffered, perhaps, from Philippinitis.

The prescribed cure for the colonialist was much like the cure for Forbes's family in Boston and Shanghai: retreat. "[T]he Country Club is a potent factor in the elimination of morbid nervousness—that American bugbear—and is equally potent in the work of reconstruction," wrote an observer of the American country club in 1905. "[F]or all who submit to its healthful, pleasing regime are surely progressing toward sane, normal, active manhood and womanhood."[180] The country clubhouse and Topside as embodiments of Baguio as a pastoral retreat arguably were designed to provide the same treatment. The use of the bungalow and the appeal of Baguio—the turn from complex colonial realities to nature and the seemingly simple life—suggest that top colonial officials and their clerks as much as their professional-managerial brothers back at home craved reprieve from their work, therapeutic solutions, and a less strenuous life.

One way some colonialists withdrew from the daily round was by playing Igorot. Forbes threw ball for a baseball team called the "Gee Strings," and he convinced a chief of the Kalingas, an ethnolinguistic group of the Cordillera, to build him a Kalinga-style house.[181] In Manila, colonialists also played Igorot. Forbes described a lecture he gave on the "non-Christian tribes": "We had lantern slides prepared with most absurd pictures, Worcester seated on Mayon Volcano, the Governor's and Worcester's faces photographed on to children and Igorots and other savages. I made my story just as far fetched and absurd as it could be. . . . We all came as Igorots and took as our motif Worcester's law prohibiting the sale of intoxicating liquor to 'non-Christians' and hailed him as the great missionary, proclaimed ourselves converts on the strength of the promise of whisky in this world and salvation in the next. . . . Each had one of the little basket hats, a shield, a spear, a red gee string over our trousers."[182] Forbes enjoyed the joke on Worcester perhaps because he himself had been the object of a similar jab. Tasked with auditing the US colonial administration early in his colonial tenure, he was asked to institute greater efficiencies in the administration.[183] As if fearing the censure of the modern-day management consultant, "[r]ecords are being straightened, clerks removed, forces diminished, elaborate forms abandoned, methods simplified everywhere just in anticipation of our visit," Forbes wrote in his journal with satisfaction.[184] The committee's work of streamlining the colonial bureaucracy brought about the "discharge of many clerks." And for this work, he wrote, the committee became known as "The Head Hunters."[185]

Worcester and Forbes served as the butts of these jokes; Igorots were the medium. And the humor sprang from the imagined distance between Philippine natives and these American administrators. The game supplied entertainment precisely because its conceit was recognized as ridiculous and "far-fetched" among its players. Perhaps in these moments, colonialists alleviated some of the symptoms of Philippinitis. But while Forbes relayed the sobriquet "head hunter" with the knowing chuckle of an insider, his good humor could well have been tinged with disquiet. The name associated his committee chiefs—those tasked with rationalizing the colonial bureaucracy—with their converse: a people imagined as barbaric and uncivilized. Implicitly, the name-calling spoke to a problem for "civilization." As Jackson Lears has suggested, "[i]mperial primitivism," playing Igorot or Indian, for example, "embodied the widespread suspicion that dark-skinned people might have characteristics worth knowing, enjoying, and even appropriating."[186] And colonialists did appropriate Ibaloi ways. Americans took their land and even adopted practices like the *cañao*. Only they then crafted a picturesque landscape of "peace, leisure, and economic self-sufficiency" that helped obscure these dispossessions and even naturalize the US occupation—an American imperial pastoral.

# Epilogue

Carlos Bulosan's semi-autobiographical novel *America Is in the Heart* (1946) was published just after the Second World War and the year the Philippines secured its independence. Born in the Philippines in 1911, Bulosan told a story of a young Filipino named Allos.[1] It was an account of the deprivations suffered by his family in the Philippines, Allos's efforts to escape from poverty, and the daily cruelties he encountered in the United States as an itinerant laborer and person of color. Gradually dispossessed of land, Allos's parents see him and his brothers leave to pursue opportunities hoped to become the family's redemption. In one of these forays before he leaves for the United States, a thirteen year-old Allos finds his way to Baguio. "Baguio is a small city in the heart of tall mountains where the weather is always temperate," Allos tells the reader. "There are no rains nor heavy winds. But in the morning there is a light mist in the air and when you walk through it you feel as though you are walking through silk." "The roads are asphalt and the most modern and beautiful in the Philippines," he continues. "The houses and theaters are built in Western fashion. Tall pine trees cover the mountains and at night one can hear the leaves singing in the slight wind from the deep canyons beyond the city that comes up with the sweet tang of fragrant vegetation from the surrounding valleys."[2] Allos ascends from the lowlands to a Baguio that he describes as almost celestial. The leaves sing, the sweet-smelling air has the touch of silk in the early morning hours, even the roads are objects of beauty. The landscape that Allos sketches is Baguio as pastoral, an image that William Cameron Forbes surely would have recognized.

The familiarity between American colonialists' and Allos's Baguio, however, ends there. For, like Virgil's eclogues, Bulosan sketches the dispossessions underwriting the resort, offering what Leo Marx might call a "complex" pastoral and contrast to "sentimental" and nostalgic invocations of the country.[3] Bulosan stopped at sites that we, too, have considered. Benguet Road was one. Allos labors on the road, nearly losing his life in one of those rainstorms that would have maddened Major Kennon and other highway engineers. Allos's account offers a representation of danger, heroism, and solidarity that counters those traded by American administrators, eager to assert engineering prowess and managerial ingenuity as the central narrative of the road. Allos's efforts to earn money continue in Baguio's marketplace, which he describes as "teeming with European and American tourists." "It was at this market that I first landed," he explains. "Europeans of affluence, Americans with big businesses in the islands, and rich Filipinos lived in Baguio. Their beautiful white houses dotted the hills." Here, in a Baguio that sounds much like Forbes's, Allos seeks work but at first finds nothing. "My clothes began to wear out. I was sick from eating what the traders discarded." His fortunes take a turn when "[o]ne day an American lady tourist asked me to undress before her camera, and gave me ten centavos for doing it. I had found a simple way to make a living. Whenever I saw a white person in the market with a camera, I made myself conspicuously ugly, hoping to earn ten centavos." This strategy has limited success, for "what interested the tourists most were the naked Igorot women and their children. Sometimes they took pictures of the old men with G-strings. They were not interested in Christian Filipinos like me. They seemed to take a particular delight in photographing young Igorot girls with large breasts and robust mountain men whose genitals were nearly exposed, their G-strings bulging large and alive."[4]

Unable to sell his labor in the marketplace, Allos comes to sell himself just outside of it. This is not panhandling or begging, to be sure, but trade. Allos sells a sight of deprivation to those white tourists who also picked through Philippine wares in the Baguio marketplace. In buying Allos's performance of the indigent native, they believe, perhaps, that they have found the real-life version of the Filipino—one they had come to anticipate from earlier tourists' tales. Bulosan implies that they sought the show-Filipino, a pitiable sight that affirmed the white viewer's superior status and justified the US presence. Allos's entrepreneurship, however, is limited by his identity as a Christian Filipino. An image of an "ugly," brown brother was not nearly as desirable as that of an Igorot. This latter figure stood to affirm lessons taught by World's Fair exhibitions, magazines like *National Geographic* and *LIFE*, and American leaders who created and recreated the Igorots as a subject people through the US occupation. The Igorot that emerged was a signifier of the difference that colonialists like Forbes and Dean Worcester had worked hard to promote. Colonialists conjured the Igorot as the United States' noble savages, writing a fable of their historic vulnerability to abusive lowlanders. They also crafted images of dog eaters and child-slaves that were set in motion across the Pacific. These expressions were accompanied by laws that advanced the alienation and commodification of Ibaloi land and the resources it held.

Bulosan's fiction hints at the costs of US colonial policy toward the Igorots. In Benguet, over time, the *baknang* lost power as Americans appropriated the basis of their authority, Cordillera lands, from which they had secured reserves of gold and on which they had grazed cattle for generations. While Igorot elites could still enjoy positions of prominence as presidents of municipalities as the Cariños had, their "power was effectively debased."[5] An Igorot township president was accountable to a provincial governor, a figure appointed by the governor general, and it was this person who held the power to distribute the surplus of jobs in government programs and public works.[6] Positions within a branch of the colonial state became the new sources of power. The Episcopal minister Bishop Charles Brent wrote around 1916 that "[t]he American nation was responsible for dragging the Igorot into the market place of the world."[7] While Igorots, especially the *baknang*, had never been wholly removed from the global market, as their trade in gold suggests, as a division of labor accompanied the state's division and commodification of land, the market displaced ritual as the basis on which the Ibaloi themselves "allocated" forms of capital (power).[8] While ordinary Ibaloi may have enjoyed greater autonomy from the *baknang* as a result, Americans' ownership and transformation of Baguio lands still circumscribed that liberty when one form of paternalism replaced another.

The Spanish had created and made meaningful the distinction between "Christian" and "non-Christian." Americans pushed the difference further, and paradoxically, by categorizing these lands and peoples as exceptional—as a "special province"—Americans brought this space, formerly one of limited colonial state authority, more securely into the bounds of the colonial state. Patricio Abinales and Donna Amoroso write that "it was under US colonial rule that the Philippine 'geo-body' was fully realized. American military power was able to achieve a substance of governance that the Spanish state and the Malolos Republic were not."[9] To a greater extent than the Spanish, Americans had made the reach of the state consistent with the territory of the Philippines and the economic practices of its inhabitants.[10]

The Igorots of Benguet were not the only Philippine peoples affected by Americans' appropriation of land and the creation of a capital in the Cordillera. Indeed, as chapter 4 shows, Americans' conjuring of the Igorot as noble savage served as justification for continued paternalist care of the entire archipelago. Paul Kramer has made this point, too; by Woodrow Wilson's election, "Struggles over the rule of non-Christians and over the rule of the islands as a whole would become inseparable."[11] Once Americans seized Ibaloi pasture and appropriated the material bases and symbolic dimensions of headmen's authority, they came to use them as their own "political resources," and Baguio became a location of U.S. power.[12] By extending access to state resources through political appointments, through tax breaks, and arguably by bringing Filipinos into relations of social credit and debt in Baguio, a "show place" and American enclave, colonialists attempted to attract Filipino elites and consolidate a class of political collaborators. In 1907, Forbes wrote that former revolutionary general Emilio Aguinaldo presented him with a "short sword" wrapped in a silk handkerchief. The handkerchief was embroidered with Aguinaldo's initials, and the sword bore a silver sheath on which was written *Kalayaan*.[13] Aguinaldo, who had once taken his revolutionary government into the Cordillera to evade US forces, had made a trade, casting his lot and liberty with the Americans.

One can understand why elites might return American favors and hospitality with gifts of their own, even one as seemingly significant as Aguinaldo's sword. While the revolutionary Malolos Republic moved to empower provincial and municipal governments in its constitution, US colonial policy like the 1904 Internal Revenue Law tended to centralize power at the national level. Just as Americans rerouted power away from traditional leaders in the special provinces and toward appointed provincial governors, so they made access to "the meager shares of the internal revenue 'apportioned' to [local governments]" contingent on "fealty to and supplication with the authorities in Manila."[14] Once more of this state apparatus fell into Filipinos' hands, especially in the period of Filipinization, Filipino leaders "began to use [the colonial state] as an instrument of 'primitive accumulation'"—a process by which resources held in common are seized and held by the state and private capital. This Filipino "ruling class" then dealt these economic and political resources like political office to friends and supporters, who in turn backed their patrons; political leaders used state corporations like the Philippine National Bank and the Manila Railroad Company in a similar fashion.[15] John Sidel has suggested that "the subordination of the state apparatus to elected officials," rather than to a government bureaucracy, paired with conditions of "primitive accumulation" ultimately conditioned clientelism and the use of "state offices" for "capital accumulation."[16] The state forged under US rule became an instrument for producing great disparities and abuses of wealth and power and for carrying out "accumulation by dispossession."[17]

William Cameron Forbes could not have supposed such a future for the Pearl of the Orient; he believed that the state merely cleared the field for merchants and that it was the businessmen who "play ball." At the end of his tenure on the archipelago in 1913, after Woodrow Wilson replaced him with Francis Burton Harrison as governor general and promoted the Filipinization of the colonial bureaucracy, Forbes was still upholding a vision of colonial plenty based on Manila as a commercial entrêpot just as Lodge had back in 1900. "The Pacific Ocean is unquestionably the future theatre of the world's commerce," Forbes insisted. And given Manila's location relative to San Francisco, Honolulu, the Suez Canal, and of course China, it ought to "be one of the world's greatest emporiums of trade."[18] Trade with China never did materialize in the way Lodge and Forbes had forecast. And, thanks to the combined effect of US tariff laws and the limits set to land ownership by the Organic Act, the Philippines did not become an outlet for surplus American capital and a "major factor" in the United States' "economic prosperity."[19] From about 1900 to 1913, "Americans bought or leased only about 40,000 acres of public land" in the Philippines.[20] By 1943, American-owned assets in the Philippines summed to about $167,100,000 compared to $590,500,000 in Cuba and over four billion dollars in Canada.[21] In the Philippines, land largely remained in the hands of rich Filipinos, who managed to pay little in way of taxes on this wealth.[22] Internal revenue in the first decade and a half of US rule came mostly from the *cédula*, a tax of limited consequence to the wealthy but onerous to the poor, just as it had been during Spanish rule.[23] "[B]y mutual consent," Norman Owen writes, "American and Filipino leaders attempted to expand and to rationalize the economy, but not to change it."[24] Despite pledges to liberate and bring justice to Philippine peoples, they wound up preserving a primary source of social inequality—landownership enjoyed by the few.

Bulosan's *America Is in the Heart* points to these very conditions. When Allos avails tourists of his body and labors along the road to Baguio, he is working to recover something that once belonged to his family: land. Allos attributes the loss of land to "[t]he sons of the professional classes" who return to the provinces after having received an education and take to "victimizing their own people and enriching themselves as the expense of the nation."[25] Once they rise to the ranks of national government office, these leading men continue this work of "exploitation of the peasantry."[26]

What Bulosan describes in his fiction owes largely to conditions generated under the Spanish[27] and perpetuated by US colonial policy.

More than an emporium of trade, under US rule, the Philippines became a source of insecure labor, and again, Bulosan's fiction points to this history. The dispossession of land and witness to the humiliation of his father advances Allos's story of exile and migration—from Baguio through Washington State, Oregon, and California. That initial dispossession renders Allos's labor available on the market and cheap. Subjects, or "nationals," like Allos would travel those Pacific sea lanes celebrated by Albert Beveridge. Thousands would follow Burnham, Forbes, and Taft back across the Pacific for work in Hawaii, California, and in the Pacific Northwest.[28] William Cameron Forbes, incidentally, hosted Hawaiian planters up in Baguio in 1911.[29] By 1916, more than 20,000 Filipinos had traveled to Hawaii, and many of these Filipinos came from the Ilocos region where Major Kennon had also recruited workers for Benguet Road.[30]

Sourcing labor from the Philippines has continued into the twenty-first century. In 2011, the 1.8 million Filipino immigrants in the United States made them the fourth largest immigrant group in the country.[31] This was close to the number of Filipinos "deployed" in jobs overseas in 2013. In 2008, the total number of Filipino migrants working outside the archipelago was over 8,000,000, or about a tenth of the country's population in that year.[32] By the end of 2010, these workers had sent back remittances totaling more than 20.5 billion US dollars, close to 11 percent of the country's GDP.[33] Sociologist Robyn Magalit Rodriguez has noted that the Philippines supplies employers around the world with a large volume of temporary workers who impose few costs to receiving countries but resolve the common conflict between economic interests and domestic politics: as sojourner-workers, they "resolve the contradictory forces of labor demand and immigration restriction." Rodriguez has studied the role of the Philippine state in facilitating this incredible outflow of migrant labor, deeming it a "labor brokerage state" and the country "the world's premier 'global enterprise' of labor." As for the origins of the state's labor entrepreneurship, she identifies US colonial state formation and neocolonial relations. Besides setting a precedent for labor recruitment and supplying workers with training and favorable impressions of the United States, the occupation helped improve the power of the country's landed elite "in exchange for concessions to US economic, political, and military interests." These circumstances have conditioned Filipino political leaders' willingness to use Philippine labor as a source of state revenue.[34]

Filipinos need not leave their country to secure employment in places that are "foreign 'in a domestic sense.'"[35] They can travel to Baguio as Allos did and find employment on Cariño's former land—at the John Hay Special Economic Zone, an "eco-tourism haven."[36] Or they can seek work at a firm like Texas Instruments, which is located in the Baguio City Export Processing Zone, one of sixty-eight manufacturing economic zones in the country in 2015.[37] The Philippine government advertises the exemptions and deductions from national and local taxes and import duties available to companies who outsource their business to workers in these spaces.[38] Ibaloi land, part of which Americans turned into a "government reservation" (a sort of extra-national space as an American enclave), has been reorganized into a new kind of extraterritorial space in the late twentieth century.[39] The government promotes the "Philippine edge" that employers can secure through native labor in these zones. Filipinos are described as "literate, English-speaking," and trainable in eight weeks rather than the four to six months estimated as necessary in other countries. Testimonials on Filipino workers from employers suggest that much and that little have changed since the time of Kennon and Holmes's work on Benguet Road: "Through my past experience of working in 17 countries' people [sic]," writes one employer, "Filipinos are the best to work together with the Japanese. They are always with a bright smile, hospitable, obedient, hard working, and loyal." Another employer, in an uncanny echo of Kennon's report that Benguet Road became a "refuge for all the unemployed," claims that "Filipino engineers are among the best in the world. Most of our engineers come from less fortunate families who have no option but to excel."[40] The government also advertises the Philippines as offering the "best sporting and recreational facilities" and "many world-class golf courses."[41] Not least of these is at the Baguio Country Club.

## No Retreat

In exposing contradictions between democratic ideals and acts of imperial conquest, Carlos Bulosan's novel recovers something of the classical pastoral described by Raymond Williams and presents what Leo Marx calls a "complex pastoral." It retains a "counterforce": "Whether represented by the plight of a dispossessed herdsman or by the sound of a locomotive in the woods," Marx writes, "this feature of the [pastoral] design brings a world which is more 'real' into juxtaposition with an idyllic vision."[42] The world more real is found in Bulosan's description of Baguio—a celestial city in which Allos endures deprivation and observes the exploitation of Igorots. And it is also in his description of the US mainland, the pastoral idyll that Baguio was meant to reproduce in micro in the colony; it is in the violence Allos meets and his response to it—the radical politics in which he engages on the West coast. If the imperial pastoral helped to generate and naturalize US power, Bulosan's counter-pastoral, or "ghetto pastoral," denaturalizes it in a dramatic reversal of the immigrant success story.[43]

Bulosan, the protagonist of his novel, and Filipinos like them would continue practicing anticolonialism in their overt political action, reminding us that contrary to many pro-imperialists' assumptions, the pursuit of new markets did not ultimately quell labor insurrection and radicalism on US shores. Bulosan, who arrived in the United States in 1930, amidst another crisis in capitalism, would go on to organize Filipino fish cannery workers in California.[44] And he followed in a long line of Filipino labor insurgents like Isabelo de los Reyes and Pascual Poblete, whose Union Obrera had caused colonialists such trouble in the first decade of the twentieth century. Their union was the seed of the Philippines' Communist and Socialist parties, which formed a front against the Japanese occupation in World War II and US neocolonialism after 1946.[45] Augusto Espiritu has argued that the ethos informing Filipino revolutionaries' vision of *kalayaan* at the turn of the century—"the idea of suffering and sacrifice based upon faith in a redemptive future"—lives on in Bulosan's "popular-front allegory" and the complex pastoral he offers.[46] In that respect, the "suffering and sacrifice" that the US occupation precipitated did liberate labor but in ways that Americans may not have anticipated. "This family tragedy," says Allos of the loss of his family's land, "marked the beginning of my conscious life" and, in that way, the beginning of the writer, whom Allos, like Bulosan, becomes.[47]

Not long before the publication of *America Is in the Heart*, Baguio was transmuted into the very reverse of refuge. On December 8, 1941, a day after the attack on Pearl Harbor, the Japanese Air Force dropped bombs on Camp John Hay, making Baguio the first Philippine city struck by the Japanese. Fearing their matériel falling into the enemy's hands, Americans destroyed equipment at the military reservation and declared Baguio an open city.[48] Even as some Americans in Manila had packed up their cars and headed north, aiming to take refuge in Baguio upon hearing of the Pearl Harbor attack, those who could fled from Baguio.[49] Some Igorots would find their way south and fight Japanese forces by early 1942. After an Igorot tank unit successfully destroyed a Japanese regiment, it earned the highest compliments of General Douglas MacArthur, son of General Arthur MacArthur, who had arrived on the archipelago nearly forty years earlier to help prosecute the Philippine-American War. Douglas MacArthur, the *New York Times* reported, "is quoted as having said that although he was cognizant of many acts of heroism on battlefields all over the world, 'for sheer, breath-taking and heart-stopping desperation, I have never known the equal of those Igorots riding the tanks.'"[50] The former model American city and resort, meanwhile, became an internment center for American, English, and Chinese residents.[51] When General Yamashita arrived on the archipelago in late 1944, he set up headquarters there and led a "protracted defensive operation."[52] He knew that an American invasion was imminent and Japanese victory unlikely; he only hoped to give Japan more time to prepare for the coming assault.[53] Yamashita would ultimately wave the white flag in Ifugao Province, just east of Benguet, and he signed surrender documents in Baguio.[54] However, this was only after occupied Baguio had been all but leveled by American carpet bombing.[55] Some observers noted at the time that "It was the civilian citizens of Baguio who were killed, not the Japanese soldiers who left the city by day. Shell-shocked, impoverished and starved survivors sought routes of escape, while most of those who stayed were either the wounded or the dying."[56]

Idyllic Baguio becoming a place of Americans' confinement and the pastoral laid waste by war magnifies in extreme ways the ironies of the hill station's history. The hill station had been a response to the perceived environmental hazards of the tropics and to conditions of war and insurgency on the archipelago. But building this green zone and, more generally, the Pearl of the Orient generated new controversies and problems for Americans. Although President McKinley initially desired only Manila, to secure the port, he and those on the ground who witnessed Philippine resistance soon understood they needed to take and, in turn, remake wider parts of the archipelago. To secure Manila as a way station to the bigger prize—the China market—they believed they had to build Baguio, a model American town and enclave to offset the bodily and psychic hazards of imperial expansion, as exercises in imperial power could quickly devolve into cases of Philippinitis. Ascending the mountain and building the hill station, however, ultimately afforded Americans neither a pristine mark of progress nor a lasting refuge.

# Notes

## INTRODUCTION

1. The only monograph on Baguio as a colonial hill station is geographer Robert Reed's *City of Pines: The Origins of Baguio as a Colonial Hill Station and Regional Capital*, 2nd ed. (Baguio City, Philippines: A-Seven Publishing, 1999 [1976]). On US colonial Baguio and Manila, see also David Brody, "Building Empire: Architecture and American Imperialism in the Philippines," *Journal of Asian American Studies* 4 (June 2001): 123–45; Brody, *Visualizing American Empire: Orientalism and Imperialism in the Philippines* (Chicago: University of Chicago Press, 2010); Thomas S. Hines, "The Imperial Façade: Daniel H. Burnham and American Architectural Planning in the Philippines," *Pacific Historical Review* 41, no. 1 (1972): 33–53; Thomas S. Hines, *Burnham of Chicago, Architect and Planner* (New York: Oxford University Press, 1974); Gerald Lico, *Arkitekturang Filipino: A History of Architecture and Urbanism in the Philippines* (Diliman, Quezon City, Philippines: University of the Philippines Press, 2008); Christopher Vernon, "Daniel Burnham and the American City Imperial," *Thesis Eleven* 123 (2014): 80-105; Vernadette Vicuña Gonzalez, *Securing Paradise: Tourism and Militarism in Hawai'i and the Philippines* (Durham: Duke University Press, 2013); Warwick Anderson, *Colonial Pathologies: American Tropical Medicine, Race, and Hygiene in the Philippines* (Durham: Duke University Press, 2006).

   For social and cultural histories of Baguio, see Patricia Okubo Afable, ed., *Japanese Pioneers in the Northern Philippine Highlands: A Centennial Tribute, 1903–2003* (Baguio City: Filipino-Japanese Foundation of Northern Luzon, Inc. 2004); Carina Agnir-Paraan and Alice Buenviaje-Wilder, eds., *Camp John Hay: How It All Began . . . Where It Is Bound* (Baguio City: John Hay Poro Point Development Corporation, 2000);

Virginia Benitez Licuanan, *Filipinos and Americans: A Love-Hate Relationship* (Manila: Baguio Country Club, 1982); Joanna K. Cariño, "The Cariños and Baguio-Benguet History: Conclusion," *Folio 2*, Cordillera Studies Center Social Science Monograph Series no. 4 (Baguio City: Cordillera Studies Center, University of the Philippines College, Baguio, 1984); June Prill-Brett, "Baguio: A Multi-Ethnic City and the Development of the Ibaloy as an Ethnic Minority," CSC Working Paper 15 (Baguio City: Cordillera Studies Center, University of the Philippines College Baguio, 1990); Erlyn Ruth Alcantara, "Baguio Between Two Wars: The Creation and Destruction of a Summer Capital," in *Vestiges of War: The Philippine-American War and the Aftermath of an Imperial Dream, 1899–1999*, ed. Angel Velasco Shaw and Luis H. Francia (New York: New York University Press, 2002).

Histories of Benguet Province and northern Luzon include Anavic Bagamaspad, Zenaida Hamada-Pawid, and Bell Balangoy, *A Peoples' History of Benguet Province* (Benguet Province: Baguio City Printing and Publishing Co., 1985); Bienvenido P. Tapang, "Innovation and Social Change: The Ibaloy Cattle Enterprise in Benguet," CSC Monograph Series 5b (Baguio City: Cordillera Studies Center, University of the Philippines College Baguio, 1985); Rowena Reyes-Boquiren, "The History and Political Economy of the Vegetable Industry in Benguet," CSC Working Paper 14 (Baguio City: Cordillera Studies Center, University of the Philippines College Baguio, 1989); William Henry Scott, *The Discovery of the Igorots: Spanish Contacts with the Pagans of Northern Luzon* (Quezon City, Philippines: New Day Publishers, 1974); William Henry Scott, *On the Cordillera: A Look at the Peoples and Cultures of the Mountain Province* (Manila: MCS Enterprises, 1966); Melanie Wiber, *Politics, Property and Law in the Philippine Uplands* (Waterloo, ON: Wilfred Laurier University Press, 1993); Delfin Tolentino, ed., *Resistance and Revolution in the Cordillera* (Baguio City: University of the Philippines College Baguio, 1994); Gerard A. Finin, *The Making of the Igorot: Ramut Ti Panagkaykaysa Dagiti Taga Cordillera: Contours of Cordillera Consciousness* (Quezon City: Ateneo de Manila University Press, 2005); Felix Maxwell Keesing, *The Ethnohistory of Northern Luzon* (Stanford: Stanford University Press, 1962).

2. For ease of reading, I will use *America* and the *United States* and *Americans* and *US citizens* interchangeably throughout this work, even as I recognize that America and Americans signify far more than US national identity.
3. Monroe Woolley, "Baguio, Simla of the Philippines," *Overland Monthly* (1913): 292.
4. *Fourth Annual Report of the Philippine Commission, 1903, In Three Parts, Part 1* (Washington, DC: Government Printing Office, 1904), 58.
5. Woolley, "Baguio, Simla of the Philippines," 293.
6. Carlos Bulosan, *America Is in the Heart: A Personal History* (Seattle: University of Washington Press, 1973 [1946]), 66.
7. In addition to the studies identified in succeeding notes, scholarship that has informed my understanding of imperial and colonial formations include

Frederick Cooper, *Colonialism in Question: Theory, Knowledge, History* (Berkeley: University of California Press, 2009); Ellen Meiksins Wood, *Empire of Capital* (London; New York: Verso, 2003); Michael Hardt and Antonio Negri, *Empire* (Cambridge: Harvard University Press, 2000); Anthony Brewer, *Marxist Theories on Imperialism: A Critical Survey* (London: Routledge, 1990); David Harvey, "The 'New' Imperialism: Accumulation by Dispossession," *Socialist Register* 40, no. 40 (March 19, 2009), http://socialistregister.com/index.php/srv/article/view/5811; Neil Smith, *American Empire: Roosevelt's Geographer and the Prelude to Globalization* (Berkeley: University of California Press, 2004); Julian Go, *Patterns of Empire: The British and American Empires, 1688 to the Present* (Cambridge: Cambridge University Press, 2011); J. H. Elliott, *Empires of the Atlantic World: Britain and Spain in America, 1492–1830* (New Haven: Yale University Press, 2006); Thomas Bender, *A Nation among Nations: America's Place in World History* (New York: Hill and Wang, 2006); Patrick Wolfe, "History and Imperialism: A Century of Theory, from Marx to Postcolonialism," *American Historical Review* 102, no. 2 (1997): 388–420; Paul A. Kramer, "Power and Connection: Imperial Histories of the United States in the World," *American Historical Review* 116, no. 5 (2011): 1348–91; Robert J. C. Young, *Postcolonialism: An Historical Introduction* (Oxford, UK; Malden, MA: Blackwell Publishers, 2001); Edward W. Said, *Orientalism* (New York: Vintage Books, 1979); Homi Bhabha, *The Location of Culture* (New York: Routledge, 1994); Dipesh Chakrabarty, *Provincializing Europe: Postcolonial Thought and Historical Difference* (Princeton: Princeton University Press, 2008); Ann Laura Stoler, *Carnal Knowledge and Imperial Power: Race and the Intimate in Colonial Rule* (Berkeley: University of California Press, 2002); Nicholas B. Dirks, *Colonialism and Culture* (Ann Arbor: University of Michigan Press, 1992); Timothy Mitchell, *Colonising Egypt* (Cambridge: Cambridge University Press, 1988); Gilbert M. Joseph, Catherine LeGrand, and Ricardo Donato Salvatore, eds., *Close Encounters of Empire: Writing the Cultural History of US-Latin American Relations* (Durham: Duke University Press, 1998); Amy Kaplan and Donald E. Pease, eds., *Cultures of United States Imperialism* (Durham: Duke University Press, 1993); Walter Mignolo, *Local Histories/Global Designs: Coloniality, Subaltern Knowledges, and Border Thinking* (Princeton: Princeton University Press, 2000).

8. Alfred W. McCoy, Francisco A. Scarano and Courtney Johnson, "On the Tropic of Cancer: Transitions and Transformations in the US Imperial State," in *Colonial Crucible: Empire in the Making of the Modern American State*, ed. Alfred W. McCoy and Francisco A. Scarano (Madison: University of Wisconsin Press, 2009), 3.
9. Until the work of revisionists, the prevailing view was that the US acquisition of the Philippines was a "great aberration," as Samuel Flagg Bemis wrote in *A Diplomatic History of the United States*, 5th ed. (New York: Holt, Rinehart, and Winston, 1965 [1936]), 163.
10. On Roosevelt and the gendered dimensions of imperial war in 1898 see Gail

Bederman, *Manliness & Civilization: A Cultural History of Gender and Race in the United States, 1880–1917* (Chicago: University of Chicago Press, 1995); Kristin L. Hoganson, *Fighting for American Manhood: How Gender Politics Provoked the Spanish-American and Philippine-American Wars* (New Haven: Yale University Press, 1998).

11. Theodore Roosevelt, quoted in Walter L. Williams, "United States Indian Policy and the Debate over Philippine Annexation: Implications for the Origins of American Imperialism," *Journal of American History* 66 (March 1980): 825. Leading works that have emphasized the racial dimensions of US imperialism circa 1898 include Paul A. Kramer, *The Blood of Government: Race, Empire, the United States, & the Philippines* (Chapel Hill: University of North Carolina Press, 2006); Vicente L. Rafael, *White Love and Other Events in Filipino History* (Durham: Duke University Press, 2000); Amy Kaplan, *The Anarchy of Empire in the Making of US Culture* (Cambridge: Harvard University Press, 2002); Eric Tyrone Lowery Love, *Race over Empire: Racism and US Imperialism, 1865–1900* (Chapel Hill: University of North Carolina Press, 2004); Stuart Creighton Miller, *"Benevolent Assimilation": The American Conquest of the Philippines, 1899–1903* (New Haven: Yale University Press, 1982).
12. Albert J. Beveridge, "'Our Philippine Policy,' Senator Beveridge's first public utterance on the Philippine situation after his return from a visit to the islands. Delivered in the Senate of the United States, January 9, 1900," in Albert Jeremiah Beveridge, *The Meaning of the Times, and Other Speeches* (Indianapolis: Bobbs-Merrill Company, 1908), 80.
13. Henry Cabot Lodge, *The Retention of the Philippine Islands. Speech of Henry Cabot Lodge of Massachusetts, in the Senate of the United States, March 7, 1900* (Washington, DC, 1900), 41.
14. Luis Francia, *A History of the Philippines: From Indios Bravos to Filipinos* (New York: Overlook Press, 2010), 74. On Spanish uses of Manila as an entrêpot, see also Robert Ronald Reed, *Hispanic Urbanism in the Philippines: A Study of the Impact of Church and State* (Manila: University of Manila, 1967); Robert Ronald Reed, *Colonial Manila: The Context of Hispanic Urbanism and Process of Morphogenesis* (Berkeley: University of California Press, 1978); Benito J. Legarda, *After the Galleons: Foreign Trade, Economic Change & Entrepreneurship in the Nineteenth Century Philippines* (Quezon City: Ateneo de Manila University Press, 1999); Elliott, *Empires of the Atlantic World*; Rainer F. Buschmann, Edward R. Slack, and James B. Tueller, *Navigating the Spanish Lake: The Pacific in the Iberian World, 1521–1898* (Honolulu: University of Hawai'i Press, 2014); William Lytle Schurz, *The Manila Galleon* (New York: E. P. Dutton, 1939); O. H. K. Spate, *The Spanish Lake* (Minneapolis: University of Minnesota Press, 1979).
15. Dennis O. Flynn and Arturo Giráldez, "Born with a 'Silver Spoon': The Origin of World Trade in 1571," *Journal of World History* 6 (1995): 201.
16. Ibid., 205.
17. Reed, *Hispanic Urbanism in the Philippines*, 110–111.

18. See Francia, *A History of the Philippines*, 75; Reed, *Hispanic Urbanism in the Philippines*, 123.
19. Reed, *Hispanic Urbanism in the Philippines*, 111.
20. See chapter 2, "Passage to India: Thomas Hart Benton and Asa Whitney," in Henry Nash Smith, *Virgin Land: The American West as Symbol and Myth* (Cambridge, MA: Harvard University Press, 1970 [1950]). "Passage to India" is the title of a poem by Walt Whitman; among other things, it reflects on the opening of the Suez Canal. For American views of the Pacific in this period, see Bruce Cumings, *Dominion from Sea to Sea: Pacific Ascendancy and American Power* (New Haven: Yale University Press, 2009); John R. Eperjesi, *The Imperialist Imaginary: Visions of Asia and the Pacific in American Culture* (Hanover, NH: Dartmouth College Press / University Press of New England, 2005); David Igler, *The Great Ocean: Pacific Worlds from Captain Cook to the Gold Rush* (Oxford: Oxford University Press, 2013).
21. Walter LaFeber, *The New Empire: An Interpretation of American Expansion, 1860–1898* (Ithaca: Cornell University Press, 1998 [1963]), 408.
22. For scholarship on early-twentieth-century US imperialism emphasizing economic interests besides LaFeber's *The New Empire*, see for examples William Appleman Williams, *The Tragedy of American Diplomacy*, 50th Anniversary Edition (New York: W. W. Norton & Company, 2009 [1959]); Thomas J. McCormick, *The China Market: America's Quest for Informal Empire, 1893–1901* (Chicago: Quandrangle Books, 1967); Emily S. Rosenberg, *Spreading the American Dream: American Economic and Cultural Expansion, 1890–1945* (New York: Hill and Wang, 1982); Thomas Schoonover, *Uncle Sam's War of 1898 and the Origins of Globalization* (Lexington: University of Kentucky Press, 2003); Cyrus Veeser, *A World Safe for Capitalism Dollar Diplomacy and America's Rise to Global Power* (New York: Columbia University, 2002); Smith, *American Empire*.
23. Smith, *American Empire*, 184.
24. Ibid., 19.
25. Ibid., 2.
26. John Gallagher and Ronald Robinson, "The Imperialism of Free Trade," *Economic History Review*, n.s., 6, no. 1 (January 1, 1953): 2.
27. See chapter 1, "Years of Preparation, 1860–1889" in LaFeber, *The New Empire*.
28. Ann Laura Stoler, "On Degrees of Imperial Sovereignty," *Public Culture* 18 (Winter 2006): 128, 136. On sovereignty as a continuous project see also Lauren A. Benton, *A Search for Sovereignty: Law and Geography in European Empires, 1400–1900* (New York: Cambridge University Press, 2010).
29. Kramer, "Power and Connection," 1375.
30. Young, *Postcolonialism*, 16–17.
31. Stoler, "On Degrees of Imperial Sovereignty;" Kramer, "Power and Connection," 1374–75.
32. Gareth Stedman Jones describes twentieth-century US empire as "invisible" in "intention at least" in "The Specificity of US Imperialism," *New Left*

*Review* 1, no. 60 (April 1970): 63. "Imperial Anticolonialism" is the title of chapter 1 of Williams's *The Tragedy of American Diplomacy*.

33. Scholarship that also grounds the US empire in place includes Catherine Lutz, *Homefront: A Military City and the American 20th Century* (Boston: Beacon, 2001); Mark L. Gillem, *America Town: Building the Outposts of Empire* (Minneapolis: University of Minnesota Press, 2007); Ron Robin, *Enclaves of America: The Rhetoric of American Political Architecture Abroad, 1900–1965* (Princeton: Princeton University Press, 1992); Andrew Friedman, *Covert Capital: Landscapes of Denial and the Making of US Empire in the Suburbs of Northern Virginia* (Berkeley: University of California Press, 2013). On the intersections of US imperialism, labor, and racial formation see Julie Greene, *The Canal Builders: Making America's Empire at the Panama Canal* (New York: Penguin Press, 2009); Greg Grandin, *Fordlandia* (New York: Picador, 2010); Jana Lipman, *Guantánamo: A Working-Class History between Empire and Revolution* (Berkeley: University of California Press, 2008); Matthew Frye Jacobson, *Barbarian Virtues: The United States Encounters Foreign Peoples at Home and Abroad, 1876–1917* (New York: Hill and Wang, 2000); Daniel E. Bender, *American Abyss: Savagery and Civilization in the Age of Industry* (Ithaca: Cornell University Press, 2009); David R. Roediger and Elizabeth D. Esch, *The Production of Difference: Race and the Management of Labor in US History* (New York: Oxford University Press, 2012); Jason M. Colby, *The Business of Empire: United Fruit, Race, and US Expansion in Central America* (Ithaca: Cornell University Press, 2013); Andrew Zimmerman, *Alabama in Africa: Booker T. Washington, the German Empire, and the Globalization of the New South* (Princeton: Princeton University Press, 2010); Daniel Bender and Jana Lipman, eds., *Making the Empire Work: Labor and United States Imperialism* (New York: New York University Press, 2015). On the intersection of technological innovations and the US civilizing mission see Michael Adas, *Dominance by Design: Technological Imperatives and America's Civilizing Mission* (Cambridge: Harvard University Press, 2006).

34. On the transnational reach of Progressives see Daniel T. Rodgers, *Atlantic Crossings: Social Politics in a Progressive Age* (Cambridge: Belknap Press of Harvard University, 2000); Ian R. Tyrrell, *Reforming the World: The Creation of America's Moral Empire* (Princeton: Princeton University Press, 2010); Emily S. Rosenberg, *Financial Missionaries to the World: The Politics and Culture of Dollar Diplomacy, 1900–1930* (Cambridge: Harvard University Press, 1999); Eileen Findlay, *Imposing Decency: The Politics of Sexuality and Race in Puerto Rico, 1870–1920* (Durham: Duke University Press, 1999); Laura Briggs, *Reproducing Empire: Race, Sex, Science, and US Imperialism in Puerto Rico* (Berkeley: University of California Press, 2002). See also William E. Leuchtenburg, "Progressivism and Imperialism: The Progressive Movement and American Foreign Policy, 1898–1916," *Mississippi Valley Historical Review* 39 (December 1, 1952): 483–504; Howard Gillette, "The Military Occupation of Cuba, 1899–1902: Workshop for American Progressivism," *American Quarterly* 25 (1973):

410–25; Joseph M. Siracusa, "Progressivism, Imperialism, and the Leuchtenburg Thesis, 1952–1974: An Historiographical Appraisal," *Australian Journal of Politics & History* 20 (December 1, 1974): 312–25.

35. Amy Chazkel and David Serlin, "Editors' Introduction," *Radical History Review* (December 2011): 1. On enclosure see also that whole issue of *Radical History Review* (December 2011); Midnight Notes Collective, "The New Enclosures," *Midnight Notes* 10 (1990), http://www.midnightnotes.org/newenclos.html; Silvia Federici, *Caliban and the Witch* (Brooklyn, NY: Autonomedia, 2004); Jim Glassman, "Primitive Accumulation, Accumulation by Dispossession, Accumulation by 'Extra-Economic' Means," *Progress in Human Geography* 30 (2006): 608–25; Harvey, "The 'New' Imperialism"; Michael Perelman, *The Invention of Capitalism: Classical Political Economy and the Secret History of Primitive Accumulation* (Durham: Duke University Press, 2000); Karl Jacoby, *Crimes against Nature: Squatters, Poachers, Thieves, and the Hidden History of American Conservation* (Berkeley: University of California Press, 2001); Greg Bankoff, *Crime, Society, and the State in the Nineteenth-Century Philippines* (Quezon City: Ateneo de Manila University Press, 1996).

36. I calculate this based on the description of the land claimed by the insular government in February 1903 in No. 636, "An Act creating a Government reservation at Baguio, in the Province of Benguet": "That parcel or tract of land in the form of a circle with its center in the house occupied by Mateo Cariño at Baguio, and with a radius of one kilometer; and also a strip of land one and one-half kilometers wide on the easterly side, and one kilometer wide on the westerly side of the Government road due east of the civil sanitarium, and extending southeasterly along said road for a distance of four kilometers." See *Philippine Islands Public Laws, with Amendments Indicated* (Manila: Bureau of Public Printing, 1904), 296. By 1907, the reservation had grown to cover 21.6 square miles. See *Eighth Annual Report of the Philippine Commission to the Secretary of War, 1907 (In Three Parts), Part 2* (Washington, DC: Government Printing Office, 1908), 206.

37. *Fifth Annual Report of the Philippine Commission, In Three Parts, Part 3* (Washington, DC: Government Printing Office, 1905), 407.

38. Finin, *The Making of the Igorot*, 8.

39. Frederick L. Wernstedt and J. E. Spencer, *The Philippine Island World: A Physical, Cultural, and Regional Geography* (Berkeley: University of California Press, 1967), 18; Otto Scheerer, "Exhibit F. The Igorrotes of Benguet (North Luzon)," in *Reports of the Taft Philippine Commission: Message from the President of the United States, Transmitting a Report of the Secretary of War, Containing the Reports of the Taft Commission, Its Several Acts of Legislation, and Other Important Information Relating to the Condition and Immediate Wants of the Philippine Islands* (Washington, DC: Government Printing Office, 1901), 149.

40. Philippine Commission and William Howard Taft, *Reports of the Taft Philippine Commission* (1901), 66. On the population of Benguet, see United States Bureau of the Census, *Census of the Philippine Islands, Taken under the Direc-*

*tion of the Philippine Commission in the Year 1903, in Four Volumes, Vol. 1, Part 1* (Washington, DC: United States Bureau of the Census, 1905), http://name.umdl.umich.edu/ajb5834.0001.001, 533.

41. Rowena Reyes-Boquiren, "Baguio's History and Cultural Heritage," *Northern Dispatch Weekly*, August 23, 2015, http://www.nordis.net/2015/08/baguios-history-and-cultural-heritage/.
42. Lazaro P. Gutierrez, *Memoirs of Baguio* (Baguio City: Summer Capital Pub. House, 1960), 11.
43. William Henry Scott, "The Word Igorot," in *On the Cordillera*, 156–57.
44. Otto Scheerer citing Pardo de Tavera, in *The Nabaloi Dialect* (Manila: Bureau of Public Printing, 1905), 97; Scott, "The Word Igorot," 164.
45. Scott, "The Word Igorot," 164.
46. Ibid., 156.
47. Benedict Anderson, *Under Three Flags: Anarchism and the Anti-Colonial Imagination* (New York: Verso, 2005), 15.
48. Michael Salman, *The Embarrassment of Slavery: Controversies over Bondage and Nationalism in the American Colonial Philippines* (Berkeley: University of California Press, 2001), 11.
49. George A. Malcolm, *The Government of the Philippine Islands: Its Development and Fundamentals* (Rochester, NY: Lawyers Co-operative Publishing Company, 1916), 287.
50. Daniel H. Burnham to Charles Moore, March 13, 1905, in Charles Moore, *Daniel H. Burnham: Architect, Planner of Cities* (New York: Da Capo Press, 1968), 245.
51. Woolley, "Baguio, Simla of the Philippines," 292.
52. On British and European hill stations see Anthony D. King, *Colonial Urban Development: Culture, Social Power, and Environment* (London: Routledge & Kegan Paul, 1976); Barbara Crossett, *The Great Hill Stations of Asia* (Boulder, CO: Westview Press, 1998); Dane Kennedy, *The Magic Mountains: Hill Stations and the British Raj* (Berkeley: University of California Press, 1996); Robert S. Aiken, *Imperial Belvederes: The Hill Stations of Malaya* (Kuala Lumpur: Oxford University Press, 1994); Thomas R. Metcalf, *An Imperial Vision: Indian Architecture and Britain's Raj* (Berkeley: University of California Press, 1989); Eric T. Jennings, *Curing the Colonizers: Hydrotherapy, Climatology, and French Colonial Spas* (Durham: Duke University Press, 2006); Eric T. Jennings, *Imperial Heights: Dalat and the Making and Undoing of French Indochina* (Berkeley: University of California Press, 2012).
53. Kennedy, *The Magic Mountains*, 12.
54. *Reports of the Taft Philippine Commission* (1901), 147.
55. William Cameron Forbes to E. H. Harriman, November 26, 1908, MS Am 1366, W. Cameron Forbes Collection, Houghton Library, Harvard University (hereafter "Forbes Papers").
56. "The Vindication of the Benguet Road," March 9, 1908, bMS Am 1364.4, Forbes Papers.

57. Woolley, "Baguio, Simla of the Philippines," 292.
58. Daniel F. Doeppers, "Manila's Imperial Makeover: Security, Health, and Symbolism," in *Colonial Crucible*, ed. McCoy and Scarano, 498.
59. On the military reservation, Camp John Hay, see Gonzalez, *Securing Paradise*; Agnir-Paraan and Buenviaje-Wilder, eds., *Camp John Hay*.
60. Leo Marx, *The Machine in the Garden: Technology and the Pastoral Ideal in America, Twenty-fifth Anniversary Edition* (New York: Oxford University Press, 2000 [1964]), 22.
61. Ibid., 23.
62. Frank G. Carpenter, *Through the Philippines and Hawaii* (Garden City, NY: Doubleday, Page & Co., 1925), 85.
63. "Ho: For Baguio," *Manila Times* (re-typed clipping), March 23, 1908, bMS Am 1364.4, Forbes Papers.
64. William James, "The Philippine Tangle," *Boston Evening Transcript*, March 1, 1899.
65. Sarah Burns, *Pastoral Inventions: Rural Life in Nineteenth-Century American Art and Culture* (Philadelphia: Temple University Press, 1989), 8.
66. Raymond Williams, *The Country and the City* (New York: Oxford University Press, 1973), 23, 14. There is a wide literature on the pastoral form. Scholarship that has informed my understanding of the pastoral includes Williams, *The Country and the City*; Lawrence Buell, *The Environmental Imagination: Thoreau, Nature Writing, and the Formation of American Culture* (Cambridge: Belknap Press of Harvard University Press, 1995); Annabel M. Patterson, *Pastoral and Ideology: Virgil to Valéry* (Berkeley: University of California Press, 1987); Burns, *Pastoral Inventions*; Louise A. Mozingo, *Pastoral Capitalism: A History of Suburban Corporate Landscapes* (Cambridge: MIT Press, 2011); Meredith Martin, *Dairy Queens: The Politics of Pastoral Architecture from Catherine d'Medici to Marie-Antoinette* (Cambridge: Harvard University Press, 2011); Guy Alan Szuberla, "Urban Vistas and the Pastoral Garden: Studies in the Literature and Architecture of Chicago (1893–1909)" (PhD diss., University of Minnesota, 1971); Jeanne Boydston, *Home and Work: Housework, Wages, and the Ideology of Labor in the Early Republic* (New York: Oxford University Press, 1990).
67. Williams, *The Country and the City*, 289, 294, 23, 22, 17, 18, 18, 20.
68. Ibid., 45–46, 25–36, 46.
69. Marx, *The Machine in the Garden*, 26, 19, 6.
70. Young-tsu Wong, *A Paradise Lost: The Imperial Garden Yuanming Yuan* (Honolulu: University of Hawai'i Press, 2001), 11.
71. Martin, *Dairy Queens*, 7–8, 11.
72. Chandra Mukerji, *Territorial Ambitions and the Gardens of Versailles* (Cambridge: Cambridge University Press, 1997), 1, 2, 35.
73. Mozingo, *Pastoral Capitalism*, 21.
74. Dean Conant Worcester, *The Philippines Past and Present*, vol. 1 (New York: Macmillan, 1914), 453.

75. Williams, *The Country and the City*, 45.
76. Ibid., 279.
77. Ibid., 294.
78. Rosalind C. Morris, "Imperial Pastoral: The Politics and Aesthetics of Translation in British Malaya," *Representations* (Summer 2007): 164, 160, 161, 159, 161.
79. My understanding and use of "space" and "place" have been shaped by the following: Gaston Bachelard, *The Poetics of Space*, translated by Maria Jolas (Boston: Beacon Press, 1964); Michel de Certeau, *The Practice of Everyday Life*, translated by Steven Rendell (Berkeley: University of California Press, 1988 [1984]); Yi-Fu Tuan, *Space and Place: The Perspective of Experience* (Minneapolis: University of Minnesota Press, 1977); David Harvey, *Spaces of Capital: Towards a Critical Geography* (New York: Routledge, 2001); Henri Lefebvre, *The Production of Space*, translated by Donald Nicholson-Smith (Malden, MA: Blackwell Publishing, 1991 [1974]); Setha M. Low and Denise Lawrence-Zúñiga, eds., *The Anthropology of Space and Place: Locating Culture* (Malden, MA: Blackwell Publishing, 1988); Edward W. Soja, *Postmodern Geographies: The Reassertion of Space in Critical Social Theory* (London: Verso, 2003 [1989]).
80. The literature on Progressivism is vast; studies that have informed my understanding of reform and reformers include Rodgers, *Atlantic Crossings*; Daniel T. Rodgers, "In Search of Progressivism," *Reviews in American History* 10, no. 4 (1982): 113–32; James T. Kloppenberg, *Uncertain Victory: Social Democracy and Progressivism in European and American Thought, 1870–1920* (New York: Oxford University Press, 1986); Robert D. Johnston, "Re-Democratizing the Progressive Era: The Politics of Progressive Era Political Historiography," *Journal of the Gilded Age and Progressive Era* 1 (2002): 68–92; Robert D. Johnston, *The Radical Middle Class: Populist Democracy and the Question of Capitalism in Progressive Era Portland, Oregon* (Princeton: Princeton University Press, 2003); Martin J. Sklar, *The Corporate Reconstruction of American Capitalism, 1890–1916: The Market, the Law, and Politics* (New York: Cambridge University Press, 1988); Michael E. McGerr, *A Fierce Discontent: The Rise and Fall of the Progressive Movement in America, 1870–1920* (New York: Free Press, 2003); Robert H. Wiebe, *The Search for Order, 1877–1920* (New York: Hill and Wang, 1967); Leuchtenburg, "Progressivism and Imperialism: The Progressive Movement and American Foreign Policy, 1898–1916"; Siracusa, "Progressivism, Imperialism, and the Leuchtenburg Thesis, 1952–1974"; Paul A. Kramer, "Reflex Actions: Colonialism, Corruption and the Politics of Technocracy in the Early 20th Century United States," in *Challenging US Foreign Policy: America and the World in the Long Twentieth Century*, ed. Bevan Sewell and Scott Lucas (New York: Palgrave Macmillan, 2011); Patricio N. Abinales, "Progressive-Machine Conflict in Early-Twentieth-Century US Politics and Colonial-State Building in the Philippines," in *The American Colonial State in the Philippines: Global Perspectives*, ed. Julian Go and Anne L. Foster (Durham: Duke University Press, 2003).

81. I borrow the concept "region of refuge" from Gonzalo Aguirre Beltrán and Deward E. Walker, *Regions of Refuge* (Washington, DC: Society for Applied Anthropology, 1979). On the hills and mountains as "regions of refuge" see also James C. Scott, *The Art of Not Being Governed: An Anarchist History of Upland Southeast Asia* (New Haven: Yale University Press, 2009).
82. Bagamaspad, Hamada-Pawid, and Balangoy, *A Peoples' History of Benguet Province*, 41. See also chapter 1 of Wiber, *Politics, Property and Law in the Philippine Uplands*. On Baguio pasture, see also Tapang, *Innovation and Social Change*, 42.
83. Nicanor G. Tiongson and Crispin Dayao, Jr., *Alpabetong Filipino* (Manila: Tahanan, 2013).
84. Karl Marx, *Capital, Volume One*, excerpted in *The Marx-Engels Reader, Second Edition*, ed. Robert C. Tucker (New York: W. W. Norton & Co.), 434–36. See also Massimo De Angelis, "Marx and Primitive Accumulation: The Continuous Character of Capital's 'Enclosures,'" *Commoner* 2 (September 2001).
85. Benedict Anderson, "Cacique Democracy and the Philippines: Origins and Dreams," *New Left Review* (May/June 1988): 3–31.
86. The quoted phrase is that of Sam Gindin and Leo Panitch in *The Making of Global Capitalism: The Political Economy of American Empire* (New York: Verso, 2012), 39.
87. David Harvey, *The Limits to Capital* (Chicago: University of Chicago Press, 1982), 233.
88. To the extent that this project concerns the state and political techniques that produced subjects of rule—as much as opportunities for resistance—it implicates another notion of the pastoral. In his study of governmentality, Michel Foucault has written on the significance of the pastoral, or, more precisely, the Christian pastorate. He identifies the Christian "pastorate as the source of a specific type of power over men, as a model and matrix of procedures for the government of men," the latter form emergent in the sixteenth and seventeenth centuries (Foucault 147). The Christian pastorate, he argues, "gave rise to an art of conducting, directing, leading, guiding, taking in hand, and manipulating men, an art of monitoring them and urging them on step by step, an art with the function of taking charge of men collectively and individually throughout their life and at every moment of their existence"(Foucault 165). This pastoral form of power at once totalizes, in caring for the flock, and also individualizes. It targets, for example, the individual conscience and, in this a way, is implicated in "the history of the subject," or subjectification (Foucault 184–85). It is this pastorate, and its "art," that Foucault understands as a "prelude" to the constellation of "institutions, procedures, analyses and reflections, calculations, and tactics" associated with the "governmentalization of the state" (Foucault 184, 108, 184). Governmentality thus becomes a way to understand the connection between "the constitution of the subject and the formation of the state" (Lemke 2).

    Foucault writes that the "transition from the pastoral of souls to the political government of men"—when the sovereign comes to undertake more than the challenge of sovereignty alone—unfolds in the "general context of resistances, revolts, and insurrections of conduct," not least of which was the Protestant Reformation (Foucault 228). In this regard, the significance of Foucault's notion of the pastoral for this project is in part theoretical; it is a component part of his conception of governmentality, and analyzing colonial practices of rule is a concern of this study. The significance of Foucault's notion of pastoral power has historical bearing, too. The period with which he identifies the transition to the "government of men" was the period in which the Spanish undertook the conquest of the Philippines and Spanish missionaries attempted to convert Philippine peoples, including the Igorots, into tractable, faithful subjects of the Crown. Vicente Rafael has written on the problem of language and translation in these attempts at religious conversion, noting the relationship between pastoral power and colonial governmentality. "Catholicism not only provided Spain's colonial enterprise with its ideological frame," he writes, "it also embedded the structure of colonial rule within the practice of religious conversion. Indeed, it was precisely the stated priority of conversion that provided the perspective from which to criticize or affirm colonial rule" (Rafael, *Contracting Colonialism*, 17). Moving forward in time, Rafael has reflected on the combination of "love and discipline" in US colonialism, too, identifying in American colonial rhetoric "the existence of a dominant desire informing the state: that of creating a continuum between an ideology of benevolence, disciplinary practices, and networks of supervision—in short, a desire to consolidate the relay between knowledge and power" (Rafael, *White Love*, 23–24). In these respects, Foucault's notion of pastoral power—a sort of "affective" power—formed a constituent part of the colonial rule in question here. Michel Foucault, *Security, Territory, Population: Lectures at the Collège De France, 1977–78*, ed. Michel Senellart, trans. Graham Burchell (New York: Picador, 2009); Thomas Lemke, "Foucault, Governmentality, and Critique," *Rethinking Marxism* 14 (2002): 49–64; Vicente Rafael, *Contracting Colonialism: Translation and Christian Conversion in Tagalog Society under Early Spanish Rule* (Ithaca: Cornell University Press, 1988); Rafael, *White Love*.

89. In the Philippine context, see for examples Hoganson, *Fighting for American Manhood*; Kramer, *The Blood of Government*; Rafael, *White Love*; Anderson, *Colonial Pathologies*; Love, *Race over Empire*; Kaplan, *The Anarchy of Empire in the Making of US Culture*; Kaplan and Pease, *Cultures of United States Imperialism*.
90. For recent examples see Alfred W. McCoy, *Policing America's Empire: The United States, the Philippines, and the Rise of the Surveillance State* (Madison: University of Wisconsin Press, 2009); McCoy and Scarano, eds., *Colonial Crucible*; Go and Foster, eds., *The American Colonial State in the Philippines*; P. N. Abinales and Donna J. Amoroso, *State and Society in the Philippines* (Lanham, MD: Rowman & Littlefield Publishers, 2005).

91. McCoy and Scarano, eds., *Colonial Crucible*, 13, 5.
92. Fernando Coronil, *The Magical State: Nature, Money, and Modernity in Venezuela* (Chicago: University of Chicago Press, 1997), 15. My attempt at a "cultural materialist" analysis is also informed by work like Mukerji's *Territorial Ambitions and the Gardens of Versailles*; her book entails "unearthing fundamental cultural dimensions of material relations whose consequences are not so much mediated through thought or language as located in an ordering of the material world itself" (36). Other works that have helped me think through ways of weaving together material, social, cultural, and political analyses of imperial and colonial power include Rosenberg, *Financial Missionaries to the World*; Jacobson, *Barbarian Virtues*; Eperjesi, *The Imperialist Imaginary*; Aihwa Ong, *Flexible Citizenship: The Cultural Logics of Transnationality* (Durham: Duke University Press, 1999).
93. Keith H. Basso, *Wisdom Sits in Places: Landscape and Language among the Western Apache* (Albuquerque: University of New Mexico Press, 1996), 106.
94. Ibid., 7, 5–6, 7.
95. Jeffrey Sklansky, "The Elusive Sovereign: New Intellectual and Social Histories of Capitalism," *Modern Intellectual History* 9 (2012): 246.
96. Thanks to Jean-Christophe Agnew who made this observation of my project's narrative.

CHAPTER ONE

1. "Philippinitis," Mudcat, http://mudcat.org/@displaysong.cfm?SongID=4667. See also Thomas P. Walsh, *Tin Pan Alley and the Philippines: American Songs of War and Love, 1898–1946, A Resource Guide* (Lanham, MD: Scarecrow Press, 2013).
2. H. Kemlein, *Kemlein & Johnson's Guide and Map of Manila and Vicinity: A Hand Book Devoted to the Interests of the Traveling Public* (Manila: Kemlein & Johnson, 1908), https://archive.org/details/kemleinjohnsons00kemlgoog, 37–38.
3. Mary Kilcline Cody, "A Paler Shade of White," in *Lost Times and Untold Tales from the Malay World*, ed. Jan Van Der Putten and Mary Kilcline Cody (Singapore: NUS Press, 2009), 91. Ann Laura Stoler discusses fears of white colonial "degeneracy" in chapter 3 of *Carnal Knowledge and Imperial Power*. In the US colonial Philippine context, see Anderson, *Colonial Pathologies*.
4. Liah Greenfeld, *Mind, Modernity, Madness: The Impact of Culture on Human Experience* (Cambridge: Harvard University Press), 606.
5. Louis H. Fales, MD, "Tropical Neurasthenia and Its Relation to Tropical Acclimation," *American Journal of Medical Sciences* 133 (April 1907): 583–84, 583, 591.
6. Anderson, *Colonial Pathologies*, 14, 76, 78.
7. Frederick L. Wernstedt and J. E. Spencer, *The Philippine Island World: A Physical, Cultural, and Regional Geography* (Berkeley: University of California Press, 1967), 18.

8. Don Enrique Hore et al., "Exhibit H. The Benguet Sanitarium," in *Reports of the Taft Philippine Commission* (1901), 162.
9. Otto Scheerer, "Exhibit F. The Igorrotes of Benguet (North Luzon)," in *Reports of the Taft Philippine Commission* (1901), 155, 152.
10. John Leddy Phelan, *The Hispanization of the Philippines: Spanish Aims and Filipino Responses, 1565–1700* (Madison: University of Wisconsin Press, 1959), 44.
11. Scheerer, "Exhibit F. The Igorrotes of Benguet (North Luzon)," in *Reports of the Taft Philippine Commission* (1901), 152.
12. On urbanism as a signal of civilization to the Spanish, see Phelan, *The Hispanization of the Philippines*, 44.
13. See Scott, *The Discovery of the Igorots*, 240; and Keesing, *The Ethnohistory of Northern Luzon*, 2.
14. Scheerer, *The Nabaloi Dialect*, 97.
15. Keesing, *The Ethnohistory of Northern Luzon*, 90–91; see also Scott, *The Art of Not Being Governed*, 25.
16. Bagamaspad, Hamada-Pawid, and Balangoy, *A Peoples' History of Benguet Province*, 34.
17. Benton, *A Search for Sovereignty*, 3, 228.
18. "Report of the Civil Governor," November 1, 1902, in *Reports of the Philippine Commission, the Civil Governor, and the Heads of the Executive Departments of the Civil Government of the Philippines Islands (1900–1903)* (Washington, DC: Government Printing Office, 1904), 299, 297.
19. Scott, *The Art of Not Being Governed*, 22, 54.
20. On Benguet's elites see Wiber, *Politics, Property and Law in the Philippine Uplands*.
21. On the Spanish sanitarium in La Trinidad see Reed, *City of Pines*, 31–48.
22. "Philippinitis," Mudcat, http://mudcat.org/@displaysong.cfm?SongID=4667.
23. On changing Western cultural attitudes toward mountains see Marjorie Hope Nicolson, *Mountain Gloom and Mountain Glory: The Development of the Aesthetics of the Infinite* (Ithaca: Cornell University Press, 1959).
24. Robert R. Rudd Papers, Schomburg Center for Research in Black Culture, New York Public Library (hereafter "Rudd Papers").
25. Brian McAllister Linn, *The Philippine War, 1899–1902* (Lawrence: University Press of Kansas, 2000), 9.
26. Ibid., 10.
27. "President's Instructions and Proclamation of General Otis," in *Elihu Root Collection of United States Documents Relating to the Philippine Islands* (Washington, DC: Government Printing Office, 1902), 776.
28. See chapter 1 of Linn, *The Philippine War, 1899–1902*, on the early US military occupation.
29. Ibid., 261.
30. Willard B. Gatewood, *"Smoked Yankees" and the Struggle for Empire: Letters from Negro Soldiers, 1898–1902* (Urbana: University of Illinois Press, 1971), 241.

31. On La Trinidad see Reed, *City of Pines*.
32. Brian McAllister Linn, *The US Army and Counterinsurgency in the Philippine War, 1899–1902* (Chapel Hill: University of North Carolina Press, 1989), 30.
33. Linn, *The Philippine War, 1899–1902*, 261.
34. Ibid.
35. Letter from William Penn Duvall to Rudd, May 18, 1900, Reel 1, Rudd Papers.
36. Linn, *The Philippine War, 1899–1902*, 146–55.
37. Letter from Duvall to Rudd, May 18, 1900, Reel 1, Rudd Papers.
38. Daniel R. Williams, *The Odyssey of the Philippine Commission* (Chicago: A. C. McClurg & Co., 1913), 260.
39. Horace Higgins, interviewed by the Philippine Commission July 19, 1899, General Records of the Department of State, Records of the U.S. Commission to the Philippine Islands, 1898–1909, Minutes and Stenographers' Notes of Meetings of the Commission, National Archives (hereafter NARA), RG 59, Entry 1027, 250: 48/28/01–02, Box 2, College Park, MD.
40. F. H. Donaldson-Sim, interviewed by the Philippine Commission July 25, 1899, General Records of the Department of State, Records of the U.S. Commission to the Philippine Islands, 1898–1909, Minutes and Stenographers' Notes of Meetings of the Commission, NARA, RG 59, Entry 1027, 250: 48/28/01–02, Box 2.
41. Ibid.
42. The population figure comes from Scheerer, *The Nabaloi Dialect*, 99.
43. Jose Camps, interviewed by the Philippine Commission July 26, 1899, General Records of the Department of State, Records of the U.S. Commission to the Philippine Islands, 1898–1909, Minutes and Stenographers' Notes of Meetings of the Commission, NARA, RG 59, Entry 1027, 250: 48/28/01–02, Box 2.
44. Ibid.
45. F. H. Donaldson-Sim, interviewed by the Philippine Commission July 25, 1899.
46. Jose Camps, interviewed by the Philippine Commission July 26, 1899.
47. "Exhibit D. Report of the committee appointed to investigate the suitableness of the township of Baguio, province of Benguet, for a health resort," in *Reports of the Taft Philippine Commission* (1901).
48. Ibid., 123.
49. Williams, *The Odyssey of the Philippine Commission*, 261.
50. "Exhibit D," in *Reports of the Taft Philippine Commission* (1901), 123–24.
51. Letter from Duvall to Robert Rudd, May 18, 1900, Reel 1, Rudd Papers.
52. "Exhibit D," in *Reports of the Taft Philippine Commission* (1901), 147.
53. *Reports of the Taft Philippine Commission* (1901), 42.
54. "Exhibit D," in *Reports of the Taft Philippine Commission* (1901), 146–47, 124, 125.
55. Otto Scheerer, *On Baguio's Past: Chapters from Local History and Tradition* (Manila: Otto Scheerer, 1933), 27.

56. Scott, *The Discovery of the Igorots*, 213.
57. "'The Ibaloi Igorot Seventy-Five Years Ago,' Account of a Spanish Expedition to Benguet in the Year 1829" [Translated from "Informe Sobre el Estado de las Islas Filipinas en 1842" (by S. Mas), Madrid, 1842: Diary of Don G. Galvey, in command of the forces for the suppression of contraband trade], in Scheerer, *The Nabaloi Dialect*, 176.
58. Manuel Scheidnagel, *Distrito de Benguet: Filipinas: Memoria Descriptiva y Económica, Acompañada del Primer Plano-Cróquis del Mismo* (Madrid: Imp. de la Dir. Gral. de Infantería, 1878), 73.
59. Rafael, *Contracting Colonialism*, 151.
60. Daniel Nemser, "Primitive Accumulation, Geometric Space, and the Construction of the 'Indian,'" *Journal of Latin American Cultural Studies* 24 (2015): 335–52.
61. Foucault, *Security, Territory, Population*, 147.
62. Nemser, "Primitive Accumulation, Geometric Space, and the Construction of the 'Indian,'" 338, 337, 348, 349, 348.
63. Melanie Wiber, "The Cañao Imperative: Changes in Resource Control, Stratification and the Economy of Ritual among the Ibaloi of Northern Luzon," in *Changing Lives, Changing Rites: Ritual and Social Dynamics in Philippine and Indonesian Uplands*, eds. Susan Russell and Clark Cunningham (Ann Arbor: Michigan Papers in South and Southeast Asia Monograph Series, University of Michigan), 47.
64. Scott, *The Art of Not Being Governed*, 181.
65. Wiber, "The Cañao Imperative," 47.
66. Scheidnagel, *Distrito de Benguet*, 22.
67. Bagamaspad, Hamada-Pawid, and Balangoy, *A Peoples' History of Benguet Province*, 77.
68. William Henry Scott, "Igorot Responses to Spanish Aims: 1576–1896," in *Resistance and Revolution in the Cordillera*, ed. Tolentino, 9–10.
69. On de Morga see Timothy Brook, *The Confusions of Pleasure: Commerce and Culture in Ming China* (Berkeley: University of California Press, 1999).
70. Scott, "Igorot Responses to Spanish Aims: 1576–1896," in *Resistance and Revolution in the Cordillera*, ed. Tolentino, 10.
71. Bagamaspad, Hamada-Pawid, and Balangoy, *A Peoples' History of Benguet Province*, 77.
72. Padre Vivar quoted in Scheerer, *On Baguio's Past*, 20.
73. Finin, *The Making of the Igorot*, 10-11.
74. Francia, *A History of the Philippines*, 128.
75. Eric A. Anderson, "The Encomienda in Early Philippine Colonial History," *Asian Studies* 14 (1976): 26.
76. Abinales and Amoroso, *State and Society in the Philippines*, 62, 55, 62.
77. Scheidnagel, *Distrito de Benguet*, 12–13.
78. De Morga quoted in Scheerer, *On Baguio's Past*, 37.
79. Scheidnagel, *Distrito de Benguet*, 12–13.

80. On trade exceeding the "range" of "political integration," see Scott, *The Art of Not Being Governed*, 51.
81. Scott, *The Discovery of the Igorots*, 13–14.
82. Antonio de Morga, *Sucesos de Las Islas Filipinas por El Doctor Antonio de Morga, Obra Publicada en Méjico El Año de 1609. Nuevamente Sacada à Luz y Anotada por José Rizal y Precedida de Un Prólogo del Prof. Fernando Blumentritt* (Paris: Garnier Hermanos, 1890), http://name.umdl.umich.edu/ahz9387.0001.001, 278–79.
83. Melanie Wiber, "Levels of Property Rights, Levels of Law: A Case Study from the Northern Philippines," *Man* 26 (1991): 477; Wiber, *Politics, Property and Law in the Philippine Uplands*, 33, 124; Tapang, *Innovation and Social Change*, 7.
84. Wiber, "The Cañao Imperative," 47; Wiber, "Levels of Property Rights, Levels of Law," 477; Wiber, *Politics, Property and Law in the Philippine Uplands*, 127.
85. Wiber, "Levels of Property Rights, Levels of Law," 477.
86. Scheerer, "Exhibit F. The Igorrotes of Benguet (North Luzon)," in *Reports of the Taft Philippine Commission* (1901), 152.
87. Bagamaspad, Hamada-Pawid, and Balangoy, *A Peoples' History of Benguet Province*, 41, 137. See also Tapang, *Innovation and Social Change*.
88. Vivar quoted in Scheerer, *On Baguio's Past*, 12.
89. Wiber, *Politics, Property and Law in the Philippine Uplands*, 56, 113.
90. Ibid., 33.
91. Wiber, "Levels of Property Rights, Levels of Law," 474.
92. Scheerer, "Exhibit F. The Igorrotes of Benguet (North Luzon)," in *Reports of the Taft Philippine Commission* (1901), 152.
93. Wiber, *Politics, Property and Law in the Philippine Uplands*, 129.
94. Wiber, "Levels of Property Rights, Levels of Law," 476; Bagamaspad, Hamada-Pawid, and Balangoy, *A Peoples' History of Benguet Province*, 212.
95. Bagamaspad, Hamada-Pawid, and Balangoy, *A Peoples' History of Benguet Province*, 114. On the cañao, see also Tapang, *Innovation and Social Change*, 8; Wiber, "The 'Cañao Imperative."
96. Tapang, *Innovation and Social Change*, 25.
97. Ibid., 26.
98. Wiber, "The 'Cañao Imperative,'" 51; Wiber, *Politics, Property and Law in the Philippine Uplands*, 127.
99. John D. Blanco, *Frontier Constitutions: Christianity and Colonial Empire in the Nineteenth-Century Philippines* (Berkeley: University of California Press, 2009), 37–38.
100. Ibid., 21.
101. Abinales and Amoroso, *State and Society in the Philippines*, 95. See also Blanco, *Frontier Constitutions*.
102. William Henry Scott, *Of Igorots and Independence: Two Essays* (Baguio City: ERA, 1993), 34; Vicente L. Rafael, *The Promise of the Foreign: Nationalism and the Technics of Translation in the Spanish Philippines* (Durham: Duke University Press, 2005), 6.

103. Scott, *The Discovery of the Igorots*, 211.
104. On Spanish attempts to control the economy see Rafael, *The Promise of the Foreign*, 5-6.
105. Scott, *The Discovery of the Igorots*, 213–214.
106. Ibid., 213–214.
107. Scott, "Igorot Responses to Spanish Aims: 1576–1896," in *Resistance and Revolution in the Cordillera*, ed. Tolentino, 25–26.
108. Scott, *The Discovery of the Igorots*, 226.
109. Ibid., 224.
110. Ibid., 233.
111. Ibid., 237.
112. Cariño, "The Cariños and Baguio-Benguet History: Conclusion," in *Folio 2*, 55.
113. Scheerer, "Exhibit F. The Igorrotes of Benguet (North Luzon)," in *Reports of the Taft Philippine Commission* (1901), 154–55.
114. Scott, *The Discovery of the Igorots*, 250.
115. Scheerer, "Exhibit F. The Igorrotes of Benguet (North Luzon)," in *Reports of the Taft Philippine Commission* (1901), 155, 152.
116. Scott, *The Discovery of the Igorots*, 295.
117. Ibid., 239.
118. Wiber, *Politics, Property and Law in the Philippine Uplands*, 127.
119. Scheerer, *On Baguio's Past*, 42.
120. Ibid., 43.
121. Ibid., 41.
122. Ibid., 30, 40; Scott, *The Discovery of the Igorots*, 117.
123. Scheerer, *On Baguio's Past*, 12.
124. Scheerer quoting Vivar in ibid., 12.
125. Scheerer, *On Baguio's Past*, 27.
126. Ibid., 41.
127. Sergio Osmeña, Sr., "How Baguio City Was Born," *Free Press*, August 29, 1959.
128. David P. Barrows, "Notes on the Property Laws of the Igorots of Southern Benguet," *Teachers' Assembly Herald* 1 (May 5, 1908): 59.
129. Scheerer, *On Baguio's Past*, 39–40. See also Scott, *The Discovery of the Igorots*, 117.
130. On the palimpsest see Harvey, *The Limits to Capital*, 233. Harvey writes: "At any moment the built environment appears as a palimpsest of landscapes fashioned according to the dictates of different modes of production at different stages of their historical development."
131. Bagamaspad, Hamada-Pawid, and Balangoy, *A Peoples' History of Benguet Province*, 231.
132. Cariño, "The Cariños and Baguio-Benguet History: Conclusion," in *Folio 2*, 62.
133. Scheerer, *On Baguio's Past*, 3.
134. Tapang, *Innovation and Social Change*, 15.

135. Prill-Brett, "Baguio: A Multi-Ethnic City and the Development of the Ibaloy as an Ethnic Minority," 3.
136. Cariño, "The Cariños and Baguio-Benguet History: Conclusion," in *Folio 2*, 62; Wiber, "Levels of Property Rights, Levels of Law," 476.
137. On Cariño's use of fences see Bagamaspad, Hamada-Pawid, and Balangoy, *A Peoples' History of Benguet Province*, 63.
138. On Spanish plans see Reed, *City of Pines*; Hore et al., "Exhibit H. The Benguet Sanitarium," in *Reports of the Taft Philippine Commission* (1901).
139. Angel Pérez, *Relaciones Agustinianas de Las Razas del Norte de Luzon, Department of the Interior, Ethnological Survey Publications, Volume III, Spanish Edition* (Manila: Bureau of Public Printing, 1904), 169.
140. Hore et al., "Exhibit H. The Benguet Sanitarium," in *Reports of the Taft Philippine Commission* (1901), 163–64.
141. Scheidnagel, *Distrito de Benguet*, 28–29.
142. See Reed, *City of Pines*.
143. F. H. Donaldson-Sim, interviewed by the Philippine Commission July 25, 1899.
144. "Exhibit D. Report of the committee appointed to investigate the suitableness of the township of Baguio, province of Benguet, for a health resort," in *Reports of the Taft Philippine Commission* (1901), 123.
145. Hore et al., "Exhibit H. The Benguet Sanitarium," in *Reports of the Taft Philippine Commission* (1901), 164.
146. Pérez, *Relaciones Agustinianas de Las Razas del Norte de Luzon*, 175.
147. Rafael, *The Promise of the Foreign*, 8.
148. Francia, *A History of the Philippines*, 112–113.
149. Kennedy, *The Magic Mountains*, 14.
150. Mark Harrison quotes army surgeon Julius Jeffreys in *Climates & Constitutions: Health, Race, Environment and British Imperialism in India, 1600–1850* (Oxford: Oxford University Press, 1999), 148.
151. Ibid., 149, 147, 19.
152. Hore et al., "Exhibit H. The Benguet Sanitarium," in *Reports of the Taft Philippine Commission* (1901), 194, 197, 204.
153. Pérez, *Relaciones Agustinianas de Las Razas del Norte de Luzon*, 176–77.
154. Hore et al., "Exhibit H. The Benguet Sanitarium," in *Reports of the Taft Philippine Commission* (1901), 188.
155. Brian M. Howell, "Moving Mountains: Protestant Christianity and the Spiritual Landscape of Northern Luzon," *Anthropological Forum* 19 (2009): 256.
156. Bagamaspad, Hamada-Pawid, and Balangoy, *A Peoples' History of Benguet Province*, 32.
157. See I. Leano, "Going Home to the Spirit World," *Saint Louis University Research Journal* 18 (1987): 189–230.
158. C. R. Moss, *Nabaloi Law and Ritual* (Berkeley: University of California Press, 1920), 283.
159. Ibid., 281.

160. Leano, "Going Home to the Spirit World," 197.
161. Bagamaspad, Hamada-Pawid, and Balangoy, *A Peoples' History of Benguet Province*, 103–4; Leano, "Going Home to the Spirit World," 197.
162. William Henry Scott, "Worship in Igorot Life," paper based on a talk given at the Second Baguio Religious Acculturation Conference, January 3, 1959, in *On the Cordillera*, 152.
163. Moss, *Nabaloi Law and Ritual*, 286.
164. Hore et al., "Exhibit H. The Benguet Sanitarium," in *Reports of the Taft Philippine Commission* (1901), 196.
165. Ibid., 195.
166. Jennings, *Curing the Colonizers*, 22.
167. Ibid., 32.
168. Anderson, *Colonial Pathologies*, 8.
169. No. 2, "An Act appropriating five thousand dollars ($5000) Mexican for the purpose of making a survey to ascertain the most advantageous route for a railroad into the mountains of Benguet, island of Luzon, and the probable cost thereof," in *Reports of the Taft Philippine Commission* (1901), 245.
170. On the military's "civic action" see Linn, *The Philippine War, 1899–1902*, 200.
171. No. 48, "An Act providing for the establishing of local civil governments in the townships of the province of Benguet," *Reports of the Taft Philippine Commission* (1901), 307.
172. "Report of the United States Philippine Commission," November 30, 1900, in *Reports of the Philippine Commission, the Civil Governor and the Heads of the Executive Departments of the Civil Government of the Philippines Islands, 1900–1903* (Washington, DC: Government Printing Office, 1904), 58.
173. "Appendix Q. Report of the Chief of the Bureau of Non Christian Tribes," in *Third Annual Report of the Philippine Commission. 1902. Part 1* (Washington, DC: Government Printing Office, 1903), 685.
174. No. 48, "An Act providing for the establishing of local civil governments in the townships of the province of Benguet," in *Reports of the Taft Philippine Commission* (1901), 307.
175. Howard T. Fry, *A History of the Mountain Province* (Quezon City: New Day Publishers, 1983), 9.
176. "Memoranda, of the Conflict, Charges, Hostilities and Misunderstandings between the Military and Civil Authorities, following the appointment of a civil governor and secretary for the province of Benguet." Submitted by General Arthur MacArthur, NARA, RG 350, BIA, 2368 Entry 5, Box 274; 150: 56/9/7.
177. *Reports of the Taft Philippine Commission* (1901), 45.
178. Bagamaspad, Hamada-Pawid, and Balangoy, *A Peoples' History of Benguet Province*, 14.
179. Currency early in the US occupation was complicated. The Mexican silver dollar (the peso) circulated along with paper, Spanish-Filipino pesos, and smaller coins in silver and copper. See *Reports of the Philippine Commission, the Civil Governor and the Heads of the Executive Departments . . .* (1904), 101.

180. See No. 48, "An Act providing for the establishing of local civil governments in the townships of the province of Benguet," in *Reports of the Taft Philippine Commission* (1901), 307.
181. *Sixth Annual Report of the Philippine Commission, 1905 (In Four Parts), Part 2* (Washington, DC: Government Printing Office, 1906), 58. See also the acts themselves: *Acts No. 1396 and 1397, The Special Provincial Government Act and the Township Government Act, Enacted September 14, 1905* (Manila: Bureau of Public Printing, 1905).
182. *Sixth Annual Report of the Philippine Commission, 1905 (In Four Parts), Part 2*, 58.
183. Owen J. Lynch, *Colonial Legacies in a Fragile Republic: A History of Philippine Land Law and State Formation* (Diliman, Quezon City: University of the Philippines College of Law, 2011), 353.
184. Benton, *A Search for Sovereignty*, 36, 2.
185. Ibid., 38
186. Hore et al., "Exhibit H. The Benguet Sanitarium," in *Reports of the Taft Philippine Commission* (1901), 204.

## CHAPTER TWO

1. On the length of the journey see Maj. L. W. V. Kennon, "Appendix H. Report of Officer in Charge of Construction of Benguet Road," in *Sixth Annual Report of the Philippine Commission, 1905 (In Four Parts), Part 3* (Washington, DC: Government Printing Office, 1906), 359. On Benguet Road, see Greg Bankoff, "'These Brothers of Ours': Poblete's Obreros and the Road to Baguio, 1903–1905," *Journal of Social History* 38 (2005): 1047–72; Arturo G. Corpuz, *The Colonial Iron Horse: Railroads and Regional Development in the Philippines, 1875–1935* (Quezon City: University of the Philippines Press, 1999); Gonzalez, *Securing Paradise*; Afable, ed., *Japanese Pioneers in the Northern Philippine Highlands;* Roediger and Esch, *The Production of Difference*, 131–35.
2. "Road Building to Music: How Major Kennon Got Work Out of Filipinos and Won a Bet," *Washington Post*, February 28, 1909.
3. "Day's Auto Tour in Philippines, Roads Found as Good as Any in France, Newly Built by American Authorities," special correspondence to the *New York Times*, May 11, 1913.
4. Bulosan, *America Is in the Heart*, 66.
5. Kennon, "Appendix H. Report of Officer in Charge of Construction of Benguet Road," in *Sixth Annual Report of the Philippine Commission, 1905 (In Four Parts) Part 3*, 361, 379.
6. Bagamaspad, Hamada-Pawid, and Balangoy, *A Peoples' History of Benguet Province*, 200.
7. "The Economic Condition," *El Renacimiento*, August 2, 1905, 1. (Published in English.)
8. David J. Doherty, *Conditions in the Philippines . . .* (Washington, DC: Government Printing Office, 1904), 3.

9. On these "extra-economic" means and primitive accumulation as an ongoing phenomenon within the capitalist mode of production, see De Angelis, "Marx and Primitive Accumulation: The Continuous Character of Capital's 'Enclosures'"; Harvey, "The 'New' Imperialism;" Perelman, *The Invention of Capitalism*; Federici, *Caliban and the Witch*; Glassman, "Primitive Accumulation, Accumulation by Dispossession, Accumulation by 'Extra-Economic Means.'"
10. *Reports of the Taft Philippine Commission* (1901), 72.
11. "Report of the Department of Commerce and Police," in *Fifth Annual Report of the Philippine Commission, 1904, In Three Parts. Part 3, 7.*
12. See De Angelis, "Marx and Primitive Accumulation: The Continuous Character of Capital's 'Enclosures,'" on the conditions in which primitive accumulation becomes a "strategy" again.
13. "Nuevas Carreteras," *El Renacimiento*, March 31, 1903, 1. Newspaper articles in Spanish have been translated by author unless otherwise noted. On this urban elite see Julian Go, *American Empire and the Politics of Meaning: Elite Political Cultures in the Philippines and Puerto Rico during US Colonialism* (Durham: Duke University Press, 2008).
14. *Reports of the Taft Philippine Commission* (1901), 71.
15. A study that brings together consideration of colonial educational policy, political tutoring, and economic development is Glenn Anthony May's *Social Engineering in the Philippines: The Aims, Execution, and Impact of American Colonial Policy, 1900–1913* (Westport, CT: Greenwood Press, 1980).
16. On the Philippine-American War see Linn, *The Philippine War, 1899–1902*; Linn, *The US Army and Counterinsurgency in the Philippine War, 1899–1902*; Samuel K. Tan, *Filipino-American War, 1898–1913* (Quezon City: University of the Philippines Press, 2002); Miller, *"Benevolent Assimilation"*; Shaw and Francia, eds., *Vestiges of War*; McCoy, *Policing America's Empire*; Reynaldo Ileto, *Pasyon and Revolution: Popular Movements in the Philippines, 1840–1910* (Manila: Ateneo University Press, 1979); David Sturtevant, *Popular Uprisings in the Philippines, 1840–1940* (Ithaca: Cornell University Press, 1976); Bonnie Miller, *From Liberation to Conquest: The Visual and Popular Cultures of the Spanish-American War of 1898* (Amherst: University of Massachusetts Press, 2011); Hoganson, *Fighting for American Manhood.*
17. Abinales and Amoroso, *State and Society in the Philippines*, 126.
18. Go, *American Empire and the Politics of Meaning*, 22.
19. Ileto, *Pasyon and Revolution*, 170.
20. Ibid., 117–18.
21. Colonel W. Burt (?) to William Howard Taft, August 14, 1900, Presidential Papers Microfilm, William H. Taft Papers, Reel 31, Series 3, Library of Congress.
22. William H. Taft, "Report of the Civil Governor of the Philippine Islands for the Period Ending December 23, 1903," in *Fourth Annual Report of the Philippine Commission, 1903, In Three Parts, Part 1*, 35. On *ladrones* see Ileto, *Pasyon and Revolution*; and Sturtevant, *Popular Uprisings in the Philippines, 1840–1940.*

See also Bankoff, *Crime, Society, and the State in the Nineteenth-Century Philippines.*

23. "Exhibit G. Philippines Constabulary," in *Third Annual Report of the Philippine Commission. 1902. Part 1*, 180.
24. "Crowding of Bilibid. Ladrones with Long Sentences Are Trying Capacity of Prison. Convicts Come in Bunches," *Manila Times*, March 11, 1903. On the suppression of political insurgents in this period, see McCoy, *Policing America's Empire*, chapter 3.
25. Ignacio Villamor, *Criminality in the Philippines, 1903–1908* (Manila: Bureau of Printing, 1909), 12.
26. *Reports of the Taft Philippine Commission* (1901), 71.
27. Doherty, *Conditions in the Philippines . . .*, 3.
28. Edward Gibbon, *The History of the Decline and Fall of the Roman Empire, Volume the First and Volume the Second*, ed. David Womersley (London: Penguin Books, 2005 [1776, 1781]), 77.
29. No. 1, "An Act appropriating two million dollars ($2,000,000) Mexican to be used in the construction and repair of highways and bridges in the Philippine Islands," in *Reports of the Taft Philippine Commission* (1901), 245.
30. Linn, *The Philippine War, 1899–1902*, 202.
31. No. 2, "An Act appropriating five thousand dollars ($5000.00) Mexican for the purpose of making a survey to ascertain the most advantageous route for a railroad into the mountains of Benguet, island of Luzon, and the probable cost thereof," in *Reports of the Taft Philippine Commission* (1901), 245.
32. No. 61, "An Act authorizing the construction of a highway from the vicinity of the town of Pozorubio [sic] in the province of Pangasinan to Baguio in the province of Benguet and appropriating seventy-five thousand dollars ($75,000) money of the United States, for that purpose," in *Annual Reports of the War Department for the Fiscal Year Ended June 30, 1901, Public Laws and Resolutions Passed by the Philippine Commission* (Washington: Government Printing Office, 1901), 103.
33. Kennon, "Appendix H. Report of Officer in Charge of Construction of Benguet Road," in *Sixth Annual Report of the Philippine Commission, 1905 (In Four Parts), Part 3*, 360.
34. Ibid., 382.
35. Ibid.
36. "El Camino de Benguet," *El Renacimiento*, March 22, 1905, 1.
37. "The Delights of the Regime," *La Vanguardia*, June 20, 1912, translated and retyped for the Bureau of Insular Affairs files, NARA, RG 350, Entry 5, Box 274.
38. "Our White Elephant," *Philippines Manila Free Press*, August 19, 1911, in NARA, RG 350, Entry 5, Box 535.
39. "Labor Leaves for Benguet. Poblete's 'Ten Thousand' Start for Twin Peaks this Morning," *Manila Times*, July 13, 1903, 1.
40. Kennon, "Appendix H. Report of Officer in Charge of Construction of Ben-

guet Road," in *Sixth Annual Report of the Philippine Commission, 1905 (In Four Parts), Part 3*, 364.

41. Ibid.
42. Ibid., 365.
43. Ibid.
44. N. M. Holmes to Chairman of Committee on Appropriations, US Philippine Commission, June 10, 1902, in "Reports on the Benguet Road for the Years 1901–2–3," NARA, RG 350, Entry 5, Box 274.
45. "The Delights of the Regime," *La Vanguardia*, June 20, 1912, translated and retyped for the Bureau of Insular Affairs files, NARA, RG 350, Entry 5, Box 274.
46. Kennon, "Appendix H. Report of Officer in Charge of Construction of Benguet Road," in *Sixth Annual Report of the Philippine Commission, 1905 (In Four Parts), Part 3*, 372.
47. Bulosan, *America Is in the Heart*, 31.
48. William Cameron Forbes, journal, September 17, 1904, MS 1365, Forbes Papers.
49. Report of William F. Pack, February 1, 1903, in *Fourth Annual Report of the Philippine Commission. 1903. In Three Parts, Part 1*, 759.
50. Kennon, "Appendix H. Report of Officer in Charge of Construction of Benguet Road," in *Sixth Annual Report of the Philippine Commission, 1905 (In Four Parts), Part 3*, 360.
51. Letter of Captain Meade to the Philippine Commission, December 14, 1900, NARA, RG 350, Entry 5, Box 261.
52. N. M. Holmes, "Report of Engineer in Charge of Benguet Road," in *Fourth Annual Report of the Philippine Commission, 1903, In Three Parts, Part 3* (Washington, DC: Government Printing Office, 1904), 261.
53. Kennon, "Appendix H. Report of Officer in Charge of Construction of Benguet Road," in *Sixth Annual Report of the Philippine Commission, 1905 (In Four Parts), Part 3*, 371.
54. Bagamaspad, Zenaida Hamada-Pawid, and Bell Balangoy, *A Peoples' History of Benguet Province*, 200.
55. *Act No. 1396, The Special Provincial Government Act, Enacted September 14, 1905*, 21–22.
56. Transcribed letter from N. M. Holmes to Luke Wright, acting governor, undated, in "Reports on the Benguet Road for the Years 1901–2–3," NARA, RG 350, Entry 5, Box 274.
57. Holmes, "Report of Engineer in Charge of Benguet Road," in *Fourth Annual Report of the Philippine Commission, 1903, In Three Parts, Part 3*, 260.
58. Ibid.
59. See United States Bureau of the Census, *Census of the Philippine Islands, Taken under the Direction of the Philippine Commission in the Year 1903, in Four Volumes, Vol. 2, No. 1*, http://name.umdl.umich.edu/ajb5834.0002.001, 919–49.
60. United States Bureau of the Census, *Census of the Philippine Islands, Taken un-*

*der the Direction of the Philippine Commission in the Year 1903, in Four Volumes, Vol. 4, No. 1,* http://name.umdl.umich.edu/ajb5834.0004.001, 193, 192.

61. See Abinales and Amoroso, *State and Society in the Philippines*, 80–81 on "The Importance of Land." See Keesing, *The Ethnohistory of Northern Luzon*, on the subsistence practices of Pangasinan peoples, Ilocanos, and Ibaloi.
62. Holmes, "Report of Engineer in Charge of Benguet Road," in *Fourth Annual Report of the Philippine Commission, 1903, In Three Parts, Part 3*, 260–61.
63. Syed Hussein Alatas, *The Myth of the Lazy Native: A Study of the Image of the Malays, Filipinos and Javanese from the 16*th *to the 20*th *Century and Its Function in the Ideology of Colonial Capitalism* (London: Frank Cass, 1977), 126.
64. Holmes, "Report of Engineer in Charge of Benguet Road," in *Fourth Annual Report of the Philippine Commission, 1903, In Three Parts, Part 3*, 262.
65. Ibid., 261.
66. Kennon, "Appendix H. Report of Officer in Charge of Construction of Benguet Road," in *Sixth Annual Report of the Philippine Commission, 1905 (In Four Parts), Part 3*, 376, 393.
67. *Third Annual Report of the Philippine Commission, 1902, Part 1*, 4.
68. Abinales and Amoroso, *State and Society in the Philippines*, 62.
69. *Third Annual Report of the Philippine Commission, 1902, Part 1*, 4.
70. Forbes to Secretary of War Henry Stimson, June 13, 1911, NARA, RG 350, Entry 5A, Box 535.
71. Kennon, "Appendix H. Report of Officer in Charge of Construction of Benguet Road," in *Sixth Annual Report of the Philippine Commission, 1905 (In Four Parts), Part 3*, 375.
72. Maj. L. W. V. Kennon to William Howard Taft, August 30, 1903, in "Reports on the Benguet Road for the Years 1901-2-3," NARA, RG 350, Entry 5, Box 274.
73. Holmes, "Report of Engineer in Charge of Benguet Road," in *Fourth Annual Report of the Philippine Commission, 1903, In Three Parts, Part 3*, 262.
74. Grace Helen Bailey, "Road Making in the Philippines: Construction of the Benguet Mountain Road," *Overland Monthly* 53 (January 1909): 31. Stoler discusses the problem of "poor whites" in European colonies in chapter 2 of *Carnal Knowledge*.
75. Roediger and Esch, *The Production of Difference*, 15.
76. Kennon, "Appendix H. Report of Officer in Charge of Construction of Benguet Road," in *Sixth Annual Report of the Philippine Commission, 1905 (In Four Parts), Part 3*, 380.
77. Holmes, "Report of Engineer in Charge of Benguet Road," in *Fourth Annual Report of the Philippine Commission, 1903, In Three Parts, Part 3*, 262.
78. Kennon, "Appendix H. Report of Officer in Charge of Construction of Benguet Road," in *Sixth Annual Report of the Philippine Commission, 1905 (In Four Parts), Part 3*, 393.
79. For labor and race management in the United States and its colonies see Roediger and Esch, *The Production of Difference*.

80. Lisa Lowe, *Immigrant Acts: On Asian American Cultural Politics* (Durham: Duke University Press, 1999 [1996]), 27–28. See also Alatas, *The Myth of the Lazy Native.*
81. "Amarillos y Negros," *El Renacimiento*, February 3, 1903.
82. "Report of the Civil Governor," November 1, 1902 in *Reports of the Philippine Commission, the Civil Governor, and the Heads of the Executive Departments . . .* (1904), 303.
83. Willard B. Gatewood, Jr., *Black Americans and the White Man's Burden, 1898–1903* (Urbana: University of Illinois Press, 1975), 306. On Fortune see also John Cullen Gruesser, *The Empire Abroad and the Empire at Home: African American Literature and the Era of Overseas Expansion* (Athens: University of Georgia Press, 2012).
84. Roediger and Esch, *The Production of Difference*, 130.
85. Gatewood, Jr., *Black Americans and the White Man's Burden, 1898–1903*, 315.
86. "Amarillos y Negros," *El Renacimiento*, February 3, 1903. See also Gatewood, Jr., *Black Americans and the White Man's Burden, 1898–1903.*
87. "Professor Moses Back from Manila," *San Francisco Chronicle*, August 12, 1903.
88. "Chinese Immigration," *Manila Times*, September 2, 1899.
89. Anderson calls the union a "free-wheeling central" in *Under Three Flags*, 228. On the union and labor in the late nineteenth and twentieth centuries see also William Henry Scott, *The Union Obrera Democratica: First Filipino Labor Union* (Quezon City: New Day Publishers, 1992); Francia, *A History of the Philippines*, 125-126; Dante G. Guevarra, *History of the Philippine Labor Movement* (Santa Mesa, Manila: Institute of Labor and Industrial Relations Polytec, 1991), 14–17.
90. "De Benguet," *El Grito del Pueblo*, August 21, 1903.
91. Scott, *The Union Obrera Democratica*; Guevarra, *History of the Philippine Labor Movement*, 14–17.
92. Edward Rosenberg, "Los Filipinos como Trabajadores," *El Grito del Pueblo*, November 22, 1903.
93. *Third Annual Report of the Philippine Commission, 1902, Part 1*, 159–176.
94. Ibid., 173.
95. "An Opportunity to Test Native Labor," *Manila Times*, July 16, 1903, 1.
96. On Poblete's radical politics, see McCoy, *Policing America's Empire*, 106. On Poblete and the Benguet Road see Bankoff, "'These Brothers of Ours.'"
97. *Fourth Annual Report of the Philippine Commission, 1903, In Three Parts, Part 3* (Washington, DC: Government Printing Office, 1904), 39–42.
98. "Lo de Los Obreros de Benguet. Inauditos Sufrimientos. Lo Que Dice El Sr. Poblete," *El Renacimiento*, July 27, 1903, 2.
99. "Nuevas Carreteras," *El Renacimiento*, March 31, 1903, 1.
100. On the investigation, see James Ross, "Exhibit N. Report of an Investigation Made by James Ross, Supervisor of Fiscals, Concerning Alleged Sufferings and Deaths Among Certain Laborers Sent from Manila During the Month of July,

1903, to Work on the Benguet Road," in *Annual Reports of the War Department for the Fiscal Year Ended June 30, 1903*, vol. 5, *Report of the Philippine Commission* (Washington, DC: Government Printing Office, 1903). See also Bankoff, "'These Brothers of Ours.'"

101. "Lo de Los Obreros de Benguet. Inauditos Sufrimientos. Lo Que Dice El Sr. Poblete," *El Renacimiento*, July 27, 1903, 2.
102. "Los de Los Obreros de Benguet. Inauditos Sufrimientos. Lo Que Dice El Sr. Poblete," *El Renacimiento*, July 28, 1903.
103. Ibid.
104. "Dissatisfaction in Labor Camp. Poblete's Brigade Strike in a Body and all Are Returning—No Accommodation Provided," *Manila Times*, July 28, 1903, 1. See also Ross, "Exhibit N. Report of an Investigation Made by James Ross . . . ," in *Annual Reports of the War Department for the Fiscal Year Ended June 30, 1903*, vol. 5 (1903); and Bankoff, "'These Brothers of Ours.'"
105. "Lo de Los Obreros de Benguet. Inauditos Sufrimientos. Lo Que Dice El Sr. Poblete," *El Renacimiento*, July 28, 1903.
106. "Poblete's Filipino Labor Bluff" (editorial), *Manila Times*, January 26, 1903, 4; "'Rake Off' On Laborers. Poblete Would Have Received a Nice Percentage Had His 1000 Taos Labored," *Manila Times*, August 1, 1903, 1.
107. See Bankoff, "'These Brothers of Ours.'"
108. McCoy, *Policing America's Empire*, 106.
109. Bankoff, "'These Brothers of Ours,'" 1064.
110. Scott, *The Union Obrera Democratica*, 187;Taft, "Report of the Civil Governor of the Philippine Islands for the Period Ending December 23, 1903," *Fourth Annual Report of the Philippine Commission, 1903, Part 1*, 36–37.
111. "Lo de Los Obreros de Benguet. Inauditos Sufrimientos. Lo Que Dice El Sr. Poblete," *El Renacimiento*, July 28, 1903.
112. Bankoff, "'These Brothers of Ours,'" 1058.
113. Kennon, "Appendix H. Report of Officer in Charge of Construction of Benguet Road," in *Sixth Annual Report of the Philippine Commission, 1905 (In Four Parts), Part 3*, 377.
114. Linn, *The Philippine War, 1899–1902*, 260.
115. Kennon, "Appendix H. Report of Officer in Charge of Construction of Benguet Road," in *Sixth Annual Report of the Philippine Commission, 1905 (In Four Parts), Part 3*, 377.
116. "Road Building to Music: How Major Kennon Got Work out of Filipinos and Won a Bet," *Washington Post*, February 28, 1909. See also Kennon, "Appendix H. Report of Officer in Charge of Construction of Benguet Road," in *Sixth Annual Report of the Philippine Commission, 1905 (In Four Parts), Part 3*, 378.
117. Bankoff, "'These Brothers of Ours,'" 1063.
118. Forbes, journal, September 17, 1904, MS Am 1365, Forbes Papers.
119. *Reports of the Philippine Commission, the Civil Governor and the Heads of the Executive Departments* . . . (1904), 513.

120. Forbes, journal, June 10, 1909, MS Am 1365, Forbes Papers.
121. Kennon, "Appendix H. Report of Officer in Charge of Construction of Benguet Road," in *Sixth Annual Report of the Philippine Commission, 1905 (In Four Parts), Part 3*, 379.
122. Ibid., 378.
123. Ibid., 379.
124. Ibid., 378.
125. Paul Samuel Reinsch, *World Politics: At the End of the Nineteenth Century as Influenced by the Oriental Situation* (New York: The MacMillan Company, 1904 [1900]), 321–22.
126. No. 518, "An Act Defining Highway Robbery or Brigandage, and Providing for the Punishment Therefore," enacted November 12, 1902, NARA, RG 350, Entry 5, Box 279; or see *Acts of the Philippine Commission [Acts Nos. 425–949 Inclusive]* (Washington, DC: Government Printing Office, 1904).
127. Taft to Anne G. Roelker, July 10, 1900, Presidential Papers Microfilm, William H. Taft Papers, Reel 31, Series 3, Library of Congress; Forbes to Theodore Roosevelt, February 18, 1905, bMS Am 1364, Forbes Papers.
128. Isabelo de los Reyes, "El Problema del Bandolerismo. Pueblo contento no se subleva," *El Renacimiento*, April 24, 1905, 1; and de los Reyes, "El Problema del Bandolerismo," *El Renacimiento*, April 19, 1905, 1.
129. Ileto, *Pasyon and Revolution*, 193.
130. Renato Constantino, *A History of the Philippines: From the Spanish Colonization to the Second World War* (New York: Monthly Review Press, 1975), 246. On the seditious plays see Rafael, *White Love*. On the United States' counterinsurgency operations see McCoy, *Policing America's Empire*.
131. *Fourth Annual Report of the Philippine Commission, 1903, In Three Parts, Part 3*, 39–42.
132. Isabelo de los Reyes, "El Problema del Bandolerismo. Pueblo contento no se subleva," *El Renacimiento*, April 24, 1905, 1.
133. Letter from Colonel W. Burt (?) to William Howard Taft, August 14, 1900, correspondence, Presidential Papers Microfilm, William H. Taft Papers, Reel 31, Series 3.
134. "La reorganización obrera de Filipinas. Un manifesto de Mr. Rosenberg," *El Renacimiento*, July 30, 1903.
135. Constantino, *A History of the Philippines*, 200.
136. Onofre D. Corpuz, *An Economic History of the Philippines* (Quezon City: University of the Philippines Press, 1997), 205.
137. *Third Annual Report of the Philippine Commission, 1902, Part 1*, 5.
138. Sturtevant, *Popular Uprisings in the Philippines*, 118; and Corpuz, *An Economic History of the Philippines*, 205.
139. Taft, "Report of the Civil Governor of the Philippine Islands for the Period Ending December 23, 1903," *Fourth Annual Report of the Philippine Commission, 1903, Part 1*, 18, 22.

140. "¡Cuantos Hambrientos! (Al Gobierno y a los Ricos)," *El Renacimiento*, May 6, 1903, 1.
141. Ibid.
142. No. 519, "An Act Defining Vagrancy and Providing for Punishment Therefore," enacted November 12, 1902, NARA, RG 350, Entry 5, Box 279; or see *Acts of the Philippine Commission [Acts Nos. 425–949 Inclusive].*
143. For a history of crime in the Spanish colonial period, see Bankoff, *Crime, Society, and the State in the Nineteenth-Century Philippines.*
144. Taft, "Report of the Civil Governor of the Philippine Islands for the Period Ending December 23, 1903," *Fourth Annual Report of the Philippine Commission, 1903, Part 1*, 37–38.
145. Villamor, *Criminality in the Philippines, 1903–1908*, 12.
146. "Report of the Civil Governor," in *Reports of the Philippine Commission, the Civil Governor, and the Heads of the Executive Departments . . .* (1904), 299, 297.
147. On US colonial forestry in the Philippines, see Greg Bankoff, "Conservation and Colonialism: Gifford Pinchot and the Birth of Tropical Forestry in the Philippines," in *Colonial Crucible*, ed. McCoy and Scarano, 479–88. A history of US forest conservation informing my account is Jacoby's *Crimes Against Nature.*
148. Lawrence Rakestraw, "George Patrick Ahern and the Philippine Bureau of Forestry, 1900–1914," *Pacific Northwest Quarterly* 53 (July 1967): 145.
149. H. N. Whitford, *The Forests of the Philippines, Part I: Forest Types and Products*, Bulletin No. 10 (Manila: Bureau of Printing, 1911), 61.
150. George P. Ahern, *A Few Pertinent Facts Concerning the Philippine Forests and Needs of the Forest Service, That Should Interest Every Filipino*, Department of the Interior, Bureau of Forestry, circular no. 3 (Manila: Bureau of Printing, 1908), 14. For Ahern's biography, see Rakestraw, "George Patrick Ahern and the Philippine Bureau of Forestry, 1900–1914."
151. Whitford, *The Forests of the Philippines*, 14.
152. On the use of *caiñgin* among Igorots see William Henry Scott, "A Preliminary Report on Upland Rice in Northern Luzon," in *On the Cordillera*. On the social implications of forest conservation, see Richard H. Grove, "Colonial Conservation, Ecological Hegemony and Popular Resistance: Towards a Global Synthesis," in *Imperialism and the Natural World*, ed. John M. MacKenzie (Manchester: Manchester University Press, 1990): 15–50. On swidden agriculture, see Michael R. Dove, "Theories of Swidden Agriculture, and the Political Economy of Ignorance," *Agroforestry Systems* 1 (1983): 85-99.
153. William Pack, "Report of the Governor of the Province of Benguet," July 1, 1905, in *Annual Reports of the War Department for the Fiscal Year Ended June 30, 1905*, vol. 10, *Report of the Philippine Commission, Part 1*. (Washington, DC: Government Printing Office, 1905), 178.
154. Worcester, *The Philippines Past and Present*, vol. 2, 848.
155. *Annual Report of the Director of Forestry of the Philippine Islands, July 1, 1905 to June 30, 1906* (Manila: Bureau of Printing, 1906) 11.

156. *Annual Report of the Director of Forestry of the Philippine Islands, July 1, 1909 to June 30, 1910* (Manila: Bureau of Printing, 1910), 7.
157. Bagamaspad, Hamada-Pawid, and Balangoy, *A Peoples' History of Benguet Province*, 206.
158. *Annual Report of the Director of Forestry of the Philippine Islands, July 1, 1909 to June 30, 1910*, 9.
159. On allotment see Stuart Banner, *How the Indians Lost Their Land: Law and Power on the Frontier* (Cambridge: Harvard University Press, 2005); Frederick E. Hoxie, *A Final Promise: The Campaign to Assimilate the Indians, 1880–1920* (Lincoln: University of Nebraska Press, 1984); C. Joseph Genetin-Pilawa, *Crooked Paths to Allotment: The Fight over Federal Indian Policy after the Civil War* (Chapel Hill: University of North Carolina Press, 2012).
160. *Annual Report of the Director of Forestry of the Philippine Islands, July 1, 1905 to June 30, 1906*, 16.
161. *Annual Report of the Director of Forestry of the Philippine Islands, July 1, 1906 to June 30, 1907* (Manila: Bureau of Printing, 1908), 7.
162. On this process in the US context see Jacoby, *Crimes Against Nature*. On state efforts to order and generate subjects in and through space, see James C. Scott, *Seeing Like a State: How Certain Schemes to Improve the Human Condition Have Failed* (New Haven: Yale University Press, 1998). See also Coronil's critique of Scott: Fernando Coronil, "Smelling Like a Market," *American Historical Review* 106, no. 1 (2001): 119–29.
163. See Jacoby, *Crimes Against Nature*.
164. Marx, *Capital, Volume One*, excerpted in *The Marx-Engels Reader*, 435. Massimo De Angelis stresses "separation" as central to Marx's understanding of primitive accumulation in "Marx and Primitive Accumulation: The Continuous Character of Capital's 'Enclosures.'" See also Harvey, "The 'New' Imperialism"; Perelman, *The Invention of Capitalism*; Glassman, "Primitive Accumulation, Accumulation by Dispossession, Accumulation by 'Extra-Economic' Means."
165. De Angelis, "Marx and Primitive Accumulation: The Continuous Character of Capital's 'Enclosures,'" 9. See also Glassman, "Primitive Accumulation, Accumulation by Dispossession, Accumulation by 'Extra-Economic' Means," 619.
166. Marx, *Capital, Volume One*, excerpted in *The Marx-Engels Reader*, 436.
167. De Angelis, "Marx and Primitive Accumulation: The Continuous Character of Capital's 'Enclosures,'" 2.
168. Harvey, "The 'New' Imperialism," 74.
169. Marx, *Capital, Volume One*, excerpted in *The Marx-Engels Reader*, 436.
170. See De Angelis, "Marx and Primitive Accumulation: The Continuous Character of Capital's 'Enclosures.'" Citing Marx, De Angelis has argued that when capital meets resistance—when, for example, the working class rejects "subordination" to capital's "'natural laws'" and the "'silent compulsion of economic relations'"—primitive accumulation becomes ongoing and assumes a

"'continuous' character"(16, 15, 16). In the US colonial Philippines, primitive accumulation persisted, by this logic, because of the conflict Filipinos, like workers along Benguet Road, presented to American road chiefs. It continued because *ladrones* and brigands, some, perhaps, mere robbers but others latter-day Robin Hoods bearing another vision of the Philippines' political economy, likewise challenged the imperial occupation.

171. Eric Foner, *A Short History of Reconstruction* (New York: Harper & Row, 1990), 93–94.
172. Ibid., 94–95.
173. Amy Dru Stanley, *From Bondage to Contract: Wage Labor, Marriage, and the Market in the Age of Slave Emancipation* (New York: Cambridge University Press, 1998), 122, 130.
174. Jane Burbank and Frederick Cooper, *Empires in World History: Power and the Politics of Difference* (Princeton: Princeton University Press, 2010), 306.
175. See Federici on primitive accumulation as "not simply an accumulation and concentration of exploitable workers and capital. It was also an accumulation of differences and divisions within the working class, whereby hierarchies built upon gender, as well as 'race' and age, became constitutive of class rule and the formation of the modern proletariat" (*Caliban and the Witch*, 63).
176. "Report of the Secretary of Commerce and Police," *Eighth Annual Report of the Philippine Commission to the Secretary of War, 1907 (In Three Parts), Part 2*, 275–77.
177. "Capable Laborers. Chinos Who Are Now on Benguet Road Are Supremely Content. Filipino Has Contract to Furnish 1000 Chinos," *Manila Times*, August 10, 1903, 8.
178. "Keep Off the Grass! Coxey's Army Invades the Nation's Capital," History Matters, http://historymatters.gmu.edu/d/5364/.
179. See Abinales and Amoroso, *State and Society in the Philippines*, 129, on different expressions of Philippine nationalism.
180. Ileto, *Pasyon and Revolution*, 117.
181. José Rizal, *The Indolence of the Filipino* (Manila: Philippine Education Co., 1913), http://archive.org/details/theindolenceofth06885gut.

## CHAPTER THREE

1. Daniel H. Burnham to Charles Moore, March 13, 1905, quoted in Moore, *Daniel H. Burnham*, 245.
2. On Burnham, see Moore, *Daniel H. Burnham*; Hines, *Burnham of Chicago, Architect and Planner*; Carl Smith, *The Plan of Chicago: Daniel Burnham and the Remaking of the American City* (Chicago: University of Chicago Press, 2007); Szuberla, "Urban Vistas and the Pastoral Garden"; Richard W. Longstreth, ed., *The Mall in Washington, 1791–1991* (Washington, DC: National Gallery of Art, 1991). On his work in the Philippines, see Thomas S. Hines, "The

Imperial Façade: Daniel H. Burnham and American Architectural Planning in the Philippines"; Brody, *Visualizing American Empire*; Brody, "Building Empire: Architecture and American Imperialism in the Philippines"; Lico, *Arkitekturang Filipino*; Vernon, "Daniel Burnham and the American City Imperial."

3. Herbert Croly, "The Promised City of San Francisco," *Architectural Record* 19 (June 1906), 427.
4. Daniel Hudson Burnham, *Plan of Chicago* (Chicago: Commercial Club, 1909), http://archive.org/details/planofchicago00burnuoft, 31. On the 1893 World's Fair, see Robert W. Rydell, *All the World's a Fair: Visions of Empire at American International Expositions, 1876–1916* (Chicago: University of Chicago Press, 1984); Alan Trachtenberg, *The Incorporation of America: Culture and Society in the Gilded Age* (New York: Hill and Wang, 1982); Paul Greenhalgh, *Ephemeral Vistas: The Expositions Universelles, Great Exhibitions, and World's Fairs, 1851–1939* (Manchester: Manchester University Press, 1988).
5. On the White City grounds see Hines, *Burnham of Chicago, Architect and Planner*, 76–79. Henry Blake Fuller called Burnham a "great impresario of architecture." See Szuberla, 22.
6. Rodgers, *Atlantic Crossings*, 12. On the City Beautiful movement see also Daniel Bluestone, *Constructing Chicago* (New Haven: Yale University Press, 1991); and William H. Wilson, *The City Beautiful Movement* (Baltimore: Johns Hopkins University Press, 1989).
7. See Bluestone, *Constructing Chicago*. Studies on Progressivism that I have drawn on include Rodgers, *Atlantic Crossings*; Rodgers, "In Search of Progressivism"; Kloppenberg, *Uncertain Victory*; Johnston, "Re-Democratizing the Progressive Era"; Johnston, *The Radical Middle Class*; Sklar, *The Corporate Reconstruction of American Capitalism, 1890–1916*; McGerr, *A Fierce Discontent*; Wiebe, *The Search for Order, 1877–1920*; Leuchtenburg, "Progressivism and Imperialism: The Progressive Movement and American Foreign Policy, 1898–1916"; Gillette, "The Military Occupation of Cuba, 1899–1902"; Siracusa, "Progressivism, Imperialism, and the Leuchtenburg Thesis, 1952–1974"; Abinales, "Progressive-Machine Conflict in Early-Twentieth-Century US Politics and Colonial-State Building in the Philippines"; Kramer, "Reflex Actions."
8. Bluestone, *Constructing Chicago*, 190.
9. On "de-commodification" as a "thread in social politics," see Rodgers, *Atlantic Crossings*, 29.
10. Louis H. Sullivan, *The Autobiography of an Idea* (Press of the American Institute of Architects, Inc., 1924), http://archive.org/details/autobiographyofa005655mbp, 325.
11. *Report of the Philippine Commission to the President*, vol. 1, *January 31, 1900* (Washington, DC: Government Printing Office, 1900), 106.
12. Elihu Root, "American Politics in the Philippines in 1902," Address of the Secretary of War at Peoria, IL, September 24, 1902, in *The Military and Co-*

*lonial Policy of the United States, Addresses and Reports*, ed. Robert Bacon and James Brown Scott (Cambridge: Harvard University Press, 1916), 76–77.

13. William Dudley Foulke (from Indiana) to William Howard Taft, February 4, 1901, Presidential Papers Microfilm, William H. Taft Papers, Reel 31, Series 3, Library of Congress.
14. Edward Gaylord Bourne, "A Trained Colonial Civil Service," *North American Review* 169 (October 1899): 529.
15. Daniel H. Burnham and Pierce Anderson, "Report on Proposed Improvements at Manila," in Moore, *Daniel H. Burnham*, 195. Szuberla also makes this observation in his "Urban Vistas and the Pastoral Garden," 38–39.
16. Burnham to Margaret Burnham, December 19, 1904, Box FF 25 (Folder 13), Daniel H. Burnham Collection, Art Institute of Chicago.
17. Burnham and Anderson, "Report on Proposed Improvements at Manila," in Moore, *Daniel H. Burnham*, 195.
18. Letter dated June 15, 1900, in Edith Moses, *Unofficial Letters of an Official's Wife* (New York: D. Appleton and Co., 1908), 9.
19. Linn, *The Philippine War, 1899–1902*, 26.
20. Doeppers, "Manila's Imperial Makeover: Security, Health, and Symbolism," in *Colonial Crucible*, ed. McCoy and Scarano, 492.
21. Linn, *The Philippine War, 1899–1902*, 31.
22. Burnham and Anderson, "Report on Proposed Improvements at Manila," in Moore, *Daniel H. Burnham*, 179.
23. Helen Herron Taft, *Recollections of Full Years* (New York: Dodd, Mead & Co., 1914), 99.
24. Doeppers, "Manila's Imperial Makeover: Security, Health, and Symbolism," in *Colonial Crucible*, ed. McCoy and Scarano, 492.
25. Linn, *The Philippine War, 1899–1902*, 57–58.
26. Ibid., 59.
27. Kramer describes Manila as the place "where fears of insurrection were most palpable" in *The Blood of Government*, 327–28.
28. "Report of the Civil Governor," in *Reports of the Philippine Commission, the Civil Governor, and the Heads of the Executive Departments* . . . (1904), 297.
29. Taft, *Recollections of Full Years*, 109.
30. Burnham Diary, in Moore, *Daniel H. Burnham*, 239–40.
31. No. 636, "An Act creating a Government reservation at Baguio, in the Province of Benguet," *Philippine Islands Public Laws, with Amendments Indicated* (Manila: Bureau of Public Printing, 1904), 296.
32. "Excerpt from Minutes of the Commission, June 1, 1903," in *Second Annual Report of the Philippine Civil Service Board to the Civil Governor of the Philippine Islands for the Year Ended September 30, 1902* (Manila: Bureau of Public Printing, 1903), 62.
33. Woolley, "Baguio, Simla of the Philippines," 292.
34. "Baguio and Commission. Summer Seat of Government Ready for Occupancy. Good Prospects in Benguet," *Manila Times*, January 16, 1903.

35. Scheerer, *On Baguio's Past*, 27.
36. Moss, *Nabaloi Law and Ritual*, 218. See also Lico, *Arkitekturang Filipino* for a history of building styles in northern Luzon.
37. Hore et al., "Exhibit H. The Benguet Sanitarium," in *Reports of the Taft Philippine Commission* (1901), 191
38. Moss, *Nabaloi Law and Ritual*, 219.
39. Hore et al., "Exhibit H. The Benguet Sanitarium," in *Reports of the Taft Philippine Commission* (1901), 191.
40. Moses, *Unofficial Letters of an Official's Wife*, 247.
41. Moss, *Nabaloi Law and Ritual*, 219.
42. Ibid., 220.
43. William Henry Scott quoting Don Alonso Martin Quirante in Scott, *The Discovery of the Igorots*, 45.
44. On the Cariños see Cariño, "The Cariños and Baguio-Benguet History: Conclusion," in *Folio 2*; Scott, *The Discovery of the Igorots*; Bagamaspad, Hamada-Pawid, and Balangoy, *A Peoples' History of Benguet Province*; Scheerer, *On Baguio's Past*; Scheerer, *The Nabaloi Dialect*.
45. Moses, *Unofficial Letters of an Official's Wife*, 247, 252–53.
46. Pérez, *Relaciones Agustinianas de Las Razas del Norte de Luzon*, 173.
47. *Transcript of Record. Supreme Court of the United States. October Term, 1907. No. 298. Mateo Cariño, Plaintiff in Error, vs. The Insular Government of the Philippine Islands. In Error to the Supreme Court of the Philippine Islands. Filed May 2, 1907*, NARA, RG 350, Entry 5, Box 781, 6–8.
48. No. 636, "An Act creating a Government reservation at Baguio, in the Province of Benguet." See *Philippine Islands Public Laws, with Amendments Indicated*, 296.
49. Bagamaspad, Hamada-Pawid, and Balangoy, *A Peoples' History of Benguet Province*, 14.
50. No. 297, "An Act appropriating eleven thousand (11,000) pesos for the purchase of land and buildings in Baguio, Benguet, for the use of the insular government and of the provincial government of Benguet." See *Annual Reports of the War Department for the Fiscal Year Ended June 30, 1902*, vol. 11, *Acts of the Philippine Commission* (Washington, DC: Government Printing Office, 1902), 59. The official in charge of the Benguet civil sanitarium in 1902 noted a purchase of land from Otto Scheerer "[a]s a nucleus of the institution." See Appendix H, "Report of the Attending Physician and Surgeon in Charge of the Civil Sanitarium at Baguio, Benguet for the Year Ending August 31, 1902," in *Third Annual Report of the Philippine Commission, 1902, Part 1*, 443
51. *Supreme Court of the United States. October Term, 1907. No. 298. Mateo Cariño, Plaintiff in Error, vs. The Insular Government of the Philippine Islands. In Error to the Supreme Court of the Philippine Islands. Brief for Plaintiff in Error*, NARA, RG 350, Entry 5, Box 781, 4. On the military's claim to some 530 acres of land as a "reservation," see Lewis W. Call, *United States Military Reservations, National*

*Cemeteries, and Military Parks: Title, Jurisdiction Etc., Revised Edition* (Washington, DC: Government Printing Office, 1907), 406.

52. *Transcript of Record. Supreme Court of the United States. October Term, 1907. No. 298. Mateo Cariño, Plaintiff in Error, vs. The Insular Government of the Philippine Islands . . .*, 22.
53. Moss, *Nabaloi Law and Ritual*, 253. On land ownership and ritual see also Wiber, "The Cañao Imperative"; Wiber, *Politics, Property and Law in the Philippine Uplands*; Tapang, *Innovation and Social Change*; Bagamaspad, Hamada-Pawid, and Balangoy, *A Peoples' History of Benguet Province.*
54. *Transcript of Record. Supreme Court of the United States. October Term, 1907. No. 298. Mateo Cariño, Plaintiff in Error, vs. The Insular Government of the Philippine Islands . . .*, 7. See also *Supreme Court of the United States. October Term, 1907. No. 298. Mateo Cariño, Plaintiff in Error, vs. The Insular Government of the Philippine Islands. In Error to the Supreme Court of the Philippine Islands. Brief for Plaintiff in Error*, 3. See Bagamaspad, Hamada-Pawid, and Balangoy, *A Peoples' History of Benguet Province*, 212, on fencing of pastureland.
55. *Transcript of Record. Supreme Court of the United States. October Term, 1907. No. 298. Mateo Cariño, Plaintiff in Error, vs. The Insular Government of the Philippine Islands . . .*, 21.
56. Ibid., 137.
57. *Transcript of Record. Supreme Court of the United States. October Term, 1907. No. 298. Mateo Cariño, Plaintiff in Error, vs. The Insular Government of the Philippine Islands . . .*, 137–39.
58. Ibid., 8.
59. *Supreme Court of the United States. October Term, 1907. No. 298. Mateo Cariño, Plaintiff in Error, vs. The Insular Government of the Philippine Islands. In Error to the Supreme Court of the Philippine Islands. Brief for Plaintiff in Error*, 3.
60. *Transcript of Record. Supreme Court of the United States. October Term, 1907. No. 298. Mateo Cariño, Plaintiff in Error, vs. The Insular Government of the Philippine Islands . . .*, 59 and 63–65.
61. Cariño, "The Cariños and Baguio-Benguet History: Conclusion," in *Folio 2*, 55.
62. Lynch, *Colonial Legacies in a Fragile Republic*, 162.
63. Ibid., 170, 135.
64. Ibid., 170
65. Ibid., 374.
66. Wiber, *Politics, Property and Law in the Philippine Uplands*, 131–32.
67. Ibid., 115.
68. "Report of the Governor of the Province of Benguet," in *Annual Reports of the War Department for the Fiscal Year Ended June 30, 1905*, vol. 10, *Report of the Philippine Commission. Part 1*, 179.
69. Ibid.
70. "Report of the Governor of Benguet," in *Eighth Annual Report of the Philippine Commission to the Secretary of War, 1907 (In Three Parts), Part 1*, 278.

71. Ibid., 278–79.
72. Ibid., 282.
73. *Transcript of Record. Supreme Court of the United States. October Term, 1907. No. 298. Mateo Cariño, Plaintiff in Error, vs. The Insular Government of the Philippine Islands* . . . , 182.
74. Ibid., 180.
75. Ibid., 48, 47.
76. "*Carino v. Insular Government* 212 U.S. 449 (1909)," Justia Law, https://supreme.justia.com/cases/federal/us/212/449/case.html.
77. *Transcript of Record. Supreme Court of the United States. October Term, 1907. No. 298. Mateo Cariño, Plaintiff in Error, vs. The Insular Government of the Philippine Islands* . . . , 26.
78. Ibid.
79. Ibid., 21–22.
80. Phelps Whitmarsh, Provincial Governor of Benguet, to the Philippine Commission, January 16, 1901, in "Correspondence in re conflict between military and civil authorities in the Province of Benguet, Islands of Luzon, Philippine Islands, Office of the Military Governor, Manila, May 14, 1901," NARA, RG 350, General Classified Files, 1898–1945, Entry 5, Box 274.
81. "Benguet Commercial Company," *Manila Times*, February 24, 1904, 4.
82. Otto Scheerer to General MacArthur, October 9, 1900, in "Correspondence in re conflict between military and civil authorities," NARA, RG 350, General Classified Files, 1898–1945, Entry 5, Box 274.
83. "Memoranda, of the Conflict, Charges, Hostilities and Misunderstandings between the Military and Civil Authorities, following the appointment of a civil governor and secretary for the province of Benguet, Submitted by General Arthur MacArthur, Military Governor to the Adjutant General," NARA, RG 350, BIA, 2368, Entry 5, Box 274.
84. Ibid.
85. Duvall to Robert Rudd, May 18, 1900, Reel 1, Rudd Papers.
86. "Memoranda, of the Conflict, Charges, Hostilities and Misunderstandings between the Military and Civil Authorities," NARA, RG 350, BIA, 2368 Entry 5, Box 274.
87. William Howard Taft to Arthur MacArthur, January 9, 1901, "Correspondence in re conflict between military and civil authorities," NARA, RG 350, General Classified Files, 1898–1945, Entry 5, Box 274.
88. Duvall to John Green Ballance, Chief Assistant, Headquarters, First District, January 24, 1904, "Correspondence in re conflict between military and civil authorities," NARA, RG 350, General Classified Files, 1898–1945, Entry 5, Box 274.
89. Cariño, "The Cariños and Baguio-Benguet History: Conclusion," in *Folio 2*, 14; Col. Duvall to Adjutant General, Department of Northern Luzon, December 28, 1900, "Correspondence in re conflict between military and

civil authorities," NARA, RG 350, General Classified Files, 1898–1945, Entry 5, Box 274.

90. Phelps Whitmarsh (no date) in "Correspondence in re conflict between military and civil authorities," NARA, RG 350, General Classified Files, 1898–1945, Entry 5, Box 274.
91. *Transcript of Record. Supreme Court of the United States. October Term, 1907. No. 298. Mateo Cariño, Plaintiff in Error, vs. The Insular Government of the Philippine Islands . . .*, 35, 177–78.
92. "The Benguet Consolidated Mining Co.," *Far Eastern Review* (June 1907): 21. See also Olivia M. Habana, "Gold Mining in Benguet: 1900–1941," *Philippine Studies* 49 (2000): 3–41; Benguet Consolidated, Inc., and Ed. C. de Jesus, *Benguet Consolidated, Inc., 1903–1978: A Brief History* (Benguet, 1978).
93. "The Philippines," *LIFE Magazine*, February 13, 1939, 58; Fry, *A History of the Mountain Province*, 183.
94. Fry, *A History of the Mountain Province*, 183.
95. "The Benguet Consolidated Mining Co.," 21.
96. "*Carino v. Insular Government* 212 U.S. 449 (1909)."
97. Banner, *How the Indians Lost Their Land* , 39–41. See also T. Alexander Aleinikoff, *Semblances of Sovereignty: The Constitution, the State, and American Citizenship* (Cambridge: Harvard University Press, 2002).
98. *No. 72. In the Supreme Court of the United States. October Term, 1908. Mateo Cariño, Plaintiff in Error, v. The Insular Government of the Philippine Islands. In Error to the Supreme Court of the Philippine Islands. Brief for the United States and the Insular Government*, NARA, RG 350, Entry 5, Box 781, 56–57.
99. On the insular cases see Bartholomew H. Sparrow, *The Insular Cases and the Emergence of American Empire* (Lawrence: University Press of Kansas, 2006), 55. See also Christina Duffy Burnett and Burke Marshall, *Foreign in a Domestic Sense: Puerto Rico, American Expansion, and the Constitution* (Durham: Duke University Press, 2001); Aleinikoff, *Semblances of Sovereignty*.
100. *Supreme Court of the United States. October Term, 1907. No. 298. Mateo Cariño, Plaintiff in Error, vs. The Insular Government of the Philippine Islands. In Error to the Supreme Court of the Philippine Islands. Brief for Plaintiff in Error*, 9.
101. Ibid., 13.
102. Ibid., 20.
103. Ibid., 19.
104. On the distinction between ownership and sovereignty see Banner, *How the Indians Lost Their Land*, 6–7.
105. William Franklin Willoughby, *Territories and Dependencies of the United States: Their Government and Administration* (New York: Century Co., 1905), 79.
106. "*Carino v. Insular Government* 212 U.S. 449 (1909)."
107. Burnett and Marshall, *Foreign in a Domestic Sense*; John A. Garraty, "Holmes's Appointment to the US Supreme Court," *New England Quarterly* 22 (September 1949): 294.

108. "*Carino v. Insular Government* 212 U.S. 449 (1909)."
109. Ibid.
110. Ibid.
111. Ibid.
112. Amy Blitz, *The Contested State: American Foreign Policy and Regime Change in the Philippines* (Lanham, MD: Rowman & Littlefield Publishers, 2000), 45–47.
113. *No. 72. In the Supreme Court of the United States. October Term, 1908. Mateo Cariño, Plaintiff in Error, v. The Insular Government of the Philippine Islands. In Error to the Supreme Court of the Philippine Islands. Brief for the United States and the Insular Government*, 49–50.
114. Frank Pommersheim, *Broken Landscape: Indians, Indian Tribes, and the Constitution* (Oxford: Oxford University Press, 2009), 125.
115. David A. Chang, "Enclosures of Land and Sovereignty: The Allotment of American Indian Lands," *Radical History Review* no. 109 (2011): 116–17.
116. See Pommersheim, *Broken Landscape*; Kaplan, *The Anarchy of Empire in the Making of US Culture*; Burnett and Marshall, *Foreign in a Domestic Sense*.
117. "City in the Skylands," *Chronicle Magazine*, April 7, 1962, 34.
118. Bagamaspad, Hamada-Pawid, and Balangoy, *A Peoples' History of Benguet Province*, 213.
119. Owen Lynch, "Concepts and Strategies for Promoting Legal Recognition of Community-Based Property Rights: Insights from the Philippines and Other Nations," in *Communities and Conservation: Histories and Politics of Community-Based Natural Resource Management*, ed. Peter J. Brosius, Anna Lowenhaupt Tsing, and Charles Zerner (Walnut Creek, CA: AltaMira Press, 2005), 401. See also Lynch, "Land Rights, Land Laws, and Land Usurpation: The Spanish Era (1565–1898)," *Philippine Law Journal* 63 (1988): 82–111.
120. On the Regalian Doctrine see Peter Causay, "Indigenizing Law or Legalizing Governmentality? The Indigenous Peoples Rights Act and the Philippine Supreme Court," http://dlc.dlib.indiana.edu/dlc/bitstream/handle/10535/1451/Peter_Causay.pdf?sequence=1, 2–3; Lynch, "Concepts and Strategies for Promoting Legal Recognition of Community-Based Property Rights: Insights from the Philippines and Other Nations"; Lynch, *Colonial Legacies in a Fragile Republic*.
121. Burnham to William Cameron Forbes, October 6, 1905, Burnham letterpress copybooks, 1890–1912, on microfilm, v. 14, Reel 6, XV, Art Institute of Chicago.
122. Daniel Burnham and Pierce Anderson, "Report on the Proposed Plan of the City of Baguio, Province of Benguet, P.I.," in Moore, *Daniel H. Burnham*, 197.
123. Burnham, *Plan of Chicago*, 29.
124. Ibid.
125. Peter Hall, *Cities of Tomorrow: An Intellectual History of Urban Planning and Design in the Twentieth Century*, 3rd edition (Malden, MA: Blackwell Publishing, 2002), 191.

126. Burnham and Anderson, "Report on Proposed Improvements at Manila," in Moore, *Daniel H. Burnham*, 188.
127. Ibid., 198.
128. "Exhibit B. Report of the Chief of the Bureau of Architecture and Construction of Public Buildings," in *Fourth Annual Report of the Philippine Commission, 1903, Part 3*, 933.
129. Reed, *Hispanic Urbanism in the Philippines*, 166–67.
130. See Doeppers, "Manila's Imperial Makeover: Security, Health, and Symbolism," in *Colonial Crucible*, ed. McCoy and Scarano.
131. Hines, "The Imperial Mall: The City Beautiful Movement and the Washington Plan of 1901–1902," 81.
132. Bluestone, *Constructing Chicago*, 185.
133. Ibid., 190, 187.
134. Wilson, *The City Beautiful Movement*, 79.
135. Burnham and Anderson, "Report on Proposed Improvements at Manila," in Moore, *Daniel H. Burnham*, 195.
136. Burnham and Anderson, "Report on the Proposed Plan of the City of Baguio, Province of Benguet, P.I.," in Moore, *Daniel H. Burnham*, 197.
137. Burnham and Anderson, "Report on Proposed Improvements at Manila," in Moore, *Daniel H. Burnham*, 188.
138. Ibid., 181–82.
139. Daniel H. Burnham and Pierce Anderson, "Appendix I. Preliminary Plan of Baguio, Province of Benguet, P.I., June 27, 1905," in *Sixth Annual Report of the Philippine Commission, 1905 (In Four Parts), Part 3*, 405.
140. Burnham and Anderson, "Report on the Proposed Plan of the City of Baguio, Province of Benguet, P.I.," in Moore, *Daniel H. Burnham*, 198.
141. Burnham, *Plan of Chicago*, 90.
142. Burnham and Anderson, "Report on the Proposed Plan of the City of Baguio, Province of Benguet, P.I.," in Moore, *Daniel H. Burnham*, 198, 199.
143. For Rudyard Kipling on Indian hill station society, see his *Plain Tales from the Hills* (New York: Penguin Books, 1990 [1890]).
144. Kennedy, *The Magic Mountains*, 14.
145. Burnham and Anderson, "Report on Proposed Improvements at Manila," in Moore, *Daniel H. Burnham*, 179.
146. Ibid., 195.
147. Gwendolyn Wright, *The Politics of Design in French Colonial Urbanism* (Chicago: University of Chicago Press, 1991), 190, 166.
148. Reed, *Hispanic Urbanism in the Philippines*, 59. On the influence of Italian Renaissance architects, see also Daniel F. Doeppers, "The Development of Philippine Cities Before 1900," *Journal of Asian Studies* 31 (August 1972): 769–92.
149. Reed, *Hispanic Urbanism in the Philippines*, 60, 61–62; Doeppers, "The Development of Philippine Cities Before 1900," 776.

150. Reed, *Hispanic Urbanism in the Philippines*, 64.
151. Ibid., 166.
152. Kennedy, *The Magic Mountains*, 4.
153. Doeppers, *Manila, 1900–1941*, 776.
154. Burnham and Anderson, "Report on the Proposed Plan of the City of Baguio, Province of Benguet, P.I.," in Moore, *Daniel H. Burnham*, 199.
155. Burnham and Anderson, "Report on Proposed Improvements at Manila," in Moore, *Daniel H. Burnham*, 183, 188, 195.
156. Wilson, *The City Beautiful Movement*, 87–89.
157. Burnham, *Plan of Chicago*, 18.
158. Ibid.
159. A. N. Rebori, "The Work of William E. Parsons in the Philippine Islands," Part I, *Architectural Record* 41 (April 1917): 309. See also Lico on Parsons in *Arkitekturang Filipino*.
160. Rodgers, *Atlantic Crossings*, 160.
161. Szuberla makes this observation, too, in "Urban Vistas and the Pastoral Garden," 38–39.
162. Burnham and Anderson, "Report on the Proposed Plan of the City of Baguio, Province of Benguet, P.I.," in Moore, *Daniel H. Burnham*, 198.
163. Kennedy, *The Magic Mountains*, 3.
164. Burnham and Anderson, "Report on the Proposed Plan of the City of Baguio, Province of Benguet, P.I.," in Moore, *Daniel H. Burnham*, 201.
165. Hines, *Burnham of Chicago, Architect and Planner*, 184; ibid., 201.
166. Burnham and Anderson, "Report on the Proposed Plan of the City of Baguio, Province of Benguet, P.I.," in Moore, *Daniel H. Burnham*, 202.
167. "Exhibit C. Report of the Consulting Architect," in *Seventh Annual Report of the Philippine Commission, 1906, (In Three Parts), Part 2* (Washington, DC: Government Printing Office, 1907), 368.
168. Burnham and Anderson, "Report on Proposed Improvements at Manila," in Moore, *Daniel H. Burnham*, 180.
169. Ibid., 183.
170. Ibid., 182.
171. Burnham, *Plan of Chicago*, 47–48.
172. Burnham and Anderson, "Report on Proposed Improvements at Manila," in Moore, *Daniel H. Burnham*, 182.
173. Ibid.
174. Roy Rosenzweig and Elizabeth Blackmar, *The Park and the People: A History of Central Park* (Ithaca: Cornell University Press, 1992), 136.
175. Burnham, *Plan of Chicago*, 53.
176. Sullivan, *The Autobiography of an Idea*, 313.
177. Bluestone, *Constructing Chicago*, 194.
178. Williams, *The Country and the City*, 46.
179. Burnham and Anderson, "Report on Proposed Improvements at Manila," in Moore, *Daniel H. Burnham*, 188–89.

180. To City Beautiful advocates, the coexistence of commerce and culture in "a single, appropriate landscape" was doubtful, writes Daniel Bluestone in *Constructing Chicago*, 194.
181. Burnham and Anderson, "Appendix I. Preliminary Plan of Baguio, Province of Benguet, P.I., June 27, 1905," in *Sixth Annual Report of the Philippine Commission, 1905 (In Four Parts), Part 3*, 405.
182. W. J. T. Mitchell, *Landscape and Power* (Chicago: University of Chicago Press, 1994), 10. See also Rafael, *White Love* on the "colonial picturesque" in chapter 2.
183. Mitchell, *Landscape and Power*, 10.
184. Marx, *The Machine in the Garden*, 23.
185. Burnham, *Plan of Chicago*, 43.
186. Szuberla, "Urban Vistas and the Pastoral Garden," 24, 39.
187. Scheerer, *On Baguio's Past*, 3.
188. Pérez, *Relaciones Agustinianas de Las Razas del Norte de Luzon*, 198.
189. "Exhibit C. Report of the Consulting Architect," in *Seventh Annual Report of the Philippine Commission, 1906 (In Three Parts), Part 2*, 365.
190. See Reed, *City of Pines*, maps on pages 104–9.
191. Hines, *Burnham of Chicago, Architect and Planner*, 213.
192. Frederic R. Coudert, Jr., "Our New Peoples: Citizens, Subjects, Nationals or Aliens," *Columbia Law Review* (January 1, 1903): 13–32. On the category of the "national" see also Christina Duffy Burnett, "Empire and the Transformation of Citizenship," in *Colonial Crucible*, ed. McCoy and Scarano; Veta R. Schlimgen, "Neither Citizens Nor Aliens: Filipino 'American Nationals' in the US Empire, 1900–1946" (PhD diss., University of Oregon, 2010).
193. Coudert, "Our New Peoples: Citizens, Subjects, Nationals or Aliens," 13, 14, 13.
194. Ibid., 17, 29, 32, 17.
195. Malcolm, *The Government of the Philippine Islands*, 392.
196. "The attached article was prepared for the Secretary of War by General McIntyre and Major Hunt, in response to a request of Mr. George Marvin, 1718 H. Street, for an article from the Secretary on the War Department as a Colonial Government. The original of this was handed to the Secretary of War, March 14, 1914." Found in NARA, RG 350, Entry 5, Box 535.
197. *Annual Reports of the War Department for the Fiscal Year Ended June 30, 1901, Report of the Secretary of War*, 87.
198. Paper read by Col. Clarence R. Edwards, Chief of the Bureau of Insular Affairs, before the National Geographic Society, at the Cosmos Club, Washington, DC, January 22, 1904, 6, NARA, RG 350, Entry 5, Box 15.
199. Mitchell, *Landscape and Power*, 16.

## CHAPTER FOUR

1. William Cameron Forbes, journal, June 13, 1905, MS 1365, Forbes Papers.
2. Forbes, journal, August 9, 1904, MS 1365, Forbes Papers.

3. On the "cargo cult" see Lamont Lindstrom, *Cargo Cult: Strange Stories of Desire from Melanesia and Beyond* (Honolulu: University of Hawai'i Press, 1993); Holger Jebens, ed., *Cargo, Cult, and Culture Critique* (Honolulu: University of Hawai'i Press, 2004); Ton Otto, "What Happened to Cargo Cults? Material Religions in Melanesia and the West," *Social Analysis* 53 (Spring 2009): 82–102.
4. Lamont Lindstrom, "Cargo Cult at the Third Millennium," in *Cargo, Cult, and Culture Critique*, ed. Jebens, 15.
5. Otto, "What Happened to Cargo Cults? Material Religions in Melanesia and the West," 83.
6. *Report of Maj.-Gen. E. S. Otis, United States Army Commanding Division of the Philippines, Military Governor. September 1, 1899 to May 5, 1900* (Washington, DC: Government Printing Office, 1900), 142; John R. M. Taylor, *The Philippine Insurrection against the United States: A Compilation of Documents with Notes and Introduction* (Pasay City, Philippines: Eugenio Lopez Foundation, 1971), 263.
7. Ileto, *Pasyon and Revolution*, 8, 117, 118, 189–90. See also Sturtevant, *Popular Uprisings in the Philippines, 1840–1940.*
8. See Rafael, *Contracting Colonialism.*
9. Henry L. Stimson to Forbes, September 1, 1911, Box 6, bMS Am 1364, Forbes Papers.
10. Kramer, *The Blood of Government*, 287–88.
11. Jacobson, *Barbarian Virtues*, 50–51.
12. I use "marketplace" to define the site of exchange and "market" to refer to a greater field of trade that traversed constructed spaces. I'm indebted to the scholarship of Jean-Christophe Agnew for cultural-historical approaches to markets and marketplaces, especially "The Threshold of Exchange: Speculations on the Market," *Radical History Review* 21 (Fall 1979): 99–118; and *Worlds Apart: The Market and the Theater in Anglo-American Thought, 1550–1750* (New York: Cambridge University Press, 1986).
13. "Exhibit I. The Preliminary Report of the Commission," *Report of the Philippine Commission to the President*, vol. 1, *January 31, 1900*, 182.
14. Karl Marx could have foreseen this; as David Harvey summarizes it, "The brilliance of Marx's dialectical method is to show that market liberalization—the credo of the liberals and the neo-liberals—will not produce a harmonious state in which everyone is better off. It will instead produce ever greater levels of social inequality." David Harvey, "The 'New' Imperialism," 73.
15. William D. Boyce, *The Philippine Islands. Illustrated* (Chicago: Rand McNally & Co., 1914), 91.
16. *Eighth Annual Report of the Philippine Commission to the Secretary of War, 1907 (In Three Parts), Part 2*, 281.
17. "Report of the Governor of Benguet," in *Eighth Annual Report of the Philippine Commission to the Secretary War, 1907 (In Three Parts), Part 1*, 278.
18. "Report of the Governor of the Province of Benguet," in *Seventh Annual Re-*

*port of the Philippine Commission, 1906, In Three Parts, Part 1* (Washington, DC: Government Printing Office, 1907), 198.

19. "Report of the Governor of Benguet," in *Report of the Philippine Commission to the Secretary of War, 1908 (In Two Parts), Part 1* (Washington, DC: Government Printing Office, 1909), 245. See also Reyes-Boquiren, "The History and Political Economy of the Vegetable Industry in Benguet," CSC Working Paper 14.
20. Tapang, *Innovation and Social Change*, 16.
21. Marshall A. Barber, "An Igorrote Dog Market at Baguio," *Graduate Magazine of the University of Kansas* 10 (June 1912): 329.
22. Moss, *Nabaloi Law and Ritual*, 227.
23. Lee Donald Warren, *Isles of Opportunity: Progress and Possibilities in the Philippine Islands* (Washington, DC: Review and Herald Pub. Association, 1928), 135.
24. Ibid.
25. Barber, "An Igorrote Dog Market at Baguio," 329.
26. Lodge, *The Retention of the Philippine Islands. Speech of Henry Cabot Lodge of Massachusetts, in the Senate of the United States, March 7, 1900*, 41.
27. Ibid., 38. For the year 1913, the Philippine Commission reported imports from the United States at $25,646,875; see *Report of the Philippine Commission to the Secretary of War, 1913 (In One Part)* (Washington, DC: Government Printing Office, 1914), 228. Goods shipped from the United States to Puerto Rico in that year totaled $33,155,005. See *Report of the Governor of Porto Rico to the Secretary of War* (Washington, DC: Government Printing Office, 1920), 7.
28. *Reports of the Taft Philippine Commission to the President* (1901), 58.
29. *Fourth Annual Report of the Philippine Commission, 1903, In Three Parts, Part 1*, 50.
30. Charles Arthur Conant, *A History of Modern Banks of Issue, with an Account of the Economic Crises of the Present Century*, Fourth Impression (New York: G. P. Putnam's Sons, 1902 [1896]), 463.
31. Charles Arthur Conant, "The Economic Basis of 'Imperialism,'" *North American Review* 167 (September 1898): 328, 330. See also Julian Go, *Patterns of Empire: The British and American Empires, 1688 to the Present* (New York: Cambridge University Press, 2011), 221.
32. Conant, "The Economic Basis of 'Imperialism,'" 338.
33. Ibid., 339.
34. Ibid., 338.
35. Charles A. Conant, *A Special Report: Coinage and Banking in the Philippine Islands Made to the Secretary of War [from the "Annual Report of the Secretary of War," Appendix G]* (Washington, DC: Government Printing Office, 1901), 16.
36. *Reports of the Philippine Commission, the Civil Governor and the Heads of the Executive Departments . . .* (1904), 101.
37. Conant, *A Special Report: Coinage and Banking in the Philippine Islands . . .*, 16.
38. Ibid., 39–42.
39. Frederick S. Allis, Jr., and Alexander W. Williams, "Forbes Family Papers,"

Massachusetts Historical Society, http://www.masshist.org/collection-guides/view/fa0269.

40. "A Chronicle of the China Trade," Harvard Business School, Historical Collections Exhibit, http://www.library.hbs.edu/hc/heard/canton-trade.html.
41. Robert B. Forbes, *Personal Reminiscences, Second Edition, Revised, to which is added Rambling Recollections Connected with China* (Boston: Little, Brown, and Company, 1882), 369, 371.
42. Henry Greenleaf Pearson, *An American Railroad Builder* (Boston: Houghton Mifflin Company, 1911), 5–6.
43. See Betty Farrell, *Elite Families: Class and Power in Nineteenth-Century Boston* (Albany: State University of New York, 1993), 154.
44. See Forbes, journal, February 1904, MS Am 1365, Forbes Papers.
45. "Hidden Dangers," *El Renacimiento*, August 26, 1905, 1. (Published in English.)
46. United States Bureau of the Census, *Census of the Philippine Islands, Taken under the Direction of the Philippine Commission in the Year 1903, in Four Volumes, Vol. 4, No. 1*, http://name.umdl.umich.edu/ajb5834.0004.001, 541–3.
47. Ibid., 544.
48. Forbes, journal, May 25, 1906, MS Am 1365, Forbes Papers; Edwin Walter Kemmerer, *Postal Savings: An Historical and Critical Study of the Postal Savings Bank System of the United States* (Princeton: Princeton University Press, 1917).
49. Forbes to C. E. Perkins, April 20, 1906, MS Am 1366, Forbes Papers.
50. Census Office of the Philippine Islands, *Census of the Philippine Islands Taken under the Direction of the Philippine Legislature in the Year 1918, Vol. 4, Part 2* (Manila: Bureau of Printing, 1921), http://name.umdl.umich.edu/AJB5835.0004.002, 730.
51. John R. Arnold, "Municipal Market Buildings in the Philippines. The Civic Betterment Brought About by the Program of Modern Market-Building of the Philippines," *American City* 8 (June 1913): 595.
52. Bureau of Public Works, "Market Construction Program," *Quarterly Bulletin* 2 (October 1913): 9–10.
53. Arnold, "Municipal Market Buildings in the Philippines," 594.
54. Bureau of Public Works, "Market Construction Program," 10.
55. Ibid., 11, 13.
56. Ibid., 12–13.
57. Ibid., 13.
58. Arnold, "Municipal Market Buildings in the Philippines," 594.
59. Bureau of Public Works, "Market Construction Program," 13.
60. Arnold, "Municipal Market Buildings in the Philippines," 591, 595, 591.
61. Ordinance No. 103. "An ordinance to establish and regulate public markets in the city of Manila, and prohibit their establishment by any other person, association, or corporation other than the city of Manila," April 30, 1908, NARA, RG 350, Entry 5, Box 305.

62. "Report of the Governor of Benguet," in *Report of the Philippine Commission to the Secretary of War, 1908 (In Two Parts), Part 1*, 246.
63. Bureau of Public Works, *Quarterly Bulletin* 2 (January 1914): 46.
64. Corpuz, *An Economic History of the Philippines*, 235.
65. Lodge, *The Retention of the Philippine Islands*, 36.
66. J. L. Harrison, "Public Works—Past and Present," *Quarterly Bulletin* 4 (April 1915): 18.
67. See Agnew, "The Threshold of Exchange: Speculations on the Market," 106, on the way marketplaces accorded "small producers" a measure of power.
68. Forbes to Henry Stimson, June 13, 1911, NARA, RG 350, Entry 5A, Box 535.
69. "William Cameron Forbes, 'Our Dependencies,' May 27–28, 1914. For discussion at the National Foreign Trade Convention, Washington, DC, under the auspices of the American Manufacturers Export Association, The Pan-American Society of the United States, and The American Asiatic Association," bMS Am 1364.4, Forbes Papers.
70. Major John P. Finley, "The Commercial Awakening of the Moro and Pagan," *North American Review* 197 (March 1913): 330.
71. On slavery as a marker of unreadiness for independence see Salman, *The Embarrassment of Slavery*; on the problem of freedom in relation to wage labor and contract ideology see Stanley, *From Bondage to Contract*.
72. Jeffrey P. Sklansky, *The Soul's Economy: Market Society and Selfhood in American Thought, 1820–1920* (Chapel Hill: University of North Carolina Press, 2002), 3.
73. Forbes to Secretary of War William Howard Taft, May 27, 1908, MS Am 1366, Forbes Papers.
74. William Howard Taft, "The Duty of Americans in the Philippines," address before Union Reading College, Manila, December 17, 1903, in John B. Devins, *An Observer in the Philippines or, Life in Our New Possessions* (New York: American Tract Society, 1905), 391–92.
75. Finley, "The Commercial Awakening of the Moro and Pagan," 328, 331.
76. W. Cameron Forbes, *The Romance of Business* (Boston: Houghton Mifflin, 1921), 135–36.
77. Norman Angell, *The Great Illusion: A Study of the Relation of Military Power in Nations to Their Economic and Social Advantage* (New York: Putnam, 1910), ix, xi.
78. Sklansky, *The Soul's Economy*, 229.
79. Richard T. Ely, *Studies in the Evolution of Industrial Society* (New York: Macmillan Company, 1903), 434.
80. Ibid., 435.
81. Gustav Peebles, "The Anthropology of Credit and Debt," *Annual Review of Anthropology* 39 (2010): 227. Relatedly, Thomas Haskell has written of the relationship between humanitarianism and the cognitive effects of capitalism in attempting to reconcile abolitionists' moral commitment with their

economic interests in ending slavery. He suggests that the very mental revolutions and practices that fostered the rise of capitalism—a sense of consequences for remote acts, a sense of mastery over circumstances, a "growing reliance on mutual promises, or contractual relations, in lieu of relations based on status, custom, or traditional authority"—may have been the same "recipe knowledge" for producing a humanitarian sensibility. See Thomas L. Haskell, "Capitalism and the Origins of the Humanitarian Sensibility, Part 2," *American Historical Review* 90 (1985): 553.

82. Renato Rosaldo, "Imperialist Nostalgia," in "Memory and Counter-Memory," special issue, *Representations* 26 (Spring 1989): 107–8.
83. Emma Sarepta Yule, "The Dog Market at Baguio," *Overland Monthly* (November 1913): 441–42.
84. Forbes, *The Romance of Business*, 136. See Frederic Jameson, "Modernism and Imperialism," in *Nationalism, Colonialism, and Literature*, ed. Terry Eagleton, Frederic Jameson, and Edward Said (Minneapolis: University of Minnesota Press, 1990); on the "representational effects" of the "imperial world system," see 156.
85. Arnold, "Municipal Market Buildings in the Philippines," 591.
86. Anderson, *Colonial Pathologies*, 114.
87. Isabel Anderson, *The Spell of the Hawaiian Islands and the Philippines, Being an Account of the Historical and Political Condition of Our Pacific Possessions, together with Descriptions of the Natural Charm and Beauty of the Countries and the Strange and Interesting Customs of their Peoples* (Boston: The Page Company, 1916), 225–26.
88. Anderson, *Colonial Pathologies*, 68. See also Reynaldo C. Ileto, "Cholera and the Origins of the American Sanitary Order in the Philippines," in *Discrepant Histories: Translocal Essays on Filipino Culture*, ed. Vicente L. Rafael (Philadelphia: Temple University Press, 1995).
89. Ileto, "Cholera and the Origins of the American Sanitary Order in the Philippines," in *Discrepant Histories*, 64.
90. Anderson, *The Spell of the Hawaiian Islands and the Philippines*, 226.
91. Anderson, *Colonial Pathologies*, 115.
92. On ritual purity see, for example, Mary Douglas, *Purity and Danger: An Analysis of Concepts of Pollution and Taboo* (New York: Routledge, 1966). The task that the colonial government assumed here—the safety of city dwellers from the dangers of unsanitary marketplaces—calls to mind Foucault's notion of pastoral power: how in its police powers, the modern state absorbed responsibilities that the Christian pastorate had once assumed in the care of the spiritual flock. See Foucault, *Security, Territory, Population*.
93. Tamara Plakins Thornton, "Capitalist Aesthetics: Americans Look at the London and Liverpool Docks," in *Capitalism Takes Command: The Social Transformation of Nineteenth-Century America*, ed. Michael Zakim and Gary John Kornblith (Chicago: University of Chicago Press, 2012), 193–94.
94. Kennedy, *The Magic Mountains*, 9.

95. Ibid.
96. Yule, "The Dog Market at Baguio," 436.
97. Barber, "An Igorrote Dog Market at Baguio," 329–31.
98. James J. Halsema, *E. J. Halsema, Colonial Engineer: A Biography* (Quezon City, Philippines: New Day Publishers, 1991), 197; Moss, *Nabaloi Law and Ritual*, 224.
99. Dean Worcester, "The Non-Christian Peoples of the Philippine Islands," *National Geographic Magazine* 24 (November 1913): 1200.
100. On race as a "visual technology" see Deborah Poole, *Vision, Race, and Modernity: A Visual Economy of the Andean Image World* (Princeton: Princeton University Press, 1997), 16.
101. Yule, "The Dog Market at Baguio," 436.
102. Marx, *Capital, Volume One*, excerpted in *The Marx-Engels Reader*, 319.
103. Yule, "The Dog Market at Baguio," 436.
104. Amit S. Rai, *Rule of Sympathy: Sentiment, Race, and Power, 1750–1850* (New York: Palgrave, 2002).
105. Ibid., 15.
106. James Blount, *The American Occupation of the Philippines, 1898–1912, with an introductory essay by Renato Constantino* (Metro Manila, Philippines: Solar Publishing Company, 1968 [1913]), 578, 573.
107. Dean C. Worcester, *Slavery and Peonage in the Philippine Islands* (Manila: Bureau of Printing, 1913), 81.
108. Ibid., 65, 52. On the case of slavery in the Philippines and its colonial politics, see Salman, *The Embarrassment of Slavery*.
109. Worcester, *Slavery and Peonage in the Philippine Islands*, 52.
110. Salman, *The Embarrassment of Slavery*, 168–69.
111. Ibid., 144.
112. Worcester, *Slavery and Peonage in the Philippine Islands*, 36.
113. Salman, *The Embarrassment of Slavery*, 164.
114. Wiber, "Levels of Property Rights, Levels of Law," 476–77
115. Paul Kramer has also discussed this in *The Blood of Government* in his chapter on the 1904 St. Louis World's Fair. See also Finin, *The Making of the Igorot*; Mark Rice, "His Name Was Don Francisco Muro: Reconstructing an Image of American Imperialism," *American Quarterly* 62 (2010): 49–76; Mark Rice, *Dean Worcester's Fantasy Islands: Photography, Film, and the Colonial Philippines* (Ann Arbor: University of Michigan Press, 2014).
116. Worcester to John H. Blair, November 6, 1912; William Dinwiddie to Dean Worcester, August 23, 1911, Box 1, Dean C. Worcester Papers, Bentley Historical Library, University of Michigan (hereafter "Worcester Papers").
117. Worcester to William Dinwiddie, June 12, 1912, Box 1, Worcester Papers.
118. G. Magie to William Dinwiddie, June 23, 1911, Box 1, Worcester Papers.
119. Worcester to M. Douglas Flattery, November 13, 1912, Box 1, Worcester Papers.
120. Worcester, "The Non-Christian Peoples of the Philippine Islands," 1255.

121. Ibid.
122. Kramer, *The Blood of Government*, 342.
123. "Report of the Governor of Benguet," in *Eighth Annual Report of the Philippine Commission to the Secretary of War, 1907 (In Three Parts), Part 1*, 277.
124. Bagamaspad, Hamada-Pawid, and Balangoy, *A Peoples' History of Benguet Province*, 227.
125. Anderson, *Under Three Flags*, 17.
126. Kramer, *The Blood of Government*, 325.
127. "'Non Christian Tribes,'" *El Renacimiento*, April 26, 1905, 1.
128. See Kramer, *The Blood of Government*, chapter 4; Rydell, *All the World's a Fair*.
129. "Peor que los Igorrotes, Hunt Ladrón, y Estafador," *El Renacimiento*, August 30, 1906, 1. See also Claire Prentice, *The Lost Tribe of Coney Island: Headhunters, Luna Park, and the Man Who Pulled off the Spectacle of the Century* (Boston: New Harvest, 2014).
130. "Advertising National Resources," *El Renacimiento*, January 3, 1906, 1. (Published in English.)
131. Maximo Manguiat Kalaw, *The Case for the Filipinos* (New York: The Century Co., 1916), http://archive.org/details/caseforfilipino00kalagoog, 183.
132. "Girls Sold in Islands," *Washington Post*, August 26, 1913; "Philippines Rife with Slave Trade," *Chicago Daily Tribune*, September 21, 1913; "Sell Girls for $100: Filipino Parents Do Not Think Such Slavery is Wrong," *Washington Post*, September 21, 1913.
133. Salman, *The Embarrassment of Slavery*, 216, 245–56.
134. The Jones Law of 1916, No. 240, "An Act to declare the purpose of the people of the United States as to the future political status of the people of the Philippine Islands, and to provide a more autonomous government for those islands," http://www.gov.ph/constitutions/the-jones-law-of-1916/. See also Julius William Pratt, *America's Colonial Experiment: How the United States Gained, Governed, and in Part Gave Away a Colonial Empire* (New York: Prentice-Hall, 1950), 205.
135. William Cameron Forbes, journal, October 24, 1907, MS 1365, Forbes Papers.
136. See Salman, *The Embarrassment of Slavery*, especially 177–79.
137. See Poole, *Vision, Race, and Modernity*. On photography and US colonial power in the Philippines see also Benito M. Vergara, Jr., *Displaying Filipinos: Photography and Colonialism in Early 20th Century Philippines* (Quezon City: University of the Philippines Press, 1995); Laura Wexler, *Tender Violence: Domestic Visions in an Age of US Imperialism* (Chapel Hill: University of North Carolina Press, 2000).
138. José S. Reyes, *Legislative History of America's Economic Policy toward the Philippines* (New York: Columbia University, 1923), 72.
139. Ibid., 117.
140. Corpuz, *An Economic History of the Philippines*, 229; "Report of the Secretary of Finance and Justice," in *Report of the Philippine Commission to the Secretary*

*of War, July 1, 1913 to December 31, 1914* (Washington, DC: Government Printing Office, 1915), 225.

141. See Corpuz, *An Economic History of the Philippines*; and Norman G. Owen, "Philippine Economic Development and American Policy: A Reappraisal," in *Compadre Colonialism: Studies in the Philippines Under American Rule*, ed. Norman G. Owen, Michigan Papers on South and Southeast Asia, no. 4 (Ann Arbor: Center for South and Southeast Asian Studies, University of Michigan, 1971).
142. Corpuz, *An Economic History of the Philippines*, 231.
143. "William Cameron Forbes, 'Our Dependencies,' For discussion at the National Foreign Trade Convention, Washington, DC," Forbes Papers.
144. Reed, *Colonial Manila*, 33.
145. Reed, *Hispanic Urbanism in the Philippines*, 174–75, 12.
146. Rafael, *The Promise of the Foreign*, 8.
147. Ibid., 10–11.
148. "Manila as Trade Depot," *Manila Times*, April 3, 1911, 4.
149. On Buencamino see Michael Cullinane, *Ilustrado Politics: Filipino Elite Responses to American Rule, 1898–1908* (Quezon City: Ateneo de Manila University Press, 2003).
150. Felipe Buencamino, "Eminently Practical Questions," *El Renacimiento*, November 4, 1905, 1.
151. Buencamino, "Eminently Practical Questions," *El Renacimiento*, November 7, 1905, 1–2.
152. Fitzhugh Lee Minnigerode, "The Simla of the Philippines," *New York Times*, December 5, 1920.
153. Rafael, *Contracting Colonialism*, 122.
154. Ibid., 128.
155. Ibid., 128, 127.
156. See also Julian Go on the two-way flow of power in *utang na loob* and patron-client relations in *American Empire and the Politics of Meaning*
157. See Rafael on the conjunctions of trade and translation in *Contracting Colonialism*, 21.
158. On this dynamic in a quite different context, see Agnew writing on Herman Melville's *The Confidence Man*: "For Melville, confidence, credibility, and credit are not the unconditional bases for social transactions, as evangelical faith and political economy would have it, but rather the unattainable objects of those transactions, for it is those same transactions, he implies, that place the ideal of mutual transparency farther out of reach." See *Worlds Apart*, 201.
159. Yule, "The Dog Market at Baguio," 442.
160. James LeRoy, *Philippine Life in Town and Country* (New York: G. P. Putnam's Sons, 1906), 292.
161. See Agnew, "The Threshold of Exchange: Speculations on the Market."

162. James A. LeRoy, "Race Prejudice in the Philippines," *Atlantic Monthly*, July 1902, 106.
163. Doherty, *Conditions in the Philippines . . .*, 15. See also Ileto, *Pasyon and Revolution*, 187.
164. Ileto, *Pasyon and Revolution*, 187.
165. Ibid., 3.
166. Vicente L. Rafael, "Welcoming What Comes: Sovereignty and Revolution in the Colonial Philippines," *Comparative Studies in Society and History* 52 (2010): 176.

## CHAPTER FIVE

1. "561 Will Take Trip to Baguio," *Manila Times*, February 25, 1911, 1.
2. "News Items," *The Teachers' Assembly Herald* 4 (1911): 17–18.
3. "Government Employes Housed at Baguio in Comfortable Buildings," *Manila Times*, March 20, 1911, 1.
4. "Report of the Secretary of Commerce and Police," in *Report of the Philippine Commission to the Secretary of War, 1911 (In One Part)* (Washington, DC: Government Printing Office, 1912), 114.
5. "The Plan of Baguio," *Teachers' Assembly Herald* 5 (May 12, 1912).
6. "Government Employes Housed at Baguio in Comfortable Buildings," *Manila Times*, March 20, 1911, 1.
7. "Many New Buildings at Baguio. Mountain Capital Has Had Building Boom During Past Year," *Manila Times*, March 15, 1911, 3.
8. Juan Claros Orendain, *Philippine Wonderland* (Baguio, 1940), 36.
9. Frank G. Carpenter, "A Switzerland in the Heart of the Tropics," *Los Angeles Times*, July 20, 1924, I13; "Report of the Secretary of Commerce and Police," in *Report of the Philippine Commission to the Secretary of War, 1911 (In One Part)*, 114.
10. George Malcolm, "Baguio, Yesterday, Today, And Tomorrow," in Baguio Golden Anniversary Supplement, *Baguio Midland Courier*, September 6, 1959, 11–12; "Mayors of the City of Baguio," City of Baguio, http://www.cityofpines.com/mayors.html.
11. Malcolm, "Baguio, Yesterday, Today, And Tomorrow," 11–12.
12. Orendain, *Philippine Wonderland*, 14.
13. Advertisement in the *Manila Times*, January 11, 1911, 3.
14. Advertisements in the *Manila Times*, January 18, 1911, 3; *Manila Times*, March 4, 1911, 10.
15. Carpenter, *Through the Philippines and Hawaii*, 86.
16. Ibid., 85.
17. Ibid., 86.
18. Richard V. Oulahan, "Lieut. Wood Won't Discuss the Details of His Stock Deals, Private Affair, He Declares, So Long as He Did Nothing to Involve His Father," *New York Times*, December 30, 1923.

19. Marx, *The Machine in the Garden*, 23.
20. On the formation of this "professional-managerial class" see Richard M. Ohmann, *Selling Culture: Magazines, Markets, and Class at the Turn of the Century* (New York: Verso, 1996).
21. Along similar lines Paul Kramer writes about Filipinos' use of "fiesta politics" as a means of "incorporating Americans into networks of obligation" in *The Blood of Government*, 186.
22. A parallel case of using nature as a "political resource" is in Mukerji's *Territorial Ambitions and the Gardens of Versailles*. Mukerji shows how in attempting to consolidate state power and limit the nobility's and emergent merchant elite's, the Sun King drew on material practices, namely engineering and the landscaping of non-human nature, to create "an uncontested and identifiable political *location* [Mukerji's emphasis] for accumulating and exercising power" (1). At the gardens of Versailles, Louis XIV forged an association between state power—his power—and nature itself.
23. Gindin and Panitch, *The Making of Global Capitalism*, 39. On creating such a "social infrastructure," see Harvey, *The Limits to Capital*.
24. "Commercial Needs of the Philippines," address by Honorable W. Cameron Forbes, *Merchants' Association Review* 1 (August 1911), bMS Am 1364.4, Forbes Papers.
25. Mukerji, *Territorial Ambitions and the Gardens of Versailles*, 1.
26. "Our Message to Our Guests," *El Renacimiento*, August 5, 1905, 12. (Published in English.)
27. Statement of Forbes addressed to "Sir", bMS Am 1364.4, Forbes Papers.
28. Burnham, *Plan of Chicago*, 51.
29. Forbes to William Haube, January 6, 1908, MS Am 1366, Forbes Papers.
30. David Paine, "Prominent Country Clubs," *New England Magazine* 32 (1905), 322–26.
31. Richard J. Moss, *Golf and the American Country Club* (Urbana: University of Illinois Press, 2001), 19.
32. James M. Mayo, *The American Country Club: Its Origins and Development* (New Brunswick, NJ: Rutgers University Press, 1998), 34–39.
33. Arnold Wright and H. A. Cartwright, *Twentieth Century Impressions of Hongkong, Shanghai, and Other Treaty Ports of China: Their History, People, Commerce, Industries, and Resources* (London: Lloyds Greater Britain Publishing Company, 1908), 388. For a summary of the extended Forbes family, see Allis, Jr., and Williams, "Forbes Family Papers," Massachusetts Historical Society, http://www.masshist.org/collection-guides/view/fa0269.
34. Elmer Osgood Cappers, *Centennial History of the Country Club, 1882–1982* (Brookline, MA: Country Club, 1981), 7.
35. Wright and Cartwright, *Twentieth Century Impressions of Hongkong, Shanghai, and Other Treaty Ports of China*, 388.
36. Randy Roberts, *The Rock, the Curse, and the Hub: A Random History of Boston Sports* (Cambridge: Harvard University Press, 2005), 136.

37. Wright and Cartwright, *Twentieth Century Impressions of Hongkong, Shanghai, and Other Treaty Ports of China*, 388.
38. Roberts, *The Rock, the Curse, and the Hub*, 136.
39. Ibid., 135.
40. Moss, *Golf and the American Country Club*, 12.
41. Peter C. Perdue, "Merchants West & East," in "Rise & Fall of the Canton Trade System—I," MIT Visualizing Cultures, http://ocw.mit.edu/ans7870/21f/21f.027/rise_fall_canton_01/cw_essay03.html.
42. Wong, *A Paradise Lost*, 10–11.
43. Ibid., 12.
44. Ralph Waldo Emerson, "The Young American" (1844), in *Emerson: Essays & Poems* (New York: Library Classics of the United States, 1996), 214, 215.
45. *A Brief History of the Baguio Country Club* (Baguio, Philippines: The Club, 1910), 9, NARA, RG 350, Entry 5, Box 535.
46. For the list of members see ibid., 16. Backgrounds on some of these figures can be found in Lewis E. Gleeck, *American Business and Philippine Economic Development* (Manila: Carmelo & Bauermann, 1975), 22, 21.
47. "List of Members of the Baguio Country Club," August 1911 bMS Am 1364.4, Forbes Papers. In his 1961 autobiography *I Walked with Heroes*, Filipino diplomat and journalist Carlos Romulo describes an effort in the late nineteen-teens by a group of "carefully chosen, presentable young men" to gain entry to the Baguio Country Club. They formed a "test group" challenging the restricted entry of Filipinos. He writes that with Manuel Quezon's support, the "ban against Filipinos was withdrawn." "It was our first victory in a race war in the Philippines." His story suggests the limits of Filipino participation in the Baguio Country Club. See Carlos S. Romulo, *I Walked with Heroes* (New York: Holt, Rinehart and Winston, 1961), 117–18.
48. *A Brief History of the Baguio Country Club*, 8–9.
49. Statement of Forbes addressed to "Sir", bMS Am 1364.4, Forbes Papers.
50. "Report of the Secretary of Commerce and Police," in *Eighth Annual Report of the Philippine Commission to the Secretary of War, 1907 (In Three Parts), Part 2*, 280-281.
51. *A Brief History of the Baguio Country Club*, 12.
52. *Reports of the Taft Philippine Commission* (1901), 148.
53. Worcester, *The Philippines Past and Present*, vol. 1, 477.
54. Anderson, *Spell of the Hawaiian Islands and the Philippines*, 249.
55. *A Brief History of the Baguio Country Club*, 7, 11.
56. "Golf," *Encyclopædia Britannica: A Dictionary of Arts, Sciences, and General Literature*, 9th ed., vol. 10 (New York: Henry G. Allen Company, 1890), 766.
57. "Prospectus of the Baguio Country Club," in *A Brief History of the Baguio Country Club, 46*.
58. Moss, *Golf and the American Country Club*, 21, 37, 49, 54.
59. "Prospectus of the Baguio Country Club," in *A Brief History of the Baguio Country Club*, 46.

60. Ibid., 46–47.
61. "The Plan of Baguio," *Teachers' Assembly Herald* 5 (May 12, 1912).
62. "Many New Buildings at Baguio. Mountain Capital Has Had Building Boom During Past Year," *Manila Times*, March 15, 1911, 3.
63. "Prospectus of the Baguio Country Club," in *A Brief History of the Baguio Country Club*, 47.
64. "Baguio," *Manila Daily Bulletin*, November 3, 1907, 3–4.
65. "The Plan of Baguio," *The Teachers' Assembly Herald* 5 (May 12, 1912).
66. Worcester to Kittie Worcester, February 5, 1908, Box 1, Worcester Papers.
67. See Williams, *The Country and the City*, on travel and seeing "objects of conspicuous aesthetic consumption" as a "form of fashionable society" (128). Williams describes the discursive formation of the landscape: "It is in the act of observing that this landscape forms; the river 'conducts the eye'; the sloping land 'displays' its grace; the stream 'inlays' the vale. It is a beautiful picture, in the strict sense. Its sense of possession, from a separate vantage-point, is a genuinely abstract aesthetic. . . . The order was being projected while it was also being composed" (126).
68. *A Brief History of the Baguio Country Club*, 43.
69. "Baguio, Philippines Summer Report," Manila Railroad brochure, 1928, American Historical Collection, Rizal Library, Ateneo de Manila University.
70. Minnigerode, "The Simla of the Philippines."
71. Carpenter, "A Switzerland in the Heart of the Tropics."
72. On the 1904 World's Fair, see Kramer, *The Blood of Government*; Jose D. Fermin, *1904 World's Fair: The Filipino Experience* (Diliman, Quezon City: University of the Philippines Press, 2004); Rydell, *All the World's a Fair.*
73. Williams, *The Country and the City*, 20.
74. William Pack, report of the provincial governor of Benguet, February 1, 1903, NARA, RG 350, Entry 150, Entry 5, Box 274; or see his report in *Annual Reports of the War Department for the Fiscal Year Ended June 30, 1903*, vol. 5, *Report of the Philippine Commission*, 759–60.
75. "Report of the Governor of Benguet," in *Annual Reports of the War Department* (Washington, DC: Government Printing Office, 1908), 280.
76. Forbes, journal, November 24, 1906, and June 19, 1906, MS 1365, Forbes Papers.
77. Thorstein Veblen, *The Theory of the Leisure Class: An Economic Study of Institutions* (New York: Macmillan and Co., 1912), 61.
78. Forbes to William Howard Taft, April 3, 1907, NARA, RG 350, Entry 5, Box 534.
79. Moss, *Nabaloi Law and Ritual*, 227.
80. Minnigerode, "The Simla of the Philippines."
81. These descriptions would also seem to signify a worldview and ontology that Timothy Mitchell terms "world-as-exhibition"—"the world conceived and grasped as though it were an exhibition." See Mitchell, *Colonising Egypt*, 222.
82. "Exhibit B. Report of the Director of Public Works," in *Seventh Annual Report of the Philippine Commission, 1906 (In Three Parts), Part 2*, 340.

83. "Exhibit C. Report of the Consulting Architect," in *Seventh Annual Report of the Philippine Commission, 1906 (In Three Parts), Part 2*, 368.
84. "Report of the Governor of the Province of Benguet," in *Seventh Annual Report of the Philippine Commission, 1906 (In Three Parts), Part 1*, 197–98.
85. "Baguio Town Site—Authorizing Sale of Lots at Public Auction" [Excerpt from minutes of the Philippine Commission of July 13, 1906], in *Annual Reports of the War Department for the Fiscal Year Ended June 30, 1906*, vol. 10, *Acts of the Philippine Commission (Nos. 1408–1538, inclusive) and Public Resolutions, Etc. From October 19, 1905, to September 15, 1906* (Washington, DC: Government Printing Office. 1906), 382.
86. Forbes to Luke E. Wright, May 28, 1906, MS Am 1366, Forbes Papers.
87. "Appendix E. Report of the Director of Lands," in *Report of the Philippine Commission to the Secretary of War, 1908 (In Two Parts), Part 2* (Washington, DC: Government Printing Office, 1909), 252.
88. "Report of the Governor of Benguet," in *Eighth Annual Report of the Philippine Commission to the Secretary of War, 1907, (In Three Parts) Part 1*, 278–79.
89. Ibid.
90. Bagamaspad, Hamada-Pawid, and Balangoy, *A Peoples' History of Benguet Province*, 232.
91. Prill-Brett, "Baguio: A Multi-Ethnic City and the Development of the Ibaloy as an Ethnic Minority," CSC Working Paper 15, 8.
92. Forbes to Secretary of War General Dickinson, March 20, 1910, NARA, RG 350, Entry 5, Box 535.
93. William T. Comstock, *Bungalows, Camps and Mountain Houses; Consisting of a Large Variety of Designs by a Number of Architects, Showing Buildings That Have Been Erected in All Parts of the Country . . . Elaborately Illustrated, Accompanied by Full Descriptive Text* (New York: W. T. Comstock, 1908), 5.
94. Anthony D. King, *The Bungalow: The Production of a Global Culture* (London: Routledge & Kegan Paul, 1984), 66, 70–71.
95. *A Brief History of the Baguio Country Club*, 12.
96. King, *The Bungalow*, 134.
97. Henry H. Saylor, *Bungalows; Their Design, Construction and Furnishing, with Suggestions Also for Camps, Summer Homes and Cottages of Similar Character* (Philadelphia: Winston, 1911), 48.
98. Forbes, journal, May 4, 1908, MS Am 1365, Forbes Papers.
99. Williams, *The Country and the City*, 124, 125.
100. "Teachers' Vacation Assembly, Baguio, Benguet," *Philippine Education* 4, no. 9 (February 1908): 6–7.
101. Forbes, journal, April 15, 1911, MS Am 1365, Forbes Papers.
102. *Tenth Annual Report of the Bureau of Civil Service to the Governor-General of the Philippine Islands for the Year Ended June 30, 1910* (Manila: Bureau of Printing, 1910), 9.
103. "Ho: For Baguio," *Manila Times*, March 23, 1908 (re-typed clipping), bMS Am 1364.4 (Box 7), Forbes Papers.

104. "Baguio: Philippines Summer Resort," brochure published by the Manila-Baguio Train Service, Manila Railroad Company, F 31, 12, American Historical Collection, Rizal Library, Ateneo de Manila University.
105. See Ohmann, *Selling Culture* on the formation of this class.
106. Woolley, "Baguio, Simla of the Philippines," 294.
107. Evan Lampe, *Work, Class, and Power in the Borderlands of the Early American Pacific: The Labors of Empire* (Lanham, MD: Lexington Books, 2014), 137. See also Paul A. Kramer, "Empires, Exceptions, and Anglo-Saxons: Race and Rule between the British and United States Empires, 1880–1910," *Journal of American History* 88 (2002): 1315–53.
108. Woolley, "Baguio, Simla of the Philippines," 294.
109. Minnigerode, "The Simla of the Philippines."
110. Forbes, journal, May 28, 1906, MS Am 1365, Forbes Papers.
111. Forbes, journal, May 19, 1906, MS Am 1365, Forbes Papers.
112. Poem in Topside guestbook, January 16, 1911, MS 1365.4, Forbes Papers.
113. Forbes to Ralph Forbes, September 1, 1906, MS Am 1366, Forbes Papers.
114. Topside guestbook, January 16, 1911, MS 1365.4, Forbes Papers.
115. "Baguio," *Manila Daily Bulletin*, November 3, 1907.
116. Forbes, journal, May 5, 1908, MS 1365, Forbes Papers.
117. See Reed, *Hispanic Urbanism in the Philippines*, 68.
118. Forbes, journal, May 19, 1906, MS 1365, Forbes Papers.
119. Forbes to Tommy Perkins, May 13, 1906, MS Am 1366, Forbes Papers.
120. King, *The Bungalow*, 49.
121. Leroy, "Race Prejudice in the Philippines," 109.
122. Brody, *Visualizing American Empire*, 167–69.
123. Patricia J. Fanning, *Norwood: A History* (Charlestown, SC: Acadia, 2002), 59–60.
124. "Forbes Taking PI Mementoes," *Manila Times*, February 26, 1912, 13.
125. Forbes to N. H. Stone, August 27, 1906, MS 1366, Forbes Papers.
126. Forbes, journal, May 5, 1908, MS Am 1365, Forbes Papers.
127. Forbes to Pierce Anderson, September 4, 1906, MS Am 1366, Forbes Papers.
128. Forbes to William James, April 18, 1907, MS Am 1366, Forbes Papers.
129. Forbes, journal, May 5, 1906, MS Am 1365, Forbes Papers.
130. Anderson, *Spell of the Hawaiian Islands and the Philippines*, 254.
131. Mukerji, *Territorial Ambitions and the Gardens of Versailles*, 203.
132. Forbes, journal, April 2, 1911, MS 1365, Forbes Papers.
133. Ibid.
134. Forbes recorded in his journal, "The Secretary was delighted with the progress, and told me that the building of my house was the thing that had saved the situation, that otherwise Baguio would have had a harder row." Forbes, journal, October 24, 1907, MS Am 1365, Forbes Papers.
135. Forbes, journal, April 2, 1911, MS 1365, Forbes Papers.
136. Forbes, journal, April 2, 1911, and April 4, 1910, MS Am 1365, Forbes Papers; Sergio Osmeña, Sr., "How Baguio City Was Born," *Free Press*, August 29, 1959.

137. "Los Terrenos en Bagyo," *El Renacimiento*, July 21, 1906, 2.
138. "Aguinaldo Off to Meet Forbes," *Christian Science Monitor*, May 24, 1909, 4.
139. "List of members of the Baguio Country Club," August 1911, bMS Am 1364.4, Forbes Papers.
140. "Big Building Boom in Baguio," *Manila Daily Bulletin*, November 29, 1912, clipping in MS Am 1192.2, vol. 5, Forbes Papers.
141. John Thayer Sidel, *Capital, Coercion, and Crime: Bossism in the Philippines* (Stanford: Stanford University Press, 1999), 7. On patron-client relations see also Anderson, "Cacique Democracy and the Philippines: Origins and Dreams."
142. Go, *American Empire and the Politics of Meaning*, 95–96.
143. Ibid.
144. Abinales and Amoroso, *State and Society in the Philippines*, 126, 141.
145. Thomas Harrington, "The Philippines on the Eve of Change," in *South East Asia Colonial History*, ed. Paul H. Kratoska (London: Routledge, 2001), 143.
146. Cullinane, *Ilustrado Politics*, 69.
147. Lynch, *Colonial Legacies in a Fragile Republic*, 239.
148. Rafael, *Contracting Colonialism*, 160–61.
149. On social and cultural capital see Pierre Bourdieu, "The Forms of Capital," in *Cultural Theory: An Anthology* (Malden, MA: Wiley-Blackwell, 2011); and Pierre Bourdieu, "Social Space and Symbolic Power," *Sociological Theory* 7 (1989): 14–25.
150. Forbes to Rudyard Kipling, January 24, 1912, MS Am 1366, Forbes Papers.
151. Lynch, *Colonial Legacies in a Fragile Republic*, 229.
152. On the political functions of balls and galas see, Lynch, *Colonial Legacies in a Fragile Republic*, 230; and Kramer, *The Blood of Government*, 185-86.
153. Kramer, *The Blood of Government*, 185-86.
154. Kramer, "Power and Connection," 1381. See also Go on Americans' use of the "power of signs" to maintain their power in *American Empire and the Politics of Meaning*, 94.
155. Woolley, "Baguio, Simla of the Philippines," 294.
156. Forbes to Ralph E. Forbes, March 19, 1907, MS Am 1366, Forbes Papers.
157. Abinales and Amoroso, *State and Society in the Philippines*, 127.
158. Sergio Osmeña to William Cameron Forbes, June 13, 1944, bMS Am 1364, Forbes Papers.
159. See Cullinane, *Ilustrado Politics*, 82, on the Partido Nacionalista and its leaders.
160. Cullinane, *Ilustrado Politics*, 169, 259–263.
161. Forbes to Ralph E. Forbes, March 19, 1907, MS Am 1366, Forbes Papers.
162. Forbes, journal, March 31, 1910, MS Am 1365, Forbes Papers.
163. Osmeña, "How Baguio City Was Born." For a photograph of Osmeña's Baguio house, see Norma Alarcón, *The Imperial Tapestry: American Colonial Architecture in the Philippines* (Manila: University of Santo Tomas Publishing House, 2008), 110.

164. Cullinane quotes Forbes's November 15–19, 1907, journal entry in *Ilustrado Politics*, 284.
165. Forbes, journal, October 16, 1907, MS Am 1365, Forbes Papers.
166. Taft to Dean Worcester, November 30, 1907, Worcester Papers.
167. Abinales and Amoroso, *State and Society in the Philippines*, 135–38.
168. "¿Tendencia Centralizadora?" *El Renacimiento*, September 18, 1906, 1.
169. "Territorio Infieles. La Asamblea Debe Tener Jurisdicción sobre Las Tribus 'Non Cristianas,'" *El Renacimiento*, February 17, 1908, 3.
170. "Las Delicias del Veraneo," *El Renacimiento*, March 13, 1908, 3.
171. This process of domestication and translation of political ideas is largely the subject of Go's *American Empire and the Politics of Meaning*.
172. "Emerson," *El Renacimiento*, April 27, 1904, 2.
173. "Our White Elephant," *Philippines Manila Free Press*, August 19, 1911, clipping in NARA, RG 350, Entry 5, Box 535.
174. Editor, "Baguio: The World's Finest Mountain Resort," *American Chamber of Commerce Journal* (April 1922): 10.
175. Anderson, *Colonial Pathologies*, 6.
176. Worcester to William Howard Taft, January 27, 1908, Box 1, Worcester Papers; Worcester to Kittie Worcester, February 5, 1908, Box 1, Worcester Papers.
177. Worcester to Kittie Worcester, February 5, 1908, Box 1, Worcester Papers.
178. "Remarks of Dean Worcester in Introducing the Speaker of the Evening, Honorable Jacob M. Dickinson, to the Members of the Mountain Club at Bontoc," August 3, 1910, NARA, RG 350, Entry 5, Box 368.
179. Carpenter, *Through the Philippines and Hawaii*, 89.
180. Paine, "Prominent Country Clubs," 37.
181. Forbes, journal, March 29, 1910, and May 5, 1911, MS Am 1365, Forbes Papers.
182. Forbes, journal, March 6, 1909, MS 1365, Forbes Papers. On "playing Indian" see Philip Joseph Deloria, *Playing Indian* (New Haven: Yale University Press, 1998).
183. Forbes to Daniel Burnham, May 14, 1905, MS Am 1366, Forbes Papers.
184. Forbes, journal, June 13, 1905, MS 1365, Forbes Papers.
185. Forbes to Daniel Burnham, May 14, 1905, MS Am 1366; Forbes, journal, June 13, 1905, MS 1365, Forbes Papers.
186. T. J. Jackson Lears, *Rebirth of a Nation: The Making of Modern America, 1877–1920* (New York: HarperCollins, 2009), 289.

## EPILOGUE

1. On Bulosan see E. San Juan, *On Becoming Filipino: Selected Writings of Carlos Bulosan* (Philadelphia: Temple University Press, 1995); E. San Juan, *Carlos Bulosan and the Imagination of the Class Struggle* (Quezon City: University of the Philippines Press, 1972); Augusto Fauni Espiritu, *Five Faces of Exile: The Nation and Filipino American Intellectuals* (Stanford: Stanford University Press, 2005);

Michael Denning, *The Cultural Front: The Laboring of American Culture in the Twentieth Century* (London; New York: Verso, 1998).

2. Bulosan, *America Is in the Heart*, 66.
3. Marx, *The Machine in the Garden*, 15, 5.
4. Bulosan, *America Is in the Heart*, 66, 67.
5. Tapang, *Innovation and Social Change*, 39.
6. Bagamaspad, Hamada-Pawid, and Balangoy, *A Peoples' History of Benguet Province*, 209, 217.
7. Charles H. Brent, "Annual Report of the Bishop of the Philippine Islands, 1916–1917," Box 34, Charles Henry Brent Papers, Manuscript Division, Library of Congress, Washington, DC.
8. Tapang, *Innovation and Social Change*, 40.
9. Abinales and Amoroso, *State and Society in the Philippines*, 125.
10. Ibid. See also Finin, *The Making of the Igorot.*
11. Kramer, *The Blood of Government*, 346. See also Salman, *The Embarrassment of Slavery*, 166.
12. Mukerji, *Territorial Ambitions and the Gardens of Versailles*, 1.
13. Forbes, journal, July 14, 1907, MS Am 1365, Forbes Papers.
14. Corpuz, *An Economic History of the Philippines*, 216, 235.
15. Anderson, "Cacique Democracy and the Philippines," 11; Abinales and Amoroso, *State and Society in the Philippines*, 141. See also Go, *American Empire and the Politics of Meaning*, on political elites' use of patronage.
16. Sidel, *Capital, Coercion, and Crime*, 18.
17. Harvey, "The 'New' Imperialism."
18. "William Cameron Forbes, 'Our Dependencies,' May 27–28, 1914 For discussion at the National Foreign Trade Convention, Washington, D.C., under the auspices of the American Manufacturers Export Association, The Pan-American Society of the United States, and The American Asiatic Association," bMS Am 1364.4, Forbes Papers.
19. Pratt, *America's Colonial Experiment*, 243, 244.
20. Blitz, *The Contested State*, 45.
21. Pratt, *America's Colonial Experiment*, 244.
22. Owen, "Philippine Economic Development and American Policy: A Reappraisal," 113.
23. Harry Luton, "American Internal Revenue Policy in the Philippines to 1916," in *Compadre Colonialism*, ed. Norman G. Owen, 134.
24. Owen, "Philippine Economic Development and American Policy: A Reappraisal," 115.
25. Bulosan, *America Is in the Heart*, 24.
26. Ibid.
27. Abinales and Amoroso, *State and Society in the Philippines*, 80.
28. Miriam Sharma, "The Philippines: A Case of Migration to Hawaii, 1906–1946," in *Labor Immigration under Capitalism: Asian Workers in the United States*

*Before World War II*, ed. Lucie Cheng and Edna Bonacich (Berkeley: University of California Press, 1984). See also Dorothy B. Fujita-Rony, *American Workers, Colonial Power: Philippine Seattle and the Transpacific West, 1919–1941* (Berkeley: University of California, 2003); Catherine Ceniza Choy, *Empire of Care: Nursing and Migration in Filipino American History* (Durham: Duke University Press, 2003); Rick Baldoz, *The Third Asiatic Invasion: Empire and Migration in Filipino America, 1898–1946* (New York: New York University Press, 2011); Kornel S. Chang, *Pacific Connections: The Making of the Western US-Canadian Borderlands* (Berkeley: University of California Press, 2012).

29. "Go to Baguio to Confer on Filipino Labor," *Manila Times*, April 17, 1911, 1, 3.
30. Baldoz, *The Third Asiatic Invasion*, 48–51.
31. "Filipino Immigrants in the United States," Migrationpolicy.org, http://migrationpolicy.org/article/filipino-immigrants-united-states.
32. Robyn Magalit Rodriguez, *Migrants for Export: How the Philippine State Brokers Labor to the World* (Minneapolis: University of Minnesota Press, 2010), xii.
33. Bangko Sentral ng Pilipinas, "Overseas Filipinos' (OF) Remittances," http://www.bsp.gov.ph/statistics/keystat/ofw.htm; The World Bank, "Personal Remittances, Received (% of GDP)," http://data.worldbank.org/indicator/BX.TRF.PWKR.DT.GD.ZS?order=wbapi_data_value_2010+wbapi_data_value+wbapi_data_value-first&sort=asc.
34. Rodriguez, *Migrants for Export*, xxiii, x, xii, 18.
35. Kaplan, *The Anarchy of Empire in the Making of US Culture*, 3.
36. Bases Conversion and Development Authority, "John Hay Special Economic Zone," http://www.bcda.gov.ph/freeport_and_ecozones/show/7.
37. "List of Operating Manufacturing Economic Zones, as of 31 January 2015," uploaded to the Philippine Economic Zone Authority website, http://www.peza.gov.ph/index.php/economic-zones/list-of-economic-zones/operating-economic-zones.
38. "Guidelines on the Registration of Information Technology Enterprises and the Establishment and Operation of IT Parks/Buildings (Approved by the PEZA Board on 29 December 2000)," document available at "Activities Eligible for PEZA Registration and Incentives," Philippine Economic Zone Authority website, http://www.peza.gov.ph/index.php/eligible-activities-incentives.
39. A few select works on economic and free trade zones: Naomi Klein, *No Logo* (New York: Picador, 2000); Keller Easterling, *Extrastatecraft: The Power of Infrastructure Space* (London: Verso, 2014); Dara Orenstein, "Foreign-Trade Zones and the Cultural Logic of Frictionless Production," *Radical History Review*, no. 109 (2011): 36–61.
40. "The Philippine Edge," Philippine Economic Zone Authority website, http://www.peza.gov.ph/index.php/homepage/2-uncategorised/28-the-philippine-edge.
41. Ibid.

42. Marx, *The Machine in the Garden*, 25.
43. See San Juan, *On Becoming Filipino*, 16. See Denning, *The Cultural Front* for Bulosan as a Popular Front author of a "ghetto pastoral."
44. Carey McWilliams, introduction to Bulosan, *America Is in the Heart*, xvi.
45. Anderson, *Under Three Flags*, 229. See also Guevarra, *History of the Philippine Labor Movement*.
46. Espiritu, *Five Faces of Exile*, 66, and on the novel as a "popular-front allegory," see San Juan, *On Becoming Filipino*, 12.
47. Bulosan, *America Is in the Heart*, 29.
48. Fry, *A History of the Mountain Province*, 190.
49. Ibid.; Frances B. Cogan, *Captured: The Japanese Internment of American Civilians in the Philippines, 1941–1945* (Athens: University of Georgia Press, 2000), 42.
50. C. Brooks Peter, "Igorots Astride Tanks in Bataan Wipe Out a Japanese Regiment," *New York Times*, February 23, 1942, 3.
51. See Cogan, *Captured*.
52. Fry, *A History of the Mountain Province*, 204.
53. Ibid.; Yuma Totani, *Justice in Asia and the Pacific Region, 1945–1952: Allied War Crimes Prosecutions* (New York: Cambridge University Press, 2015), 33.
54. Bagamaspad, Hamada-Pawid, and Balangoy, *A Peoples' History of Benguet Province*, 287–88.
55. Fry, *A History of the Mountain Province*, 205.
56. Ibid.

# Bibliography

## *Archival Sources*

Art Institute of Chicago, Chicago, Illinois
Daniel H. Burnham Collection, Ryerson and Burnham Archives
Bentley Historical Library, University of Michigan, Ann Arbor
Dean C. Worcester Papers
Filipinas Heritage Library Photo Collection, Makati City, Philippines
Houghton Library, Harvard University, Cambridge, Massachusetts
William Cameron Forbes Papers
Lopez Memorial Museum, Manila, Philippines
Manuscript Division, Library of Congress, Washington, DC
Charles Brent Papers
Elihu Root Papers
William Howard Taft Papers
Museum of Anthropological Archaeology, University of Michigan, Ann Arbor
Worcester Photograph Collection
National Archives and Records Administration, College Park, Maryland
Bureau of Insular Affairs, Record Group 350
Private Collection
Erlyn Ruth Alcantara (photos)
Rizal Library, Ateneo de Manila University, Quezon City, Philippines
American Historical Collection
Schomburg Center for Research in Black Culture, New York Public Library
Robert R. Rudd Papers

## *Newspapers*

*El Grito del Pueblo*
*El Renacimiento*
*Manila Times*

## *Government-Published Primary Sources*

*Acts Nos. 1396 and 1397: The Special Provincial Government Act and the Township Government Act, Enacted September 14, 1905* . . . Manila: Bureau of Public Printing, 1905.

*Acts of the Philippine Commission*. Washington, DC: Government Printing Office, 1902.

*Acts of the Philippine Commission [Acts Nos. 425-949 Inclusive]*. Washington, DC: Government Printing Office, 1904.

*Acts of the Philippine Commission*. Washington, DC: Government Printing Office, 1901–1907.

Ahern, George P. *Compilation of Notes on the Most Important Timber Tree Species of the Philippine Islands*. Manila: Bureau of Printing, 1901.

———. *A Few Pertinent Facts Concerning the Philippine Forests and Needs of the Forest Service, That Should Interest Every Filipino*, circular no. 3. Manila: Bureau of Printing, 1908.

———. *Spanish Public Land Laws in the Philippine Islands and Their History to August 13, 1898*. Washington, DC: Government Printing Office, 1901.

*Annual Reports of the War Department*:

*Annual Reports of the War Department for the Fiscal Year Ended June 30, 1901, Public Laws and Resolutions Passed by the Philippine Commission*. Washington, DC: Government Printing Office, 1901.

*Annual Reports of the War Department for the Fiscal Year Ended June 30, 1902*. Vol. 11, *Acts of the Philippine Commission*. Washington, DC: Government Printing Office, 1902.

*Annual Reports of the War Department for the Fiscal Year Ended June 30, 1903*. Vol. 5, *Report of the Philippine Commission*. Washington, DC: Government Printing Office, 1903.

*Annual Reports of the War Department for the Fiscal Year Ended June 30, 1905*. Vol. 10, *Report of the Philippine Commission, Part 1*. Washington, DC: Government Printing Office, 1905.

*Annual Reports of the War Department for the Fiscal Year Ended June 30, 1906*.Vol. 10, *Acts of the Philippine Commission (Nos. 1408–1538, Inclusive) and Public Resolutions, etc., from October 19, 1905, to September 15, 1906*. Washington, DC: Government Printing Office. 1906.

*War Department, USA, Annual Reports, 1907 (In Ten Volumes)*.Vol. 8, *Report of the Philippine Commission*. Washington, DC: Government Printing Office, 1908.

Bureau of Forestry, Philippines. *Annual Report of the Director of Forestry of the Philippine Islands, July 1, 1905 to June 30, 1906*. Manila: Bureau of Printing. 1906.

———. *Annual Report of the Director of Forestry of the Philippine Islands, July 1, 1906 to June 30, 1907*. Manila: Bureau of Printing, 1908.

———. *Annual Report of the Director of Forestry of the Philippine Islands, July 1, 1909 to June 30, 1910*. Manila: Bureau of Printing, 1910.

Bureau of Public Works, Philippines. *Annual Report of the Director of Public Works to the Honorable the Secretary of Commerce and Police*. Manila: Bureau of Printing, 1910–1930.

———. *Quarterly Bulletin*. Manila: Bureau of Printing, 1910–1931.

Call, Lewis. W. *United States Military Reservations, National Cemeteries, and Military Parks: Title, Jurisdiction Etc.* Rev. ed. Washington, DC: Government Printing Office, 1907.

"Carino v. Insular Government 212 U.S. 449 (1909)." Justia Law. Accessed March 27, 2015. https://supreme.justia.com/cases/federal/us/212/449/case.html.

Census Office of the Philippine Islands. *Census of the Philippine Islands Taken under the Direction of the Philippine Legislature in the Year 1918, Vol. 4, Part 2*. Manila: Bureau of Printing, 1921. http://name.umdl.umich.edu/AJB5835.0004.002.

Doherty, David J. *Conditions in the Philippines . . .* [Washington, DC]: [Government Printing Office], 1904. http://catalog.hathitrust.org/api/volumes/oclc/25970795.html.

*Elihu Root Collection of United States Documents Relating to the Philippine Islands*. Washington, DC: Government Printing Office, 1902.

Jenks, Albert Ernest. *The Bontoc Igorot*. Department of Interior Ethnographical Survey Publications, vol. 1. Manila: Bureau of Public Printing, 1905.

Pérez, Angel. *Relaciones Agustinianas de Las Razas del Norte de Luzon*, Department of the Interior, Ethnological Survey Publications, vol. 3, Spanish edition. Manila: Bureau of Public Printing, 1904.

*Philippine Islands Public Laws, with Amendments Indicated*. Manila: Bureau of Public Printing, 1904.

*Preliminary Statement of the Commissioners Appointed by the President of the United States to Investigate Affairs in the Philippine Islands*. Washington, DC: Government Printing Office, 1899.

*Report of the Governor of Porto Rico to the Secretary of War*. Washington, DC: Government Printing Office, 1920.

*Report of Maj.-Gen. E. S. Otis, United States Army Commanding Division of the Philippines, Military Governor. September 1, 1899 to May 5, 1900*. Washington, DC: Government Printing Office, 1900.

*Reports of the Philippine Commission*:

*Report of the Philippine Commission to the President*. Vol. 1, *January 31, 1900*. Washington, DC: Government Printing Office, 1900.

*Reports of the Taft Philippine Commission: Message from the President of the United States, Transmitting a Report of the Secretary of War, Containing the Reports of the Taft Commission, Its Several Acts of Legislation, and Other Important Information Relating to the Condition and Immediate Wants of the Philippine Islands*. Washington, DC: Government Printing Office, 1901.

*Third Annual Report of the Philippine Commission, 1902, Part 1*. Washington, DC: Government Printing Office, 1903.

*Fourth Annual Report of the Philippine Commission, 1903, In Three Parts, Part 1*. Washington, DC: Government Printing Office, 1904.

*Fourth Annual Report of the Philippine Commission, 1903, In Three Parts, Part 3*. Washington, DC: Government Printing Office, 1904.

*Reports of the Philippine Commission, the Civil Governor, and the Heads of the Executive Departments of the Civil Government of the Philippines Islands, 1900–1903*. Washington, DC: Government Printing Office, 1904.

*Fifth Annual Report of the Philippine Commission, In Three Parts, Part 3*. Washington, DC: Government Printing Office, 1905.

*Sixth Annual Report of the Philippine Commission, 1905 (In Four Parts), Part 2*. Washington, DC: Government Printing Office, 1906.

*Sixth Annual Report of the Philippine Commission, 1905 (In Four Parts), Part 3*. Washington, DC: Government Printing Office, 1906.

*Seventh Annual Report of the Philippine Commission, 1906 (In Three Parts), Part 2*. Washington, DC: Government Printing Office, 1907.

*Eighth Annual Report of the Philippine Commission to the Secretary War, 1907 (In Three Parts), Part 1*. Washington, DC: Government Printing Office, 1908.

*Eighth Annual Report of the Philippine Commission to the Secretary of War, 1907 (In Three Parts), Part 2*. Washington, DC: Government Printing Office, 1908.

*Report of the Philippine Commission to the Secretary of War, 1908 (In Two Parts), Part 1*. Washington, DC: Government Printing Office, 1909.

*Report of the Philippine Commission to the Secretary of War, 1908 (In Two Parts), Part 2*. Washington, DC: Government Printing Office, 1909.

*Report of the Philippine Commission to the Secretary of War, 1911 (In One Part)*. Washington, DC: Government Printing Office, 1912.

*Report of the Philippine Commission to the Secretary of War, 1913 (In One Part)*. Washington, DC: Government Printing Office, 1914.

*Report of the Philippine Commission to the Secretary of War, July 1, 1913 to December 31, 1914*. Washington, DC: Government Printing Office, 1915.

Scheerer, Otto. *The Nabaloi Dialect*. Manila: Bureau of Public Printing, 1905.

*Tenth Annual Report of the Bureau of Civil Service to the Governor-General of the Philippine Islands for the Year Ended June 30, 1910*. Manila: Bureau of Printing, 1910.

United States Bureau of the Census. *Census of the Philippine Islands, Taken under the Direction of the Philippine Commission in the Year 1903, in Four Volumes . . .* Washington, DC: United States Bureau of the Census, 1905. http://name.umdl.umich.edu/ajb5834.0002.001.

Villamor, Ignacio. *Criminality in the Philippine Islands, 1903–1908*. Manila: Bureau of Printing, 1909.

Whitford, H. N. *The Forests of the Philippines, Part I: Forest Types and Products*, bulletin no. 10. Manila: Bureau of Printing, 1911.

Worcester, Dean C. *Slavery and Peonage in the Philippine Islands*. Manila: Bureau of Printing, 1913.

## *Published Primary Sources*

Anderson, Isabel. *The Spell of the Hawaiian Islands and the Philippines, Being an Account of the Historical and Political Condition of Our Pacific Possessions, together with Descriptions of the Natural Charm and Beauty of the Countries and the Strange and Interesting Customs of their Peoples*. Boston: The Page Company, 1916.

Angell, Norman. *The Great Illusion: A Study of the Relation of Military Power in Nations to Their Economic and Social Advantage*. New York: Putnam, 1910.

Arnold, John R. "Municipal Market Buildings in the Philippines. The Civic Betterment Brought About by the Program of Modern Market-Building of the Philippines." *American City* 8, no. 6 (June 1913).

Baguio Country Club. *The Baguio Country Club, Baguio, Philippine Islands*. Baguio: The Club, 1910.

*Baguio: The Mountain Resort of the Philippine Islands*. Manila: Thos. Cook & Son, 1919.

Bailey, Grace Helen. "Road Making in the Philippines: Construction of the Benguet Mountain Road." *Overland Monthly and the Out West Magazine*, 53 (January 1909): 33–38.

Barber, Marshall A. "An Igorrote Dog Market at Baguio." *Graduate Magazine of the University of Kansas* 10, no. 9 (June 1912).

"Benguet Consolidated Mining Co., The," *Far Eastern Review* (June 1907).

Beveridge, Albert J. *The Meaning of the Times and Other Speeches*. Indianapolis: Bobbs-Merrill, 1908.

Blount, James H. *American Occupation of the Philippines, 1898–1912*. Metro Manila: Solar Publishing Corporation, 1968, reprint of 1913 G. P. Putnam & Sons edition.

Bourne, Edward Gaylord. *Discovery, Conquest, and Early History of the Philippine Islands; Being a Separate Issue of the Historical Introduction to Blair & Robertson's "The Philippine Islands: 1493–1898"*. Cleveland: A. H. Clark Company, 1907.

———. "A Trained Colonial Civil Service." *North American Review* 169 (October 1899): 530–34.

Boyce, W. D. *The Philippine Islands*. Chicago: Rand McNally, 1914.

Brooks, Howard K. *Brooks' Foreign Exchange Text Book: An Elementary Treatise on Foreign Exchange and the Monetary Systems of the World*. Chicago: H. K. Brooks, 1906.

Bulosan, Carlos. *America Is in the Heart*. Seattle: University of Washington Press, 2000 [1946].

Burnham, Daniel Hudson. *Plan of Chicago*. Chicago: Commercial Club, 1909. http://archive.org/details/planofchicago00burnuoft.

Burnham, Daniel H., and Peirce Anderson. "Report on Proposed Improvements at Manila," and "Report on the Proposed Plan of the City of Baguio, Province of Benguet, P.I." Printed in *Proceedings of the Thirty-Ninth Annual Convention of the American Institute of Architects* (Washington, DC, 1906) and in Charles Moore, *Daniel H. Burnham: Architect, Planner of Cities*, vol. 2 (Boston: Houghton Mifflin Company, 1921).

———."White City and Capital City." *Century Magazine* (February 1902).

Carpenter, Frank G. "A Switzerland in the Heart of the Tropics." *Los Angeles Times*, July 20, 1924.

———. *Through the Philippines and Hawaii*. Garden City, NY: Doubleday, Page & Co., 1925.

"City in the Skylands." *Chronicle Magazine*, April 7, 1962.

Comstock, William T. *Bungalows, Camps and Mountain Houses; Consisting of a Large Variety of Designs by a Number of Architects, Showing Buildings That Have Been Erected in All Parts of the Country . . . Elaborately Illustrated, Accompanied by Full Descriptive Text*. New York: W. T. Comstock, 1908.

Conant, Charles A. "The Economic Basis of 'Imperialism.'" *North American Review* 167 (September 1898): 326–40.

———. *A History of Modern Banks of Issue, with an Account of the Economic Crises of the Present Century*, 4th impression. New York: G. P. Putnam's Sons, 1902 [1896].

———. *A Special Report: Coinage and Banking in the Philippine Islands Made to the Secretary of War [from the "Annual Report of the Secretary of War," Appendix G]*. Washington, DC: Government Printing Office, 1901.

Coudert, Frederic R. "Our New Peoples: Citizens, Subjects, Nationals or Aliens." *Columbia Law Review* (January 1, 1903): 13–32.

Croly, Herbert. "The Promised City of San Francisco." *Architectural Record* 19 (June 1906).

"Day's Auto Tour in Philippines, Roads Found as Good as Any in France, Newly Built by American Authorities." Special correspondence to the *New York Times*, May 11, 1913.

Devins, John B. *An Observer in the Philippines; Or, Life in Our New Possessions*. New York: American Tract Society, 1905.

Editor. "Baguio: The World's Finest Mountain Resort." *American Chamber of Commerce Journal* (April 1922).

Edwards, Clarence R. "Governing the Philippine Islands." *National Geographic Magazine* 15 (July 1904): 273–84.

———. "The Work of the Bureau of Insular Affairs." *National Geographic Magazine* 15, no. 6 (June 1904): 239–55.

Ely, Richard T. *Studies in the Evolution of Industrial Society*. New York: The Macmillan Company, 1903.

———. "The Past and the Present of Political Economy." *Overland Monthly* 2 (September 1883): 225–35.

Fales, Louis H., "Tropical Neurasthenia and Its Relation to Tropical Acclimation." *American Journal of Medical Sciences* (April 1907): 582–92.

Finley, John P. "The Commercial Awakening of the Moro and Pagan." *The North American Review* CXCVII (March 1913): 325–34.

Folkmar, D. "The Administration of a Philippine Province." In *American Colonial Policy and Administration*. Philadelphia: American Academy of Political and Social Science, 1907.

Forbes, Robert B. *Personal Reminiscences, Second Edition, Revised, to which is added Rambling Recollections Connected with China*. Boston: Little, Brown, and Company, 1882.

Forbes, W. Cameron. *Notes on Early History of Baguio*. Manila: Manila Daily Bulletin, 1933.

———. *The Philippine Islands*. 2 vols. Boston: Houghton Mifflin Co., 1928.

———. *The Romance of Business*. Boston: Houghton, Mifflin & Co., 1921.

———."What Had Best Be Done for the Material Advancement of the Philippines." In *Proceedings of the Twenty-Sixth Annual Meeting of the Lake Mohonk Conference of Friends of the Indian and Other Dependent Peoples*, 117–23. New York: Lake Mohonk Conference of Friends and the Indian and Other Dependent Peoples, 1908.

Forbes, W. Cameron, and William Atkinson Jones. *Reply to Jones. W. Cameron Forbes, Governor-General of the Philippine Islands, Replies to False Charges Contained in the Speeches Made by William A. Jones, of Virginia, in the House of Representatives, January 28 and February 13, 1913*. Manila, 1913.

Gibbon, Edward. *The History of the Decline and Fall of the Roman Empire, Volume the First and Volume the Second*, ed. David Womersley. London: Penguin Books, 2005 [1776, 1781].

James, William. "The Philippine Tangle." *Boston Evening Transcript*, March 1, 1899.

Jenks, Maud Huntley. *Death Stalks the Philippine Wilds: Letters of Maud Huntley Jenks*. Minneapolis: Lund Press, 1951.

Kalaw, Maximo Manguiat. *The Case for the Filipinos*. New York: The Century Co., 1916. http://archive.org/details/caseforfilipino00kalagoog.

Kane, Samuel E. *Baguio and a Guide to the Mountain Province, Philippine Islands: With Interesting Customs of the Mountain People*. Baguio: Catholic School Press, 1930.

———. *Baguio: Gateway to Wonderland, Describing 20 Trips in the Mountain Province*. Manila: Sugar News Press, 1931.

Keesing, Felix M., and Marie Keesing. *Taming Philippine Headhunters: A Study of Government and of Cultural Change in Northern Luzon*. London: George Allen & Unwin Ltd., 1934.

Kemlein, H. *Kemlein & Johnson's Guide and Map of Manila and Vicinity: A Hand Book Devoted to the Interests of the Traveling Public*. Manila: Kemlein & Johnson, 1908.

Kemmerer, Edwin Walter. *Postal Savings: An Historical and Critical Study of the Postal Savings Bank System of the United States*. Princeton: Princeton University Press, 1917.

Kipling, Rudyard. *Plain Tales from the Hills*. New York: Penguin Books, 1990.

Ireland, Alleyen. *Tropical Colonization: An Introduction to the Study of the Subject*. New York: Macmillan, 1899.

Leroy, James A. *Philippine Life in Town and Country*. New York: G. P. Putnam's Sons, 1906.

———. "Race Prejudice in the Philippines." *Atlantic Monthly* (July 1902): 100–112.

Lodge, Henry Cabot. *The Retention of the Philippine Islands: Speech of Hon. Henry Cabot Lodge, of Massachusetts, in the Senate of the United States, March 7, 1900*. Washington, DC: [Government Printing Office], 1900. http://catalog.hathitrust.org/api/volumes/oclc/14720100.html.

Malcolm, George. "Baguio, Yesterday, Today, And Tomorrow." In Baguio Golden Anniversary Supplement, *Baguio Midland Courier*, September 6, 1959.

———. *The Government of the Philippine Islands: Its Development and Fundamentals*. Rochester, NY: Lawyers Co-operative Publishing Company, 1916. http://archive.org/details/aex3186.0001.001.umich.edu.

Manila Railroad Company. *Baguio: Philippine Summer Resort*. Manila: Manila Railroad Company, 1928.

Minnigerode, Fitzhugh Lee. "The Simla of the Philippines." *New York Times*, December 5, 1920.

Morga, Antonio de, Ferdinand Blumentritt, and José Rizal. *Sucesos de las Islas Filipinas*. Paris: Garnier Hermanos, 1890.

Moses, Edith. *Unofficial Letters of an Official's Wife*. New York: D. Appleton and Co., 1908.

Moss, C. R. *Nabaloi Law and Ritual*. Berkeley: University of California Press, 1920.

Osmeña, Sergio. "How Baguio City Was Born." *Free Press*, August 29, 1959.

Paine, David. "Prominent Country Clubs." *New England Magazine* 32 (1905): 322–38.

"Philippines, The." *LIFE*, February 13, 1939.

"Professor Moses Back from Manila." *San Francisco Chronicle*, August 12, 1903.

Rebori, A. N. "The Work of William E. Parsons in the Philippine Islands," Part I. *Architectural Record* 41 (April 1917): 305–24.

———."The Work of William E. Parsons in the Philippine Islands," Part II. *Architectural Record* 41 (May 1917): 423–34.

Reinsch, Paul Samuel. *World Politics: At the End of the Nineteenth Century as Influenced by the Oriental Situation*. New York: The MacMillan Company, 1904 [1900].

Reyes, José S. *Legislative History of America's Economic Policy toward the Philippines*. New York: Columbia University, 1923.

Rizal, Jose. 1913. *The Indolence of the Filipino*. Manila: Philippine Education Co., 1913. Accessed March 26, 2015. http://archive.org/details/theindolenceofth06885gut.

"Road Building to Music: How Major Kennon Got Work Out of Filipinos and Won a Bet." *Washington Post*, February 28, 1909.

Romulo, Carlos S. *I Walked with Heroes*. New York: Holt, Rinehart and Winston, 1961.

Root, Elihu. *Addresses on Government and Citizenship*. Cambridge: Harvard University Press, 1916.

———. *The Military and Colonial Policy of the United States, Addresses and Reports*. Edited by Robert Bacon and James Brown Scott. Cambridge: Harvard University Press, 1916.

Rosenberg, Edward. "Filipinos as Workmen: Data and Conclusions on the Labor Situation and General Conditions in the Philippine Islands." *American Federationalist* 10 (1903).

Rowe, L. S. "The Supreme Court and the Insular Cases." *Annals of the American Academy of Political and Social Science* 18, no. 2 (1901): 38–62. doi:10.1177/000271620101800202.

Saylor, Henry H. *Bungalows: Their Design, Construction and Furnishing, with Suggestions Also for Camps, Summer Homes and Cottages of Similar Character*. Philadelphia: Winston, 1911.

Scheerer, Otto. *On Baguio's Past: Chapters from Local History and Tradition*. Manila: Otto Scheerer, 1933.

Scheidnagel, Manuel. *Distrito de Benguet: Filipinas; memoria descriptiva y económica, acompañada del primer plano-cróquis del mismo*. Madrid: Imp. de la Dir. Gral. de Infantería, 1878.

Sullivan, Louis H. *The Autobiography of an Idea*. New York: Press of the American Institute of Architects, Inc., 1924. http://archive.org/details/autobiographyofa005655mbp.

Sumulong, Juan. "The Philippine Problem from a Filipino Standpoint." *North American Review* 179, no. 577 (December 1904): 860–67.

Taft, Helen Herron. *Recollections of Full Years*. New York: Dodd, Mead & Company, 1914.

Taylor, John R. M, United States, and Bureau of Insular Affairs. *The Philippine Insurrection against the United States: A Compilation of Documents with Notes and Introduction*. Pasay City, Philippines: Eugenio Lopez Foundation, 1971.

Veblen, Thorstein. *The Theory of the Leisure Class: An Economic Study of Institutions*. New York: Macmillan, 1912.

Warren, Lee Donald. *Isles of Opportunity: Progress and Possibilities in the Philippine Islands*. Washington, DC: Review and Herald Pub. Association, 1928.

Whitney, Caspar W. "Evolution of the Country Club." *Harper's New Monthly Magazine*, December 1894, 17–32.

Willcox, Cornelis De Witt. *The Head Hunters of Northern Luzon, from Ifugao to Kalinga: A Ride Through the Mountains of Northern Luzon, with an Appendix on the Independence of the Philippines*. Kansas City: Franklin Hudson Publishing, Co., 1912.

Williams, Daniel R. *The Odyssey of the Philippine Commission*. Chicago: A. C. McClurg & Co., 1913.

Willoughby, William Franklin. *Territories and Dependencies of the United States: Their Government and Administration*. New York: Century Co., 1905.

Woolley, Monroe. "Baguio, Simla of the Philippines." *Overland Monthly*, September 1913, 292–95.

Worcester, Dean C. "Headhunters Under the Stars and Stripes." *New York Times*, October 25, 1908.

———. "The Non-Christian Peoples of the Philippine Islands." *National Geographic Magazine* 24, no. 11 (November 1913): 1158–256.

———. "Philippine Forest Wealth." *American Forestry* 21, no. 1 (January 1915): 1–18.

———. *The Philippine Islands*. New York: Macmillan Company, 1899.

———. *The Philippines: Past and Present*. Vols. 1 and 2. New York: Macmillan, 1914.

———. "What Would Lincoln Do? Slave Trading Under the Stars and Stripes." *Chicago Daily Tribune*, January 11, 1914.

Wright, Arnold and H. A. Cartwright. *Twentieth Century Impressions of Hongkong, Shanghai, and Other Treaty Ports of China: Their History, People, Commerce, Industries, and Resources*. London: Lloyds Greater Britain Publishing Company, 1908.

Yule, Emma Sarepta. "The Dog Market at Baguio." *Overland Monthly*, November 1913, 436–42.

## *Secondary Sources*

Abinales, P. N., and Donna J. Amoroso. *State and Society in the Philippines*. Lanham, MD: Rowman & Littlefield Publishers, 2005.

Adas, Michael. *Dominance by Design: Technological Imperatives and America's Civilizing Mission*. Cambridge: Harvard University Press, 2006.

Afable, Patricia Okubo, ed. *Japanese Pioneers in the Northern Philippine Highlands: A Centennial Tribute, 1903–2003*. Baguio City: Filipino-Japanese Foundation of Northern Luzon, Inc., 2004.

Agnew, Jean-Christophe. "The Threshold of Exchange: Speculations on the Market." *Radical History Review* 21 (Fall 1979): 99–118.

———. *Worlds Apart: The Market and the Theater in Anglo-American Thought, 1550–1750*. New York: Cambridge University Press, 1993.

Agnir-Paraan, Lucris Carina, and Alice Buenviaje-Wilder, eds. *Camp John Hay: How It All Began . . . Where It Is Bound*. Camp John Hay, Baguio City: John Hay Poro Point Development Corporation, 2000.

Aguirre Beltrán, Gonzalo, and Deward E. Walker. *Regions of Refuge*. Washington, DC: Society for Applied Anthropology, 1979.

Aiken, S. Robert. *Imperial Belvederes: The Hill Stations of Malaya*. Kuala Lumpur: Oxford University Press, 1994.

Alarcón, Norma. *The Imperial Tapestry: American Colonial Architecture in the Philippines*. Manila: University of Santo Tomas Pub. House, 2008.

Alatas, Syed Hussein. *The Myth of the Lazy Native: A Study of the Image of the Malays, Filipinos and Javanese from the 16th to the 20th Century and Its Function in the Ideology of Colonial Capitalism*. London: Frank Cass, 1977.

Alcantara, Erlyn Ruth. "Baguio Between Two Wars: The Creation and Destruction of a Summer Capital." In *Vestiges of War: The Philippine-American War and the Aftermath of an Imperial Dream, 1899–1999*. Edited by Angel Velasco Shaw and Luis H. Francia. New York: New York University Press, 2002.

Aleinikoff, T. Alexander. *Semblances of Sovereignty: The Constitution, the State, and American Citizenship*. Cambridge: Harvard University Press, 2002.

Alpers, Paul. *What Is Pastoral?* Chicago: University of Chicago Press, 1996.

Anderson, Benedict. "Cacique Democracy in the Philippines: Origins and Dreams." *New Left Review* 169 (May-June 1988): 3–31.

———. *Imagined Communities: Reflections on the Origin and Spread of Nationalism*. Rev. ed. London: Verso, 2003 [1983].

———. *Under Three Flags: Anarchism and the Anti-Colonial Imagination*. London: Verso, 2005.

Anderson, Eric A. "The Encomienda in Early Philippine Colonial History." *Asian Studies* 14 (1976): 25–36.

Anderson, Warwick. *Colonial Pathologies: American Tropical Medicine, Race, and Hygiene in the Philippines*. Durham: Duke University Press, 2006.

Bachelard, Gaston. *The Poetics of Space*. Translated by Maria Jolas. Boston: Beacon Press, 1964.

Bagamaspad, Anavic, Zenaida Hamada-Pawid, and Bell Balangoy. *A People's History of Benguet Province*. Benguet Province, 1985.

Baldoz, Rick. *The Third Asiatic Invasion: Empire and Migration in Filipino America, 1898–1946*. New York: New York University Press, 2011.

Balisacan, A. M., and Hal Hill. *The Philippine Economy: Development, Policies, and Challenges*. New York: Oxford University Press, 2003.

Banner, Stuart. *How the Indians Lost Their Land: Law and Power on the Frontier*. Cambridge: Belknap Press of Harvard University Press, 2005.

Bankoff, Greg. "Conservation and Colonialism: Gifford Pinchot and the Birth of Tropical Forestry in the Philippines." In *Colonial Crucible: Empire in the Making of the Modern American State*. Edited by Alfred W. McCoy and Francisco A. Scarano. Madison: University of Wisconsin Press, 2009.

———. *Crime, Society, and the State in the Nineteenth-Century Philippines*. Manila: Ateneo de Manila University Press, 1996.

———. "'These Brothers of Ours': Poblete's Obreros and the Road to Baguio, 1903–1905." *Journal of Social History* 38, no. 4 (Summer 2005): 1047–72.

———. "Wants, Wages, and Workers: Laboring in the American Philippines, 1899–1908." *Pacific Historical Review* 74, no. 1 (February 2005): 59–86.

Basso, Keith H. *Wisdom Sits in Places: Landscape and Language Among the Western Apache*. Albuquerque: University of New Mexico Press, 1996.

Bederman, Gail. *Manliness & Civilization: A Cultural History of Gender and Race in the United States, 1880–1917*. Chicago: University of Chicago Press, 1995.

Bee, Ooi Jin. *Depletion of the Forest Resources in the Philippines*. Field Report Series, no. 18. Singapore: ASEAN Economic Research Unit, Institute of Southeast Asian Studies, 1987.

Bemis, Samuel Flagg. *A Diplomatic History of the United States*, 5th ed. New York: Holt, Rinehart, and Winston, 1965 [1936].

Bender, Daniel E. *American Abyss: Savagery and Civilization in the Age of Industry*. Ithaca: Cornell University Press, 2009.

Bender, Daniel, and Jana Lipman, eds. *Making the Empire Work: Labor and United States Imperialism*. New York: New York University Press, 2015.

Bender, Thomas. *A Nation among Nations: America's Place in World History*. New York: Hill and Wang, 2006.

Benguet Consolidated Inc., and Ed. C. de Jesus. *Benguet Consolidated, Inc., 1903–1978: A Brief History*. Benguet, 1978.

Benton, Lauren A. *A Search for Sovereignty: Law and Geography in European Empires, 1400—1900*. Cambridge: Cambridge University Press, 2010.

Best, Jonathan. *A Philippine Album: American Era Photographs, 1900–1930*. Makati City, Philippines: Bookmark, Inc., 1998.

Bhabha, Homi. *The Location of Culture*. London: Routledge, 1993.

Blanco, John D. *Frontier Constitutions Christianity and Colonial Empire in the Nineteenth-Century Philippines*. Berkeley: University of California Press, 2009.

Blitz, Amy. *The Contested State: American Foreign Policy and Regime Change in the Philippines*. Lanham, MD: Rowman & Littlefield Publishers, 2000.

Bluestone, Daniel. *Constructing Chicago*. New Haven: Yale University Press, 1991.

Booth, Howard J., and Nigel Rigby. *Modernism and Empire*. Manchester: University of Manchester Press, 2000.

Bourdieu, Pierre. "The Forms of Capital." In *Cultural Theory: An Anthology*. Malden, MA: Wiley-Blackwell, 2011.

———. *The Logic of Practice*. Translated by Richard Nice. Stanford: Stanford University Press, 1990 [1980].

———. "Social Space and Symbolic Power." *Sociological Theory* 7, no. 1 (1989): 14–25.

Boydston, Jeanne. *Home and Work: Housework, Wages, and the Ideology of Labor in the Early Republic*. New York: Oxford University Press, 1990.

Brewer, Anthony. *Marxist Theories on Imperialism: A Critical Survey*. London: Routledge, 1990.

Briggs, Laura. *Reproducing Empire: Race, Sex, Science, and US Imperialism in Puerto Rico*. Berkeley: University of California Press, 2002.

Brody, David. "Building Empire: Architecture and American Imperialism in the Philippines." *Journal of Asian American Studies* 4, no. 2 (June 2001): 123–45.

———. *Visualizing American Empire: Orientalism and Imperialism in the Philippines.* Chicago: University of Chicago Press, 2010.

Brook, Timothy. *The Confusions of Pleasure: Commerce and Culture in Ming China.* Berkeley: University of California Press, 1999.

Brosius, J. Peter, Anna Lowenhaupt Tsing, and Charles Zerner, eds. *Communities and Conservation: Histories and Politics of Community-Based Natural Resource Management.* Lanham, MD: AltaMira Press, 2005.

Buell, Lawrence. *The Environmental Imagination: Thoreau, Nature Writing, and the Formation of American Culture.* Cambridge: Belknap Press of Harvard University Press, 1995.

Bulosan, Carlos, and E. San Juan. *On Becoming Filipino: Selected Writings of Carlos Bulosan.* Philadelphia: Temple University Press, 1995.

Burnett, Christina Duffy. "Empire and the Transformation of Citizenship." In *Colonial Crucible: Empire in the Making of the Modern American State.* Edited by Alfred W. McCoy and Francisco A. Scarano. Madison: University of Wisconsin Press, 2009.

Burnett, Christina Duffy, and Burke Marshall. *Foreign in a Domestic Sense: Puerto Rico, American Expansion, and the Constitution.* Durham: Duke University Press, 2001.

Burns, Sarah. *Pastoral Inventions: Rural Life in Nineteenth-Century American Art and Culture.* Philadelphia: Temple University Press, 1989.

Buschmann, Rainer F., Edward R. Slack, and James B. Tueller. *Navigating the Spanish Lake: The Pacific in the Iberian World, 1521–1898.* Honolulu: University of Hawai'i Press, 2014.

Campomanes, Oscar. "The New Empire's Forgetful and Forgotten Citizens: Unrepresentability and Unassimilability in Filipino-American Postcolonialities." *Critical Mass* 2, no. 2 (Spring 1995): 145–200.

———."New Formations in Asian American Studies and the Question of US Imperialism." *positions* 3 (1997): 523–50.

Cappers, Elmer Osgood. *Centennial History of the Country Club, 1882–1982.* Brookline, MA: Country Club, 1981.

Carino, Joanna K. "The Cariños and Baguio-Benguet History: Conclusion." In *Folio 2.* Social Science Monograph Series no. 4. [Baguio]: Cordillera Studies Center, University of the Philippines College Baguio, 1984.

Certeau, Michel de. *The Practice of Everyday Life.* Translated by Steven Rendell. Berkeley: University of California Press, 1988 [1984].

Chakrabarty, Dipesh. *Provincializing Europe: Postcolonial Thought and Historical Difference.* Princeton: Princeton University Press, 2008.

Chang, David A. "Enclosures of Land and Sovereignty: The Allotment of American Indian Lands." *Radical History Review,* no. 109 (2011): 108–19.

Chang, Kornel S. *Pacific Connections: The Making of the Western US-Canadian Borderlands.* Berkeley: University of California Press, 2012.

Chazkel, Amy and David Serlin. "Editors' Introduction." *Radical History Review,* no. 109 (December 2011): 1–11.

Cheng, Charles L. and Katherine V. Bersamira. *The Ethnic Chinese in Baguio and in the Cordillera Philippines: The Untold Stories*. Baguio City: Unique Printing Press, 1997.

Cheng, Lucie, and Edna Bonacich. *Labor Immigration under Capitalism: Asian Workers in the United States before World War II*. Berkeley: University of California Press, 1984.

Chibber, Vivek. *Postcolonial Theory and the Specter of Capital*. London: Verso, 2013.

Choy, Catherine Ceniza. *Empire of Care: Nursing and Migration in Filipino American History*. Durham: Duke University Press, 2003.

Ciriaco, Conrado F. *Catalogue of Philippine Picture Postcards, American Period: 1898–1941*. [Manila]: C. F. Ciriaco, 1995.

Cody, Mary Kilcline. "A Paler Shade of White." In *Lost Times and Untold Tales from the Malay World*. Edited by Jan Van Der Putten and Mary Kilcline Cody. Singapore: NUS Press, 2009.

Cogan, Frances B. *Captured: The Japanese Internment of American Civilians in the Philippines, 1941–1945*. Athens: University of Georgia Press, 2000.

Colby, Jason M. *The Business of Empire: United Fruit, Race, and US Expansion in Central America*. Ithaca: Cornell University Press, 2013.

Comaroff, John L., and Jean Comaroff. "Law and Disorder in the Postcolony: An Introduction." In *Law and Disorder in the Postcolony*. Edited by Jean Comaroff and John L. Comaroff. Chicago: University of Chicago Press, 2006.

Conklin, Harold C. *Hanunóo Agriculture: A Report on an Integral System of Shifting Cultivation in the Philippines*. Northford, CT: Elliot's Books, 1975 [1957].

Constantino, Renato, and Letizia R. Constantino. *A History of the Philippines: From the Spanish Colonization to the Second World War*. New York: Monthly Review Press, 1975.

Cooper, Frederick. *Colonialism in Question: Theory, Knowledge, History*. Berkeley: University of California Press, 2009.

Coronil, Fernando. *The Magical State: Nature, Money, and Modernity in Venezuela*. Chicago: University of Chicago Press, 1997.

———. "Smelling Like a Market." *American Historical Review* 106, no. 1 (2001): 119–29.

———. "Towards a Critique of Globalcentrism: Speculations on Capitalism's Nature." *Public Culture* 12 (Spring 2000): 351–74.

Corpuz, Arturo G. *The Colonial Iron Horse: Railroads and Regional Development in the Philippines, 1875–1935*. Quezon City: University of the Philippines, 1999.

Corpuz, Onofre D. *An Economic History of the Philippines*. Quezon City: University of the Philippines Press, 1997.

Cosgrove, Denis E., and Stephen Daniels. *The Iconography of Landscape: Essays on the Symbolic Representation, Design, and Use of Past Environments*. Cambridge: Cambridge University Press, 1988.

Cronon, William. *Changes in the Land: Indians, Colonists, and the Ecology of New England*. New York: Hill and Wang, 1983.

———. *Nature's Metropolis: Chicago and the Great West.* New York: W. W. Norton, 1991.

Crossette, Barbara. *The Great Hill Stations of Asia.* Boulder, CO: Westview Press, 1998.

Cruz, Romeo V. *America's Colonial Desk and the Philippines, 1898–1934.* Quezon City: University of the Philippines Press, 1974.

Cuasay, Peter. "Indigenizing Law or Legalizing Governmentality? The Indigenous Peoples Rights Act and the Philippine Supreme Court." Paper presented at "Politics of the Commons: Articulating Development and Strengthening Local Practices," Chiang Mai, Thailand, July 11–14, 2003. Accessed at http://dlc.dlib.indiana.edu/archive/00001146/.

Cullinane, Michael. *Ilustrado Politics: Filipino Elite Responses to American Rule, 1898–1908.* Manila: Ateneo de Manila University Press, 2003.

Cumings, Bruce. *Dominion from Sea to Sea: Pacific Ascendancy and American Power.* New Haven: Yale University Press, 2009.

Davis, David Brion. *The Problem of Slavery in the Age of Revolution.* Ithaca: Cornell University Press, 1975.

Davis, William G. *Social Relations in a Philippine Market: Self-Interest and Subjectivity.* Berkeley: University of California Press, 1973.

De Angelis, Massimo. "Marx and Primitive Accumulation: The Continuous Character of Capital's 'Enclosures.'" *Commoner* 2 (September 2001).

Deloria, Philip Joseph. *Playing Indian.* New Haven: Yale University Press, 1998.

Delmendo, Sharon. *The Star-Entangled Banner: One Hundred Years of America in the Philippines.* New Brunswick, NJ: Rutgers University Press, 2004.

Denning, Michael. *The Cultural Front: The Laboring of American Culture in the Twentieth Century.* New York: Verso, 1998.

Dirks, Nicholas B. *Colonialism and Culture.* Ann Arbor: University of Michigan Press, 1992.

Doeppers, Daniel F. "The Development of Philippine Cities Before 1900." *Journal of Asian Studies* 31 (August 1972): 769–92.

———. *Manila, 1900–1941: Social Change in a Late Colonial Metropolis.* New Haven: Yale University Southeast Asia Studies, 1984.

———. "Manila's Imperial Makeover: Security, Health, and Symbolism." In *Colonial Crucible: Empire in the Making of the Modern American State.* Edited by Alfred W. McCoy and Francisco A. Scarano. Madison: University of Wisconsin Press, 2009.

Douglas, Mary. *Purity and Danger: An Analysis of Concepts of Pollution and Taboo.* New York: Routledge, 1966.

Dove, Michael. R. "Theories of Swidden Agriculture, and the Political Economy of Ignorance." *Agroforestry Systems* 1 (1983): 85–99.

Eagleton, Terry, Fredric Jameson, and Edward W. Said. *Nationalism, Colonialism, and Literature.* Minneapolis: University of Minnesota Press, 1990.

Easterling, Keller. *Extrastatecraft: The Power of Infrastructure Space.* London: Verso, 2014.

Elliott, J. H. *Empires of the Atlantic World: Britain and Spain in America, 1492–1830.* New Haven: Yale University Press, 2006.

Eperjesi, John R. *The Imperialist Imaginary: Visions of Asia and the Pacific in American Culture.* Hanover, NH: Dartmouth College Press, University Press of New England, 2005.

Espiritu, Augusto Fauni. *Five Faces of Exile: The Nation and Filipino American Intellectuals.* Stanford: Stanford University Press, 2005.

Fanning, Patricia J. *Norwood: A History.* Charleston, SC: Acadia, 2002.

Farrell, Betty. *Elite Families: Class and Power in Nineteenth-Century Boston.* Albany: State University of New York Press, 1993.

Federici, Silvia. *Caliban and the Witch.* Brooklyn, NY: Autonomedia, 2004.

Fermin, Jose D. *1904 World's Fair: The Filipino Experience.* Diliman, Quezon City: University of the Philippines Press, 2004.

"Filipino Immigrants in the United States." *Migrationpolicy.org.* Accessed March 28, 2015. http://migrationpolicy.org/article/filipino-immigrants-united-states.

Findlay, Eileen. *Imposing Decency: The Politics of Sexuality and Race in Puerto Rico, 1870–1920.* Durham: Duke University Press, 1999.

Finin, Gerard A. *The Making of the Igorot: Contours of Cordillera Consciousness.* Honolulu: University of Hawai'i Press, 2006.

Flynn, Dennis O. and Arturo Giráldez. "Born with a 'Silver Spoon': The Origin of World Trade in 1571." *Journal of World History* 6 (1995): 201–21.

Foner, Eric. *A Short History of Reconstruction, 1863–1877.* New York: Harper & Row, 1990.

Foucault, Michel. *Security, Territory, Population: Lectures at the Collège De France, 1977–78.* Edited by Michel Senellart. Translated by Graham Burchell. New York: Picador, 2009.

Francia, Luis. *A History of the Philippines: From Indios Bravos to Filipinos.* 1st ed. New York: Overlook Press, 2010.

Friedman, Andrew. *Covert Capital: Landscapes of Denial and the Making of US Empire in the Suburbs of Northern Virginia.* Berkeley: University of California Press, 2013.

Fry, Howard T. *A History of the Mountain Province.* Quezon City: New Day Publishers, 1983.

Fujita-Rony, Dorothy B. *American Workers, Colonial Power: Philippine Seattle and the Transpacific West, 1919–1941.* Berkeley: University of California, 2003.

Gallagher, John, and Ronald Robinson. "The Imperialism of Free Trade." *Economic History Review*, n.s., 6, no. 1 (January 1, 1953): 1–15.

Garraty, John A. "Holmes's Appointment to the US Supreme Court." *New England Quarterly* 22, no. 3 (September 1949).

Gatewood, Willard B. *"Smoked Yankees" and the Struggle for Empire: Letters from Negro Soldiers, 1898–1902.* Urbana: University of Illinois Press, 1971.

Genetin-Pilawa, C. Joseph. *Crooked Paths to Allotment: The Fight over Federal Indian*

*Policy after the Civil War*. Chapel Hill: University of North Carolina Press, 2012.

Gillem, Mark L. *America Town: Building the Outposts of Empire*. Minneapolis: University of Minnesota Press, 2007.

Gillette, Howard. "The Military Occupation of Cuba, 1899–1902: Workshop for American Progressivism." *American Quarterly* 25, no. 4 (1973): 410–25.

Gilmore, Glenda Elizabeth. *Who Were the Progressives? Readings*. Boston: Bedford/St. Martin's, 2002.

Gindin, Sam, and Leo Panitch. *The Making of Global Capitalism: The Political Economy of American Empire*. London: Verso, 2012.

Glassman, Jim. "Primitive Accumulation, Accumulation by Dispossession, Accumulation by 'Extra-Economic' Means." *Progress in Human Geography* 30, no. 5 (2006): 608–25.

Gleeck, Lewis E. *American Business and Philippine Economic Development*. Manila: Carmelo & Bauermann, 1975.

———. *The Manila Americans, 1901–1964*. Manila: Carmelo & Bauermann, Inc., 1977.

Go, Julian. *American Empire and the Politics of Meaning: Elite Political Cultures in the Philippines and Puerto Rico during US Colonialism*. Durham: Duke University Press, 2008.

———. *Patterns of Empire: The British and American Empires, 1688 to Present*. Cambridge: Cambridge University Press, 2011.

Go, Julian, and Anne L. Foster. *The American Colonial State in the Philippines: Global Perspectives*. Durham: Duke University Press, 2003.

Golay, Frank. *Face of Empire: United States-Philippine Relations, 1898–1946*. Madison: Center for Southeast Asian Studies, 1997.

Gonzalez, Vernadette Vicuña. *Securing Paradise: Tourism and Militarism in Hawai'i and the Philippines*. Durham: Duke University Press, 2013.

Goswami, Manu. *Producing India: From Colonial Economy to National Space*. Chicago: University of Chicago Press, 2004.

Grandin, Greg. *Fordlandia*. New York: Picador, 2010.

Greene, Julie. *The Canal Builders: Making America's Empire at the Panama Canal*. New York: Penguin Press, 2009.

Greenhalgh, Paul. *Ephemeral Vistas: The Expositions Universelles, Great Exhibitions, and World's Fairs, 1851–1939*. Manchester: Manchester University Press/St. Martin's Press, 1988.

Grier, Katherine C. *Pets in America: A History*. Chapel Hill: University of North Carolina Press, 2006.

Grove, Richard H. "Colonial Conservation, Ecological Hegemony and Popular Resistance: Towards a Global Synthesis." In *Imperialism and the Natural World*, ed. John M. MacKenzie. Manchester: Manchester University Press, 1990.

———. *Green Imperialism: Colonial Expansion, Tropical Edens and the Origins of Environmentalism, 1600–1860*. Cambridge: Cambridge University Press, 1995.

Grove, Richard H., Vinita Damodaran, and Satpal Sangwan, eds. *Nature and the Orient: The Environmental History of South and Southeast Asia*. Delhi: Oxford University Press, 1998.

Gruesser, John Cullen. *The Empire Abroad and the Empire at Home: African American Literature and the Era of Overseas Expansion*. Athens: University of Georgia, 2012.

Guevarra, Dante G. *History of the Philippine Labor Movement*. Santa Mesa, Manila: Institute of Labor & Industrial Relations, Polytechnic University of the Philippines, 1991.

Gutierrez, Lazaro P. *Memoirs of Baguio*. Baguio City: Summer Capital Pub. House, 1960.

Habana, Olivia M. "Gold Mining in Benguet: 1900–1941." *Philippine Studies* 49, no. 1 (2000): 3–41.

Hall, Peter. *Cities of Tomorrow: An Intellectual History of Urban Planning and Design in the Twentieth Century*. 3rd edition. Malden, MA: Blackwell Publishing, 2002.

Halsema, James J. *E. J. Halsema, Colonial Engineer: A Biography*. Quezon City, Philippines: New Day Publishers, 1991.

Hardt, Michael, and Antonio Negri. *Empire*. Cambridge: Harvard University Press, 2000.

Harrington, Thomas. "The Philippines on the Eve of Change." In *South East Asia Colonial History*. Edited by Paul H. Kratoska. London: Routledge, 2001.

Harrison, Mark. *Climates & Constitutions: Health, Race, Environment and British Imperialism in India, 1600–1850*. Oxford: Oxford University Press, 1999.

Harvey, David. *The Limits to Capital*. Chicago: University of Chicago Press, 1982.

———. "The 'New' Imperialism: Accumulation by Dispossession." *Socialist Register* 40, no. 40 (March 19, 2009).

———. *Spaces of Capital: Towards a Critical Geography*. New York: Routledge, 2001.

———. *Spaces of Global Capitalism*. London: Verso, 2006.

Haskell, Thomas L. "Capitalism and the Origins of the Humanitarian Sensibility." Pts. 1 and 2. *American Historical Review* 90, no. 2 (April 1985): 339–61; 90, no. 3 (June 1985): 547–66.

Hays, Samuel P. *Conservation and the Gospel of Efficiency: The Progressive Conservation Movement, 1890–1920*. Cambridge: Harvard University Press, 1959.

Hines, Thomas S. *Burnham of Chicago: Architect and Planner*. New York: Oxford University Press, 1974.

———. "The Imperial Façade: Daniel H. Burnham and American Architectural Planning in the Philippines." *Pacific Historical Review* 41, no. 1 (February 1972): 33–53.

Hobsbawm, Eric. *The Age of Capital, 1848–1875*. London: Abacus, 2004 [1975].

———. *The Age of Empire, 1875–1914*. New York: Vintage Books, 1989 [1987].

Hoganson, Kristin L. *Consumers' Imperium: The Global Production of American Domesticity, 1865–1920*. Chapel Hill: University of North Carolina Press, 2007.

———. *Fighting for American Manhood: How Gender Politics Provoked the Spanish-American and Philippine-American Wars*. New Haven: Yale University Press, 1998.

Howell, Brian M. "Moving Mountains: Protestant Christianity and the Spiritual Landscape of Northern Luzon." *Anthropological Forum* 19, no. 3 (2009): 253–69.

Hoxie, Frederick E. *A Final Promise: The Campaign to Assimilate the Indians, 1880–1920*. Lincoln: University of Nebraska Press, 1984.

Igler, David. *The Great Ocean: Pacific Worlds from Captain Cook to the Gold Rush*. Oxford: Oxford University Press, 2013.

Ileto, Reynaldo C. *Filipinos and their Revolution: Event, Discourse, and Historiography*. Manila: Ateneo de Manila University Press, 1998.

———. *Pasyon and Revolution: Popular Movements in the Philippines, 1840–1910*. Quezon City: Ateneo de Manila University Press, 1979.

Jackson, John Brinckerhoff. *A Sense of Place, a Sense of Time*. New Haven: Yale University Press, 1994.

Jackson, John Brinckerhoff. *Landscape in Sight: Looking at America*. Edited by Helen Lefkowitz Horowitz. New Haven: Yale University Press, 1997.

Jacobson, Matthew Frye. *Barbarian Virtues: The United States Encounters Foreign Peoples at Home and Abroad, 1876–1917*. New York: Hill and Wang, 2000.

Jacoby, Karl. *Crimes Against Nature: Squatters, Poachers, Thieves, and the Hidden History of American Conservation*. Berkeley: University of California Press, 2001.

Jameson, Fredric. "Modernism and Imperialism." In *National, Colonialism and Literature*. Edited by Terry Eagleton, Frederic Jameson, and Edward Said. Minneapolis: University of Minnesota Press, 1990.

Jebens, Holger. *Cargo, Cult, and Culture Critique*. Honolulu: University of Hawai'i Press, 2004.

Jenesta, Frank. *The White Apos*. Quezon City: New Day Publishers, 1987.

Jennings, Eric Thomas. *Curing the Colonizers: Hydrotherapy, Climatology, and French Colonial Spas*. Durham: Duke University Press, 2006.

———. *Imperial Heights: Dalat and the Making and Undoing of French Indochina*. Berkeley: University of California Press, 2012.

Jessup, Philip C. *Elihu Root*. Hamden, CT: Archon Books, 1964.

Johnson, Walter. *Soul by Soul: Life Inside the Antebellum Slave Market*. Cambridge, MA: Harvard University Press, 1999.

Johnston, Robert D. *The Radical Middle Class: Populist Democracy and the Question of Capitalism in Progressive Era Portland, Oregon*. Princeton: Princeton University Press, 2003.

———. "Re-Democratizing the Progressive Era: The Politics of Progressive Era Political Historiography." *Journal of the Gilded Age and Progressive Era* 1, no. 1 (2002): 68–92.

Jones, Gareth Stedman. "The Specificity of US Imperialism." *New Left Review* 1, no. 60 (April 1970): 59–86.

Joseph, G. M., Catherine LeGrand, and Ricardo Donato Salvatore, eds. *Close Encounters of Empire: Writing the Cultural History of US-Latin American Relations.* Durham: Duke University Press, 1998.

Kaplan, Amy. *The Anarchy of Empire in the Making of US Culture.* Cambridge: Harvard University Press, 2002.

Kaplan, Amy, and Donald E. Pease. *Cultures of United States Imperialism.* Durham: Duke University Press, 1993.

Keesing, Felix Maxwell. *The Ethnohistory of Northern Luzon.* Stanford: Stanford University Press, 1962.

Kennedy, Dane Keith. *The Magic Mountains: Hill Stations and the British Raj.* Berkeley: University of California Press, 1996.

Kerkvliet, Benedict J. *The Huk Rebellion: A Study of Peasant Revolt in the Philippines.* Berkeley: University of California Press, 1977.

King, Anthony D. *The Bungalow: The Production of a Global Culture.* London: Routledge & Kegan, 1984.

———. *Colonial Urban Development: Culture, Social Power, and Environment.* London: Routledge & Kegan Paul, 1976.

Klein, Naomi. *No Logo.* New York: Picador, 2000.

Kloppenberg, James T. *Uncertain Victory: Social Democracy and Progressivism in European and American Thought, 1870–1920.* Oxford: Oxford University Press, 1986.

Kramer, Paul A. *The Blood of Government: Race, Empire, the United States, and the Philippines.* Chapel Hill: University of North Carolina Press, 2006.

———. "Empires, Exceptions, and Anglo-Saxons: Race and Rule between the British and United States Empires, 1880–1910." *Journal of American History* 88, no. 4 (March 2002): 1315–53.

———."Power and Connection: Imperial Histories of the United States in the World." *American Historical Review* 116, no. 5 (2011): 1348–91.

———. "Reflex Actions: Colonialism, Corruption and the Politics of Technocracy in the Early 20th Century United States." In *Challenging US Foreign Policy: America and the World in the Long Twentieth Century.* Edited by Bevan Sewell and Scott Lucas. New York: Palgrave Macmillan, 2011.

LaFeber, Walter. *The New Empire: An Interpretation of American Expansion, 1860–1898.* Ithaca: Cornell University Press, 1963.

Lampe, Evan. *Work, Class, and Power in the Borderlands of the Early American Pacific: The Labors of Empire.* Lanham, MD: Lexington Books, 2014.

Larkin, John. *Sugar and the Origins of Modern Philippine Society.* Berkeley: University of California Press, 1993.

Leano, I. "Going Home to the Spirit World." *Saint Louis University Research Journal* 18, no. 2 (1987): 189–230.

Lears, T. J. Jackson. *No Place of Grace: Antimodernism and the Transformation of American Culture, 1880–1920.* New York: Pantheon Books, 1981.

———. *Rebirth of a Nation: The Making of Modern America, 1877–1920.* New York: HarperCollins, 2009.

Lefebvre, Henri. *The Production of Space*. Translated by Donald Nicholson-Smith. Malden, MA: Blackwell Publishing, 1991 [1974].

Legarda, Benito J. *After the Galleons: Foreign Trade, Economic Change & Entrepreneurship in the Nineteenth Century Philippines*. Quezon City: Ateneo de Manila University Press, 1999.

Lemke, Thomas. "Foucault, Governmentality, and Critique." *Rethinking Marxism* 14 (2002): 49–64.

Leuchtenburg, William E. "Progressivism and Imperialism: The Progressive Movement and American Foreign Policy, 1898–1916." *Mississippi Valley Historical Review* 39, no. 3 (December 1, 1952): 483–504.

Lewis, Martin W. *Wagering the Land: Ritual, Capital, and Environmental Degradation in the Cordillera of Northern Luzon, 1900–1986*. Berkeley: University of California Press, 1991.

Lico, Gerald. *Arkitekturang Filipino: A History of Architecture in the Philippines*. Diliman, Quezon City, Philippines: University of the Philippines Press, 2008.

Licuanan, Virginia Benitez. *Filipinos and Americans: A Love-Hate Relationship*. Manila: Baguio Country Club, 1982.

Lindstrom, Lamont. *Cargo Cult: Strange Stories of Desire from Melanesia and Beyond*. Honolulu: University of Hawai'i Press, 1993.

Linebaugh, Peter. *The London Hanged: Crime and Civil Society in the Eighteenth Century*. New York: Cambridge University Press, 1992.

Linn, Brian McAllister. *The Philippine War, 1899–1902*. Lawrence: University Press of Kansas, 2000.

———. *The US Army and Counterinsurgency in the Philippine War, 1899–1902*. Chapel Hill: University of North Carolina Press, 1989.

Lipman, Jana. *Guantánamo: A Working-Class History between Empire and Revolution*. Berkeley: University of California Press, 2008.

Longstreth, Richard W., ed. *The Mall in Washington, 1791–1991*. Washington, DC: National Gallery of Art, 1991.

Love, Eric Tyrone Lowery. *Race over Empire: Racism and US Imperialism, 1865–1900*. Chapel Hill: University of North Carolina Press, 2004.

Low, Setha M., and Denise Lawrence-Zúñiga. *The Anthropology of Space and Place: Locating Culture*. Malden, MA: Blackwell, 2003.

Lowe, Lisa. *Immigrant Acts: On Asian American Cultural Politics*. Durham: Duke University Press, 1999 [1996].

Luton, Harry. "American Internal Revenue Policy in the Philippines to 1916." In *Compadre Colonialism: Studies on the Philippines under American Rule*. Edited by Norman G. Owen. Ann Arbor: Center for South and Southeast Asian Studies, University of Michigan, 1971.

Lutz, Catherine. *Homefront: A Military City and the American 20th Century*. Boston: Beacon, 2001.

Lynch, Owen J. *Colonial Legacies in a Fragile Republic: A History of Philippine Land Law and State Formation with Emphasis on the Early US Regime, 1898–1913*. Diliman, Quezon City: University of the Philippines, College of Law, 2011.

———. "Concepts and Strategies for Promoting Legal Recognition of Community-Based Property Rights: Insights from the Philippines and Other Nations." In *Communities and Conservation: Histories and Politics of Community-Based Natural Resource Management*. Edited by Peter J. Brosius, Anna Lowenhaupt Tsing, and Charles Zerner. Walnut Creek, CA: AltaMira Press, 2005.

Martin, Meredith. *Dairy Queens: The Politics of Pastoral Architecture from Catherine de' Medici to Marie-Antoinette*. Cambridge: Harvard University Press, 2011.

Marx, Karl. *Capital, Volume One*. Excerpted in *The Marx-Engels Reader*. 2nd ed. Edited by Robert C. Tucker. New York: W. W. Norton, 1978 [1972].

Marx, Leo. *The Machine in the Garden: Technology and the Pastoral Ideal in America*. New York: Oxford University Press, 2000 [1964].

Massey, Doreen B. *Space, Place, and Gender*. Minneapolis: University of Minnesota Press, 2005 [1994].

———. *Spatial Divisions of Labor: Social Structures and the Geography of Production*. New York: Methuen, 1984.

Matsuda, Matt K. *Pacific Worlds: A History of Seas, Peoples, and Cultures*. Cambridge: Cambridge University Press, 2012.

May, Glenn Anthony. *Social Engineering in the Philippines: The Aims, Execution, and Impact of American Colonial Policy, 1900–1913*. Westport, CT: Greenwood Press, 1980.

Mayo, James M. *The American Country Club: Its Origins and Development*. New Brunswick, NJ: Rutgers University Press, 1998.

McCormick Thomas J. *The China Market: America's Quest for Informal Empire, 1893–1901*. Chicago: Quandrangle Books, 1967.

McCoy, Alfred W. *Policing America's Empire: The United States, the Philippines, and the Rise of the Surveillance State*. Madison: University of Wisconsin Press, 2009.

McCoy, Alfred W., and Ed. C. de Jesus, eds. *Philippine Social History: Global Trade and Local Transformations*. Honolulu: University of Hawai'i Press, 1982.

McCoy, Alfred W., and Francisco A. Scarano, eds. *Colonial Crucible: Empire in the Making of the Modern American State*. Madison: University of Wisconsin Press, 2009.

McCoy, Alfred W., Francisco A. Scarano, and Courtney Johnson. "On the Tropic of Cancer: Transitions and Transformations in the US Imperial State." In *Colonial Crucible: Empire in the Making of the Modern American State*. Edited by Alfred W. McCoy and Francisco A. Scarano. Madison: University of Wisconsin Press, 2009.

McGerr, Michael E. *A Fierce Discontent: The Rise and Fall of the Progressive Movement in America, 1870–1920*. New York: Free Press, 2003.

Metcalf, Thomas R. *An Imperial Vision: Indian Architecture and Britain's Raj*. Berkeley: University of California Press, 1989.

Midnight Notes Collective. "The New Enclosures." *Midnight Notes* 10 (1990). http://www.midnightnotes.org/newenclos.html.

Mignolo, Walter. *Local Histories/Global Designs: Coloniality, Subaltern Knowledges, and Border Thinking*. Princeton: Princeton University Press, 2000.

Miller, Bonnie. *From Liberation to Conquest: The Visual and Popular Cultures of the Spanish-American War of 1898*. Amherst: University of Massachusetts, 2011.

Miller, Stuart Creighton. *"Benevolent Assimilation": The American Conquest of the Philippines, 1899–1903*. New Haven: Yale University Press, 1982.

Mitchell, Timothy. *Colonising Egypt*. Cambridge: Cambridge University Press, 1988.

Mitchell, W. J. T. *Landscape and Power*. Chicago: University of Chicago Press, 1994.

Mojares, Resil B. *Brains of the Nation: Pedro Paterno, T. H. Pardo de Tavera, Isabelo de los Reyes and the Production of Modern Knowledge*. Quezon City: Ateneo de Manila University Press, 2006.

Moore, Charles. *Daniel H. Burnham: Architect, Planner of Cities*. New York: Da Capo Press, 1968.

Morris, Rosalind C. "Imperial Pastoral: The Politics and Aesthetics of Translation in British Malaya." *Representations* (Summer 2007): 159–94.

Moss, Richard J. *Golf and the American Country Club*. Urbana: University of Illinois Press, 2001.

Mozingo, Louise A. *Pastoral Capitalism: A History of Suburban Corporate Landscapes*. Cambridge, MA: MIT Press, 2011.

Mukerji, Chandra. *Territorial Ambitions and the Gardens of Versailles*. Cambridge: Cambridge University Press, 1997.

Nemser, Daniel. "Primitive Accumulation, Geometric Space, and the Construction of the 'Indian.'" *Journal of Latin American Cultural Studies* 24 (2015): 335–52.

Ngai, Mae M. *Impossible Subjects: Illegal Aliens and the Making of Modern America*. Princeton: Princeton University Press, 2004.

Nicolson, Marjorie Hope. *Mountain Gloom and Mountain Glory: The Development of the Aesthetics of the Infinite*. Ithaca: Cornell University Press, 1959.

Ohmann, Richard M. *Selling Culture: Magazines, Markets, and Class at the Turn of the Century*. London: Verso, 1996.

Ong, Aihwa. *Flexible Citizenship: The Cultural Logics of Transnationality*. Durham: Duke University Press, 1999.

Orendain, Juan Claros. *Philippine Wonderland*. Baĝuio, 1940.

Orenstein, Dara. "Foreign-Trade Zones and the Cultural Logic of Frictionless Production." *Radical History Review*, no. 109 (2011): 36–61.

Otto, Ton. "What Happened to Cargo Cults? Material Religions in Melanesia and the West." *Social Analysis* 53, no. 1 (Spring 2009): 82–102.

Owen, Norman G. *Compadre Colonialism: Studies on the Philippines under American Rule*. Ann Arbor: Center for South and Southeast Asian Studies, University of Michigan, 1971.

———. *Prosperity without Progress: Manila Hemp and Material Life in the Colonial Philippines*. Quezon City: Ateneo de Manila University Press, 1984.

Paredes, Ruby R., ed. *Philippine Colonial Democracy*. Monograph Series, no. 32, Yale University Southeast Asia Studies. New Haven: Yale University, 1988.

Patterson, Annabel M. *Pastoral and Ideology: Virgil to Valéry*. Berkeley: University of California Press, 1987.

Pearson, Henry Greenleaf. *An American Railroad Builder, John Murray Forbes*. Boston: Houghton Mifflin Co., 1911.

Peebles, Gustav. "The Anthropology of Credit and Debt." *Annual Review of Anthropology* 39 (2010): 225–40.

Perdue, Peter C. "Merchants West & East." In "Rise & Fall of the Canton Trade System—I." *MIT Visualizing Cultures*. Accessed March 26, 2015. http://ocw.mit.edu/ans7870/21f/21f.027/rise_fall_canton_01/cw_essay03.html.

Perelman, Michael. *The Invention of Capitalism: Classical Political Economy and the Secret History of Primitive Accumulation*. Durham: Duke University Press, 2000.

Peter, C. Brooks. "Igorots Astride Tanks in Bataan Wipe Out a Japanese Regiment." *New York Times*, February 23, 1942.

Peterson, John A. *The Birth of City Planning in the United States, 1840–1917*. Baltimore: Johns Hopkins University Press, 2003.

Phelan, John Leddy. *The Hispanization of the Philippines: Spanish Aims and Filipino Responses, 1565–1700*. Madison: University of Wisconsin Press, 1959.

Pomeroy, William J. *American Neo-Colonialism: Its Emergence in the Philippines and Asia*. New York: International Publishers, 1970.

Pommersheim, Frank. *Broken Landscape: Indians, Indian Tribes, and the Constitution*. Oxford: Oxford University Press, 2009.

Poole, Deborah. *Vision, Race, and Modernity: A Visual Economy of the Andean Image World*. Princeton: Princeton University Press, 1997.

Pratt, Julius William. *America's Colonial Experiment: How the United States Gained, Governed, and in Part Gave Away a Colonial Empire*. New York: Prentice-Hall, 1950.

Pratt, Mary Louise. *Imperial Eyes: Travel Writing and Transculturation*. London: Routledge, 1992.

Prentice, Claire. *The Lost Tribe of Coney Island: Headhunters, Luna Park, and the Man Who Pulled off the Spectacle of the Century*. Boston: New Harvest, 2014.

Prill-Brett, June. "Baguio: A Multi-Ethnic City and the Development of the Ibaloy as an Ethnic Minority." CSC Working Paper no. 15. Baguio City: Cordillera Studies Center, 1990.

———. "Indigenous Experience of Autonomy in the Cordillera." CSC Working Paper no. 10. Baguio City: Cordillera Studies Center, 1989.

Rafael, Vicente L. *Contracting Colonialism: Translation and Christian Conversion in Tagalog Society under Early Spanish Rule*. Ithaca: Cornell University Press, 1988.

———, ed. *Discrepant Histories: Translocal Essays on Filipino Culture*. Philadelphia: Temple University Press, 1995.

———. *The Promise of the Foreign: Nationalism and the Technics of Translation in the Spanish Philippines*. Durham: Duke University Press, 2005.

———. "Welcoming What Comes: Sovereignty and Revolution in the Colonial Philippines." *Comparative Studies in Society and History* 52, no. 1 (2010): 157–79.

———. *White Love and Other Events in Filipino History*. Durham: Duke University Press, 2000.

Rai, Amit S. *Rule of Sympathy: Sentiment, Race, and Power, 1750–1850*. New York: Palgrave, 2002.

Rakestraw, Lawrence. "George Patrick Ahern and the Philippine Bureau of Forestry, 1900–1914." *Pacific Northwest Quarterly* 53 (July 1967): 142–50.

Reed, Robert Ronald. *City of Pines: The Origins of Baguio as a Colonial Hill Station and Regional Capital*. 2nd ed. Baguio City: A-Seven Publishing, 1999 [1976].

———. *Colonial Manila: The Context of Hispanic Urbanism and Process of Morphogenesis*. Berkeley: University of California Press, 1978.

———. *Hispanic Urbanism in the Philippines: A Study of the Impact of Church and State*. Manila: University of Manila, 1967.

Reyes-Boquiren, Rowena. "The History and Political Economy of the Vegetable Industry in Benquet." CSC Working Paper 14. Baguio City: Cordillera Studies Center, 1989.

———. "Baguio's History and Cultural Heritage." *Northern Dispatch Weekly*, August 23, 2015. http://www.nordis.net/2015/08/baguios-history-and-cultural-heritage/.

Rice, Mark. *Dean Worcester's Fantasy Islands: Photography, Film, and the Colonial Philippines*, 2014.

———. "His Name Was Don Francisco Muro: Reconstructing an Image of American Imperialism." *American Quarterly* 62, no. 1 (2010): 49–76.

Ritvo, Harriet. *The Animal Estate: The English and Other Creatures in the Victorian Age*. Cambridge: Harvard University Press, 1987.

Roberts, Randy. *The Rock, the Curse, and the Hub: A Random History of Boston Sports*. Cambridge: Harvard University Press, 2005.

Robin, Ron. *Enclaves of America: The Rhetoric of American Political Architecture Abroad, 1900–1965*. Princeton: Princeton University Press, 1992.

Rodgers, Daniel T. *Atlantic Crossings: Social Politics in a Progressive Age*. Cambridge, MA: Belknap, 2000.

. "In Search of Progressivism." *Reviews in American History* 10, no. 4 (1982): 113–32.

Rodriguez, Robyn Magalit. *Migrants for Export: How the Philippine State Brokers Labor to the World*. Minneapolis: University of Minnesota Press, 2010.

Roediger, David R., and Elizabeth D. Esch. *The Production of Difference: Race and the Management of Labor in US History*. New York: Oxford University Press, 2012.

Rosaldo, Renato. "Imperialist Nostalgia." In "Memory and Counter-Memory," special issue, *Representations* 26 (Spring 1989): 107–22.

Rosenberg, Emily S. *Financial Missionaries to the World: The Politics and Culture of Dollar Diplomacy, 1900–1930*. Cambridge: Harvard University Press, 1999.

———. *Spreading the American Dream: American Economic and Cultural Expansion, 1890–1945*. New York: Hill and Wang, 1982.

Rosenzweig, Roy and Elizabeth Blackmar. *The Park and the People: A History of Central Park*. Ithaca: Cornell University Press, 1992.

Roth, Dennis Morrow. *The Friar Estates of the Philippines*. Albuquerque: University of New Mexico Press, 1977.

Rydell, Robert W. *All the World's a Fair: Visions of Empire at American International Expositions, 1876–1916*. Chicago: University of Chicago Press, 1984.

Said, Edward W. *Orientalism*. New York: Vintage Books, 1979.

Salman, Michael. *The Embarrassment of Slavery: Controversies over Bondage and Nationalism in the American Colonial Philippines*. Berkeley: University of California Press, 2001.

San Juan, E. *After Postcolonialism: Remapping Philippines-United States Confrontations*. Lantham, MD: Rowan & Littlefield Publishers, Inc., 2000.

———. *Beyond Postcolonial Theory*. New York: St. Martin's Press, 1998.

———. *Carlos Bulosan and the Imagination of the Class Struggle*. Quezon City: University of the Philippines Press, 1972.

Schaffer, Kristen. *Daniel H. Burnham: Visionary Architect and Planner*. New York: Rizzoli, 2003.

Schlimgen, Veta R. "Neither Citizens Nor Aliens: Filipino 'American Nationals' in the US Empire, 1900–1946." PhD dissertation, University of Oregon, 2010.

Schoonover, Thomas. *Uncle Sam's War of 1898 and the Origins of Globalization*. Lexington: University of Kentucky Press, 2003.

Schumacher, John N., SJ. *The Making of a Nation: Essays on Nineteenth-Century Filipino Nationalism*. Manila: Ateneo de Manila University Press, 1996.

Schurz, William Lytle. *The Manila Galleon*. New York: E. P. Dutton, 1939.

Scott, James C. *The Art of Not Being Governed: An Anarchist History of Upland Southeast Asia*. Yale Agrarian Studies. New Haven: Yale University Press, 2009.

———. *Seeing like a State: How Certain Schemes to Improve the Human Condition Have Failed*. New Haven: Yale University Press, 1998.

Scott, William Henry. *The Discovery of the Igorots: Spanish Contacts with the Pagans of Northern Luzon*. Quezon City: New Day Publishers, 1974.

———. "Igorot Responses to Spanish Aims: 1576–1896." In *Resistance and Revolution in the Cordillera*. Edited by Delfin Tolentino. Baguio City: University of the Philippines College Baguio, 1994.

———. *Of Igorots and Independence: Two Essays*. Baguio City, Philippines: ERA, 1993.

———. *On the Cordillera: A Look at the Peoples and Cultures of the Mountain Province*. Manila: MCS Enterprises, 1966.

———. *The Union Obrera Democratica: First Filipino Labor Union*. Quezon City: New Day Publishers, 1992.

Sewell, Bevan, and Scott Lucas. *Challenging US Foreign Policy: America and the World in the Long Twentieth Century*. New York: Palgrave Macmillan, 2011.

Sharma, Miriam. "The Philippines: A Case of Migration to Hawaii, 1906–1946." In *Labor Immigration under Capitalism: Asian Workers in the United States Before World War II*. Edited by Lucie Cheng and Edna Bonacich. Berkeley: University of California Press, 1984.

Shaw, Angel Velasco, and Luis H. Francia, eds. *Vestiges of War: The Philippine-American War and the Aftermath of an Imperial Dream, 1899–1999*. New York: New York University Press, 2002.

Sidel, John Thayer. *Capital, Coercion, and Crime: Bossism in the Philippines*. Stanford: Stanford University Press, 1999.

Siracusa, Joseph M. "Progressivism, Imperialism, and the Leuchtenburg Thesis, 1952–1974: An Historiographical Appraisal." *Australian Journal of Politics & History* 20, no. 3 (December 1, 1974): 312–25.

Sklansky, Jeffrey. "The Elusive Sovereign: New Intellectual and Social Histories of Capitalism." *Modern Intellectual History* 9, no. 1 (2012): 233–48.

———. *The Soul's Economy: Market Society and Selfhood in American Thought, 1820–1920*. Chapel Hill: University of North Carolina Press, 2002.

Sklar, Martin J. *The Corporate Reconstruction of American Capitalism, 1890–1916: The Market, the Law, and Politics*. New York: Cambridge University Press, 1988.

Smith, Carl. *The Plan of Chicago: Daniel Burnham and the Remaking of the American City*. Chicago: University of Chicago Press, 2007.

Smith, Henry Nash. *Virgin Land: The American West as Symbol and Myth*. Cambridge: Harvard University Press, 2007 [1950].

Smith, Neil. *American Empire: Roosevelt's Geographer and the Prelude to Globalization*. Berkeley: University of California Press, 2004.

———. *Uneven Development: Nature, Capital, and the Production of Space*. New York: Blackwell, 1984.

Soja, Edward W. *Postmodern Geographies: The Reassertion of Space in Critical Social Theory*. London: Verso, 1989.

Sparrow, Bartholomew H. *The Insular Cases and the Emergence of American Empire*. Lawrence: University Press of Kansas, 2006.

Spate, O. H. K. *The Spanish Lake*. Minneapolis: University of Minnesota Press, 1979.

Stanley, Amy Dru. *From Bondage to Contract: Wage Labor, Marriage, and the Market in the Age of Slave Emancipation*. Cambridge: Cambridge University Press, 1998.

Steinberg, David Joel. *The Philippines, a Singular and a Plural Place*. Boulder, CO: Westview Press, 1990.

Stoler, Ann Laura. *Carnal Knowledge and Imperial Power: Race and the Intimate in Colonial Rule*. Berkeley: University of California Press, 2002.

———. "On Degrees of Imperial Sovereignty." *Public Culture* 18 (Winter 2006): 125–46.

Stoler, Ann, and Frederick Cooper, eds. *Tensions of Empire: Colonial Cultures in a Bourgeois World*. Berkeley: University of California Press, 1997.

Sturtevant, David. *Popular Uprisings in the Philippines, 1840–1940*. Ithaca: Cornell University Press, 1976.

Sullivan, Rodney J. *Exemplar of Americanism: The Philippine Career of Dean C. Worcester*. Michigan Papers on South and Southeast Asia, no. 36. Ann Arbor: Center for South and Southeast Asian Studies, 1991.

Szuberla, Guy Alan. "Urban Vistas and the Pastoral Garden: Studies in the Literature and Architecture of Chicago (1893–1909)." PhD dissertation, University of Minnesota, 1971.

Tan, Samuel K. *The Filipino-American War, 1898–1913*. Quezon City: University of the Philippines Press, 2002.

Tapang, Bienvenido P. "Innovation and Social Change: The Ibaloy Cattle Enterprise in Benguet." CSC Monograph Series 5b. Baguio: Cordillera Studies Center, 1985.

Tchen, John Kuo Wei. *New York Before Chinatown: Orientalism and the Shaping of American Culture, 1776–1882*. Baltimore: John Hopkins University Press, 1999.

Thomas, Nicholas. *Colonialism's Culture: Anthropology, Travel and Government*. Cambridge, UK: Polity Press, 1994.

———. *Entangled Objects: Exchange, Material Culture, and Colonialism in the Pacific*. Cambridge: Harvard University Press, 1991.

Thornton, Tamara Plakins. "Capitalist Aesthetics: Americans Look at the London and Liverpool Docks." In *Capitalism Takes Command: The Social Transformation of Nineteenth-Century America*. Edited by Michael Zakim and Gary John Kornblith. Chicago: University of Chicago Press, 2012.

Tolentino, Delfin, ed. *Resistance and Revolution in the Cordillera*. Baguio City: University of the Philippines College Baguio, 1994.

Totani, Yuma. *Justice in Asia and the Pacific Region, 1945–1952: Allied War Crimes Prosecutions*. New York: Cambridge University Press, 2015.

Trachtenberg, Alan. *The Incorporation of America: Culture and Society in the Gilded Age*. New York: Hill and Wang, 1982.

Tuan, Yi-Fu. *Space and Place: The Perspective of Experience*. Minneapolis: University of Minnesota Press, 1977.

Tucker, Richard C., ed. *Marx-Engels Reader*. New York: Norton, 1987.

Tucker, Richard P. *Insatiable Appetite: The United States and the Ecological Degradation of the Tropical World*. Berkeley: University of California Press, 2000.

Tyrrell, Ian R. *Reforming the World: The Creation of America's Moral Empire*. Princeton: Princeton University Press, 2010.

Vergara, Benito M., Jr. *Displaying Filipinos: Photography and Colonialism in Early 20th Century Philippines*. [Quezon City]: University of the Philippines Press, 1995.

Vernon, Christopher. "Daniel Burnham and the American City Imperial." *Thesis Eleven* 123 (2014): 80–105.

Walsh, Thomas P. *Tin Pan Alley and the Philippines: American Songs of War and Love, 1898–1946, A Resource Guide*. Lanham, MD: Scarecrow Press, 2013.

Wernstedt, Frederick L., and J. E. Spencer. *The Philippine Island World: A Physical, Cultural, and Regional Geography*. Berkeley: University of California Press, 1967.

Wexler, Laura. *Tender Violence: Domestic Visions in an Age of US Imperialism*. Chapel Hill: University of North Carolina Press, 2000.

White, G. Edward. *Justice Oliver Wendell Holmes : Law and the Inner Self*. New York: Oxford University Press, 1995.

Wiber, Melanie. "The 'Cañao Imperative': Changes in Resource Control, Stratification and the Economy of Ritual among the Ibaloi of Northern Luzon." In

*Changing Lives, Changing Rites: Ritual and Social Dynamics in Philippine and Indonesian Uplands*. Edited by Susan Russell and Clark Cunningham. Ann Arbor: Michigan Papers in South and Southeast Asia Monograph Series, University of Michigan.

———. "Levels of Property Rights, Levels of Law: A Case Study from the Northern Philippines." *Man* 26 (1991): 469–92.

———. *Politics, Property and Law in the Philippine Uplands*. Waterloo, ON: Wilfred Laurier University Press, 1993.

Wiebe, Robert H. *The Search for Order, 1877–1920*. New York: Hill and Wang, 1967.

Williams, Raymond. *The Country and the City*. New York: Oxford University Press, 1975 [1973].

Williams, Walter L. "United States Indian Policy and the Debate over Philippine Annexation: Implications for the Origins of American Imperialism." *Journal of American History* 66 (March 1980): 810–31.

Williams, William Appleman. *The Contours of American History*. Cleveland: World Pub. Co., 1961.

———. *The Tragedy of American Diplomacy*. 50th anniversary ed. New York: W. W. Norton & Company, 2009.

Wilson, William H. *The City Beautiful Movement*. Baltimore: Johns Hopkins University Press, 1989.

Wolfe, Patrick. "History and Imperialism: A Century of Theory, from Marx to Postcolonialism." *American Historical Review* 102, no. 2 (1997): 388–420.

Wong, Young-tsu. *A Paradise Lost: The Imperial Garden Yuanming Yuan*. Honolulu: University of Hawai'i Press, 2001.

Wood, Ellen Meiksins. *Empire of Capital*. London: Verso, 2003.

Wright, Gwendolyn. *The Politics of Design in French Colonial Urbanism*. Chicago: University of Chicago Press, 1991.

Young, Robert J. C. *Postcolonialism: An Historical Introduction*. Malden, MA: Blackwell Publishers, 2001.

Zakim, Michael, and Gary John Kornblith, eds. *Capitalism Takes Command: The Social Transformation of Nineteenth-Century America*. Chicago: University of Chicago Press, 2012.

Zimmerman, Andrew. *Alabama in Africa: Booker T. Washington, the German Empire, and the Globalization of the New South*. Princeton: Princeton University Press, 2010.

# Index

Page numbers followed by the letter *f* indicate a figure.